WORLD ATLAS OF HOLOCENE SEA-LEVEL CHANGES

FURTHER TITLES IN THIS SERIES
Volumes 1–7, 11, 15, 16, 18, 19, 21, 23, 29 and 32 are out of print.

8 E. LISITZIN
SEA-LEVEL CHANGES
9 R.H. PARKER
THE STUDY OF BENTHIC COMMUNITIES
10 J.C.J. NIHOUL (Editor)
MODELLING OF MARINE SYSTEMS
12 E.J. FERGUSON WOOD and R.E. JOHANNES
TROPICAL MARINE POLLUTION
13 E. STEEMANN NIELSEN
MARINE PHOTOSYNTHESIS
14 N.G. JERLOV
MARINE OPTICS
17 R.A. GEYER (Editor)
SUBMERSIBLES AND THEIR USE IN OCEANOGRAPHY AND OCEAN ENGINEERING
20 P.H. LEBLOND and L.A. MYSAK
WAVES IN THE OCEAN
22 P. DEHLINGER
MARINE GRAVITY
24 F.T. BANNER, M.B. COLLINS and K.S. MASSIE (Editors)
THE NORTH-WEST EUROPEAN SHELF SEAS: THE SEA BED AND THE SEA IN MOTION
25 J.C.J. NIHOUL (Editor)
MARINE FORECASTING
26 H.G. RAMMING and Z. KOWALIK
NUMERICAL MODELLING MARINE HYDRODYNAMICS
27 R.A. GEYER (Editor)
MARINE ENVIRONMENTAL POLLUTION
28 J.C.J. NIHOUL (Editor)
MARINE TURBULENCE
30 A. VOIPIO (Editor)
THE BALTIC SEA
31 E.K. DUURSMA and R. DAWSON (Editors)
MARINE ORGANIC CHEMISTRY
33 R. HEKINIAN
PETROLOGY OF THE OCEAN FLOOR
34 J.C.J. NIHOUL (Editor)
HYDRODYNAMICS OF SEMI-ENCLOSED SEAS
35 B. JOHNS (Editor)
PHYSICAL OCEANOGRAPHY OF COASTAL AND SHELF SEAS
36 J.C.J. NIHOUL (Editor)
HYDRODYNAMICS OF THE EQUATORIAL OCEAN
37 W. LANGERAAR
SURVEYING AND CHARTING OF THE SEAS
38 J.C.J. NIHOUL (Editor)
REMOTE SENSING OF SHELF SEA HYDRODYNAMICS
39 T. ICHIYE (Editor)
OCEAN HYDRODYNAMICS OF THE JAPAN AND EAST CHINA SEAS
40 J.C.J. NIHOUL (Editor)
COUPLED OCEAN–ATMOSPHERE MODELS
41 H. KUNZENDORF (Editor)
MARINE MINERAL EXPLORATION
42 J.C.J. NIHOUL (Editor)
MARINE INTERFACES ECOHYDRODYNAMICS
43 P. LASSERRE and J.M. MARTIN (Editors)
BIOGEOCHEMICAL PROCESSES AT THE LAND-SEA BOUNDARY
44 I.P. MARTINI (Editor)
CANADIAN INLAND SEAS
45 J.C.J. NIHOUL and B.M. JAMART (Editors)
THREE-DIMENSIONAL MODELS OF MARINE AND ESTUARIN DYNAMICS
46 J.C.J. NIHOUL and B.M. JAMART (Editors)
SMALL-SCALE TURBULENCE AND MIXING IN THE OCEAN
47 M.R. LANDRY and B.M. HICKEY (Editors)
COASTAL OCEANOGRAPHY OF WASHINGTON AND OREGON
48 S.R. MASSEL
HYDRODYNAMICS OF COASTAL ZONES
49 V.C. LAKHAN and A.S. TRENHAILE (Editors)
APPLICATIONS IN COASTAL MODELING
50 J.C.J. NIHOUL and B.M. JAMART (Editors)
MESOSCALE SYNOPTIC COHERENT STRUCTURES IN GEOPHYSICAL TURBULENCE
51 G.P. GLASBY (Editor)
ANTARCTIC SECTOR OF THE PACIFIC
52 P.W. GLYNN (Editor)
GLOBAL ECOLOGICAL CONSEQUENCES OF THE 1982–83 EL NINO–SOUTHERN OSCILLATION
53 J. DERA (Editor)
MARINE PHYSICS
54 K. TAKANO (Editor)
OCEANOGRAPHY OF ASIAN MARGINAL SEAS
55 TAN WEIYAN
SHALLOW WATER HYDRODYNAMICS
56 R. CHARLIER and J. JUSTUS
OCEAN ENERGY. RESOURCES FOR THE FUTURE
57 P.C. CHU and J.C. GASCARD (Editors)
DEEP CONVECTION AND DEEP WATER FORMATION IN THE OCEANS

Elsevier Oceanography Series, 58

WORLD ATLAS OF HOLOCENE SEA-LEVEL CHANGES

Paolo Antonio PIRAZZOLI

Assistant: Jean PLUET

CNRS, Laboratoire de Géographie Physique, 1 Place Aristide Briand, 92190 Meudon-Bellevue, France

ELSEVIER

Amsterdam — London — New York — Tokyo 1991

ELSEVIER SCIENCE PUBLISHERS B.V.
Sara Burgerhartstraat 25
P.O. Box 211, 1000 AE Amsterdam, Netherlands

Distributors for the United States and Canada:

ELSEVIER SCIENCE PUBLISHING COMPANY INC.
655, Avenue of the Americas
New York, NY 10010, U.S.A.

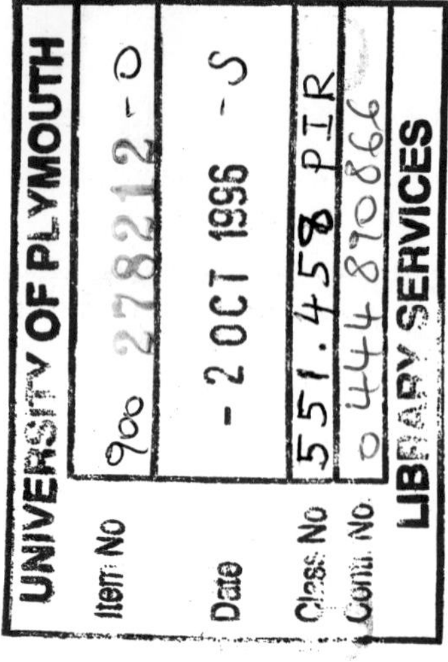

Library of Congress Cataloging-in-Publication Data

Pirazzoli, P. A. (Paolo A.)
World atlas of Holocene sea-level changes / Paolo Antonio Pirazzoli ; assistant, Jean Pluet.
p. cm. -- (Elsevier oceanography series ; 58)
Includes bibliographical references and indexes.
ISBN 0-444-89086-6
1. Sea level--Atlases. 2. Geology, Stratigraphic--Holocene--Atlases. I. Pluet, Jean. II. Title. III. Series.
GC89.P53 1991
551.4'58--dc20 91-25634
CIP

ISBN 0-444-89086-6

This book is printed on acid-free paper.

Printed in The Netherlands

PREFACE

This study is concerned with the current record of sea-level changes during the past 10,000 years, their rates, and our ability to estimate these changes accurately. It aims to:

1) give a global review of the state of the art in the field;

2) provide a visual assessment of geological trends in sea level during the Holocene; by comparing them with trends deduced from tide gauges and near-future trends predicted by climate models, this should help the assessment of near-future sea-level changes at a local scale; this comparison is essential in order to estimate the impacts of future sea-level rise;

3) give a reference compilation of Holocene sea-level curves, most useful when estimating vertical earth movements, establishing palaeogeographic maps and testing geophysical models.

Since 1974, mainly in the framework of international research programmes sponsored by Unesco and by the International Union of Geological Sciences, research on specific local sea-level histories has been undertaken in many new areas, revealing unexpected differences in behaviour from place to place, which geophysical models helped to clarify. In this Atlas, some 800 local relative sea-level curves, deduced from field data from all parts of the world, have been assembled and compared to over 100 curves predicted by geophysical models.

Part 1 of the Atlas is an extensive Introduction, intended for a scientific public with no specialized knowledge of this field. Following an historical background of the subject, the possible causes of sea-level changes and the main methods used to reconstruct former sea-level positions are briefly discussed. An attempt is made to answer questions such as: what kind of sea-level indicators should be used? What are the principal assumptions made by authors? How are former sea levels dated? What accurate are sea-level studies? What are the main assumptions reached by geophysical models? How should this Atlas be consulted?

Part 2 is the main core of the Atlas. It consists of 77 regional plates, each containing 4 to 20 relative sea-level curves, drawn to the same scale with locations, authors' names, year of publication and some indicative values of the spring tidal range in the region, as well as an accompanying text of comments. The plates are simple to consult and enable a direct comparison of relative sea-level trends to be made between sites and regions. The text, mainly intended for a specialized public, gives whenever possible, a very brief analysis for each curve, of the type of data and main assumptions made; in some cases, comments contain criticism.

Part 3 (Conclusions) summarizes the state of the art, proposes improvements in methodology, specifies which regions are uplifting and which are subsiding, and finally, mentions those which require further study.

Part 4 includes an extensive bibliography with over 750 references. Two indexes (geographical localities and authors) complete the study.

This Atlas is a contribution to the IGCP Project 274 "Coastal evolution during the Quaternary" and to the activities of the INQUA Committee on Global Change and of the INQUA Commissions on Quaternary Shorelines, on Neotectonics and on the Holocene. It can also be considered as a contribution of the Commission of European Communities (DG XII) to international research programmes such as Past Global Changes of the International Geosphere-Biosphere Programme (IGBP), the Human Dimensions of Global Change Programme in the areas of Environmental History and of Global Risk Assessment, and the International Decade for Natural Disaster Reduction.

ACKNOWLEDGEMENTS

This work would not have been possible without the financial contribution of the Commission of European Communities [contract n° EV4C-0067-F (CD)] and the research facilities provided by the Centre National de la Recherche Scientifique, first at the Laboratoire Intergéo, Paris, then at the Laboratoire de Géographie Physique, Meudon-Bellevue.

Two postgraduate students, J. Pluet and C. Vaucourt, were of help in assembling publications with sea-level curves; J. Pluet also prepared the reduction of most curves and a draft of about half of the plates. The assistance of Ms. M. Delahaye was essential in revising the English text.

Among the many persons who contributed to this Atlas in several ways, I am very grateful to all those who, since 1977, have sent their publications with sea-level curves to NIVMER, to IGCP-200 or more recently to the author of this Atlas; R.J.N. Devoy, S. Jelgersma, I. Shennan and M.J. Tooley have completed reference lists concerning their study areas at an early stage of the Atlas preparation; P. Giresse and O. van de Plassche have commented on certain data; J.T. Andrews, C. Baeteman, A.P. Belperio, J.H.A. Bosch, R.J. Fulton, K. Lambeck, N.A. Mörner, W.R. Peltier, P. Sanlaville and several others have provided "last minute" publications and F. Antonioli, C. Baroni, G. Dai Pra, N.A. Mörner, G. Orombelli, O. van de Plassche and S. Stiros have even contributed papers containing sea-level curves, still in press; W. Fjeldskaar, V. Gornitz and N.O. Svensson have sent information on sea-level databases; I should like to thank M.M. Birot for great help in reading Russian publications and M. Pirazzoli-t'Serstevens, for kindly reading publications in Chinese and Japanese.

For the illustrations, I am indebted to the following for permissions (received by 30 July 1991) to reproduce copyright material:

Fig. 1-1: reprinted with permission from *Physics and Chemistry of the Earth* vol. 4, R.W. Fairbridge, Eustatic changes in sea level, copyright (C) 1961, Pergamon Press PLC.
Fig. 1-3: reprinted by permission from *Nature* vol. 181 pp. 1518-1519, H. Godwin, R.P. Suggate and E.H. Willis, copyright (C) 1958, Macmillan Magazines Ltd.
Fig. 1-4: reprinted with permission from *Quaternary Research* vol. 11, J.A. Clark and C.S. Lingle, copyright (C) 1979, University of Washington.

Fig. 1-15: copyright (C) 1991, L.F. Montaggioni.

Fig. 1-16: reprinted from *Proc. Fifth Int. Coral Reef Congr. Tahiti*, vol. 1, L.F. Montaggioni, copyright (C) 1985, Antenne Museum-EPHE, Moorea, French Polynesia.

Fig. 1-17: reprinted from Shore-level changes in South Norway during the last 13,000 years, traced by biostratigraphical methods and radiometric datings, from *Norske Geografisk Tidsskrift* by permission of U. Hafsten and of Universitetsforlaget AS (Norwegian University Press).

Fig. 1-19: reprinted with permission of J. Laborel.

Fig. 1-25 and 2-8: reprinted with permission from *Journal of Quaternary Science*, vol. 4, I. Shennan, pp. 77-89, copyright (C) 1989, Longman Group UK Ltd.

Fig. 1-26: copyright (C) 1991, N.O. Svensson.

Fig. 2-1 and 2-2: maps compiled by H. Haila and M. Eronen for the Atlas of Finland, reprinted from Stages of the Baltic Sea and Late Quaternary Shoreline Displacement in Finland, Excursion Guide by J. Donner & M. Eronen, Helsinki, 1981.

Fig. 2-3, 2-9, 2-17, 2-23, 2-25, 2-31, 2-32 and 2-41: reprinted from *Journal of Coastal Research*, Spec. Issue 1, copyright (C) 1986, The Coastal Education and Research Foundation, Inc.

Fig. 2-4: reprinted with permission of J.B. Sissons and of the Institute of British Geographers (copyright (C) 1983) from Shorelines and Isostasy, Academic Press Inc.

Fig. 2-5: reprinted with permission of I. Shennan and of the Institute of British Geographers (copyright (C) 1987) from Sea-Level Changes, Basil Blackwell.

Fig. 2-29: reprinted from The Geomorphology of the Great Barrier Reef, D. Hopley, copyright (C) 1982, John Wiley & Sons, Inc.

Fig. 2-30: reprinted with permission from *Geophysical Journal of the Royal Astronomical Society*, vol. 96, M. Nakada and K. Lambeck, copyright (C) 1989.

Fig. 2-35: reprinted from *Proc. Fifth Int. Coral Reef Congr. Tahiti*, vol. 3, L. Martin, J.M. Flexor, D. Blitzkow and K. Suguio, copyright (C) 1985, Antenne Museum-EPHE, Moorea, French Polynesia.

Fig. 2-39: reprinted with permission from *Canadian Journal of Earth Science*, vol 18, G. Quinlan and C. Beaumont, pp. 1146-1163, copyright (C) 1981, Research Journals, National Research Council Canada.

Fig. 2-40: reprinted with permission of J.T. Andrews from Quaternary Geology of Canada and Greenland, Geological Survey of Canada.

CONTENTS

PART 1: INTRODUCTION

HISTORICAL BACKGROUND

The Holocene is the most recent period of the earth's history, comprising the last 10,000 years. It is characterized by a climate much milder than in the preceding glacial period, during which huge ice caps covered not only Antarctica and Greenland (as today), but also part of North America and of North Europe, as well as several other high latitude and high altitude areas.

When the last glaciation was at a maximum, some 21,000 years ago, sea level was much lower (-120 ± 20 m according to most estimations) than at present. With the melting of the ice caps, huge water masses which had accumulated on the continents in the form of ice, flowed into the ocean, causing a global sea level rise, at rates possibly as much as 20 mm/yr at certain periods. During this time when the sea gradually invaded continental shelves previously exposed, the warmer climate led to increased evaporation and rainfall. People living near the coast were gradually compelled to retreat inland when water inundated their territory. The virtually universal myth of the Flood probably represents the trace in human memory of these prehistoric events. This was the case especially in the early Holocene, when after a last cold episode called Younger Dryas, sea level was again rising rapidly, increased rainfall was making possible the development of forests and savanas, while wide lakes were forming even in the Sahara region (Petit-Maire, 1986). Civilisation and history started slightly later, following the end of the global sea-level rise, when deltas and coastal plains were built up with fertile sediments from rivers such as the Nile and the Euphrates, enabling large groups of people to settle down and develop a complex social organization.

During the historical period of mankind, and until quite recently, the fact that sea level could have varied was recognized only by very few, mainly in areas where sea level was creating problems. In the northern Baltic, where the relative sea-level drop could not be ignored, Celsius suggested in 1733-4 that it was due to "evaporation and the amount of water used by plants to form humus". One of his contemporaries, Linnaeus, also supported the theory that a constantly diminishing global water supply was responsible for the changes observed in sea level, whereas Runeberg, in 1765-9, ascribed them to small movements of the earth's crust (Devoy, 1987).

The first reference marks for the purpose of measuring future sea-level changes, the ancestors of modern tide-gauge datums, were established in 1682 at Amsterdam (Van Veen, 1954), in 1732 on one side of the Doge's Palace in Venice (Zendrini, 1802; Pirazzoli, 1974) (Fig. 2-7) and in 1774 at a sluice in Stockholm (Ekman, 1988).

Suess (1885), developing the ideas of Strabo (Geography, I, 3), who believed that no slope could exist on the ocean surface, introduced the concept of "eustatic" changes in sea level, i.e. of vertical displacements of the ocean surface occurring uniformly throughout the world. This concept strongly influenced sea-level

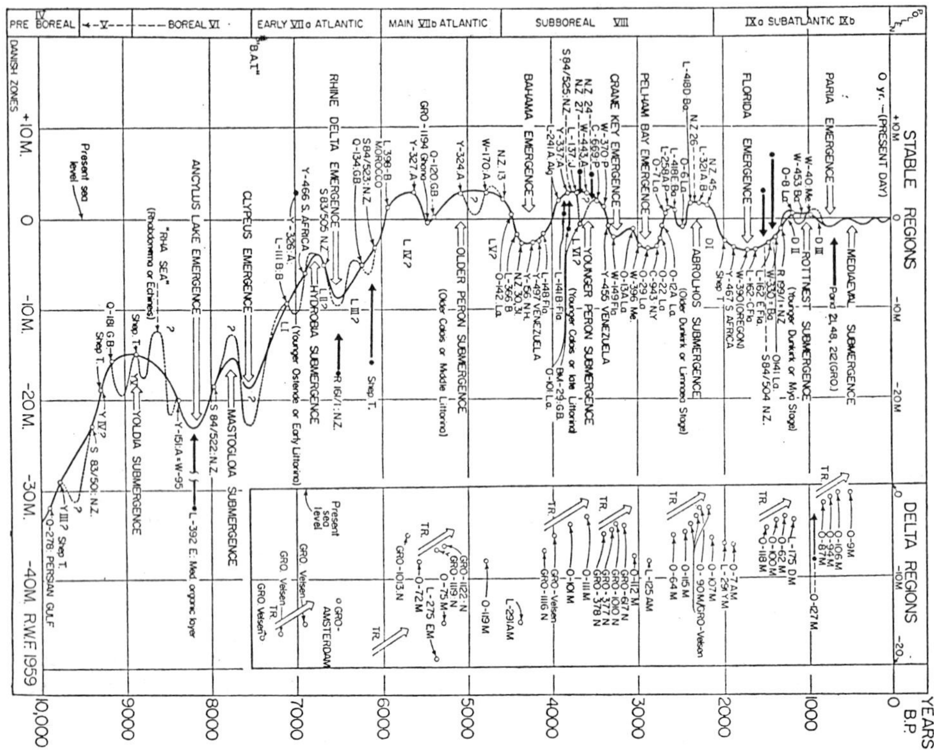

Fig. 1-1. Holocene eustatic changes in sea level according to Fairbridge (1961).

research until the late 1970s. Daly (1934) stressed the importance of changes in sea level and "glacio-isostatic" effects accompanying the last deglaciation phase, with uplift movements in areas of ice melting and subsidence movements in a wide peripheral belt.

The first examples of well-documented Holocene sea-level curves known to the present author are those reported from Sweden by Granlund (1932) and especially by Lidén (1938), who used sequences of varves as a very precise dating tool. Godwin (1940) produced a curve of sea-level changes in the Fenland dated by pollen analysis and archaeological data. It was only in the 1950s, however, with the advent of radiocarbon dating, that several significant Holocene sea-level studies were published.

Two benchmark review papers followed in the early 1960s, each of them combining in a same graph sea-level data from various regions of the world: Fairbridge (1961) deduced a fluctuating sea-level pattern (Fig. 1-1); Shepard (1963, 1964) a gradually rising one (Fig. 1-2). A third school of thought (e.g.: Fisk, 1956; Godwin et al., 1958; McFarlan, 1961) suggested the occurrence of a continuously rising sea level without important fluctuations from late-Glacial times until about 5500 yr BP, followed by sea-level

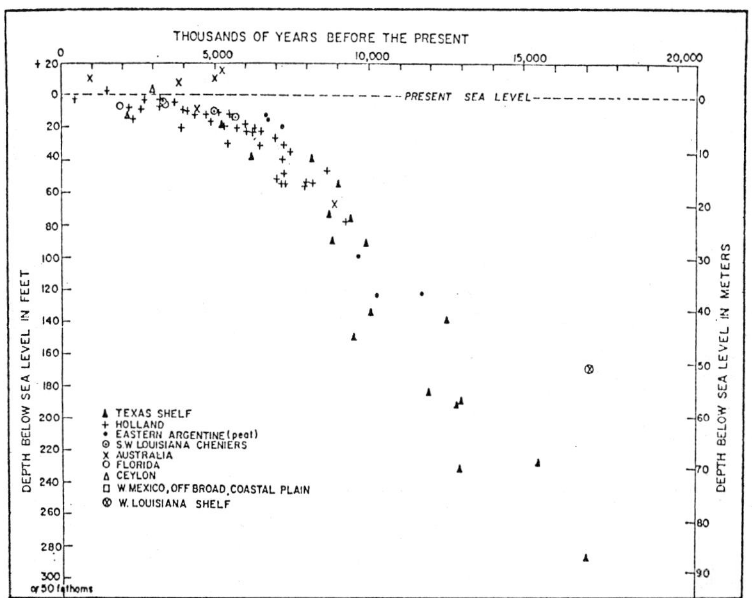

Fig. 1-2. Elevation and age plot of samples "from stable areas" used by Shepard (1963) to infer a gradual eustatic sea-level rise of about 6 m during the past 6000 years.

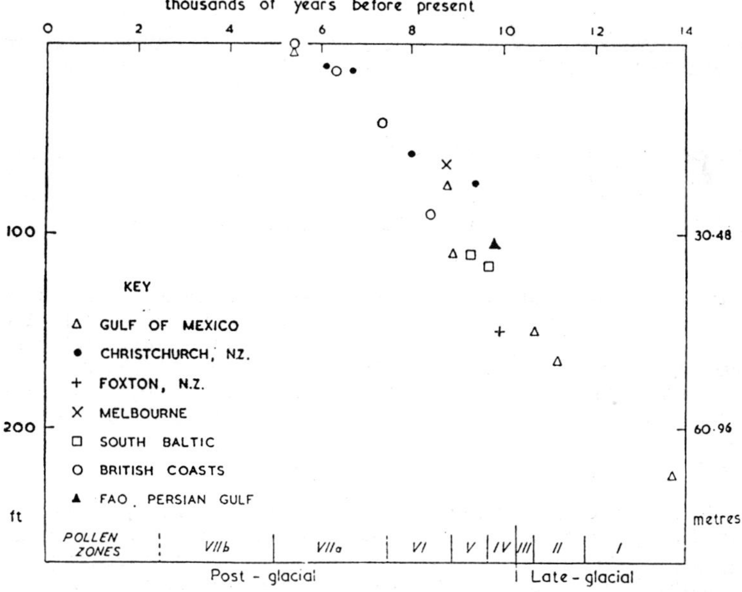

Fig. 1-3. The eustatic rise of sea level according to Godwin et al. (1958).

stability (Fig. 1-3). A commun assumption was that, since the oceans are interconnected, the "eustatic" change in sea level would be the same everywhere and therefore that "stable" regions in any area of the world would be representative of the eustatic situation. As discussed in detail by Pirazzoli (1976b), oscillations in the Fairbridge (1961) curve were the result of a hybrid mixing of data (some of them unreliable) from subsident coasts (corresponding to troughs of the oscillations of the Fairbridge's curve) with data from elevated coasts (corresponding to crests in the curve); the gradually rising trend proposed by Shepard was based mainly on the fact that the "stable" areas considered (Texas, Florida, the Netherlands, Louisiana, etc.) are really all more or less subsiding; finally the sea-level stability suggested after 5500 yr BP by Godwin et al. (1958) depended mostly on insufficient late Holocene data.

Several new sea-level curves were published in the 1960s, most of them differing one from the other. Several authors claimed that their curve was eustatic, i.e. representative of the world pattern, thus giving rise to animated discussions and even disputes. In order for everyone to agree, sea level would have had to be, at one and the same time, higher, equal, and lower than at present and remain stable, though at the same time fluctuating. The lack of mutual understanding was so great that Unesco and the International Union of Geological Sciences (IUGS) launched in 1974, in the framework of the International Geological Correlation Programme (IGCP), Project 61 "Sea level changes during the last hemicycle (c. 15,000 yr)", with the aim of detemining by 1982, the eustatic curve which could reconcile the various observers. Project 61 was headed by A.L. Bloom, who believed that the differences in elevation between contemporaneous shorelines were probably the consequence of *hydro-isostatic* effects (1967, 1971), i.e. of the differences in load caused on the sea floor by the water melting from ice caps. This idea stemmed from the suggestion by Daly (1934, p. 164) that "the plastic adjustment of the earth to the change in water-load with deglaciation" could result "in emergence of both islands and continental shores, but by amounts varying from ocean to ocean and also along the shore of any single ocean".

In 1976, three types of new results and ideas contributed to making the initial aim of IGCP Project 61 obsolete: 1) the old notion of geoidal changes was supported anew by Mörner (1976a), in a benchmark paper suggesting that displacements of the bumps and depressions revealed by satellites on the ocean surface topography could cause differences between coastal areas in the relative sea-level history; 2) the first compilations of worldwide sea-level data (Pirazzoli, 1976b, 1977; Bloom, 1977) showed clearly that late Holocene sea-level histories can vary considerably around the world; 3) in December 1976, at the annual meeting of IGCP Project 61 in Dakar, A.L. Bloom presented the first results obtained using a new global isostatic model of the sea-level response to changing earth loads (Clark et al., 1978; Peltier et al., 1978) predicting the existence of six sea-level zones, resulting from retreat of Northern Hemisphere ice sheets, each characterized by a typical regional trend (Fig. 1-4). These and other results helped to demonstrate that the determination of a single sea-level curve of global applicability was an illusory task and that, by definition, the initial goal of IGCP Project 61, too attached to obsolete theories derived from Strabo, though updated in part by Daly (1934) and Bloom (1967), was ill conceived.

With the succeeding Project 200 "Sea-level correlation and applications", led by the present writer from 1983 to 1987, the aim was to determine local sea-level histories as precisely as possible, to consider all such variations as potentially influenced by a complex of local, regional and global processes and to identify and assess internal and external effects. The ultimate purpose was to provide a basis for predicting near-future sea-level changes, especially in densely populated low-lying coastal areas. One major conclusion of IGCP Project 200 was the confirmation, by field data and models, of the spatial variability in sea level which makes the application of local field data on a worldwide eustatic scale unrealistic and misleading.

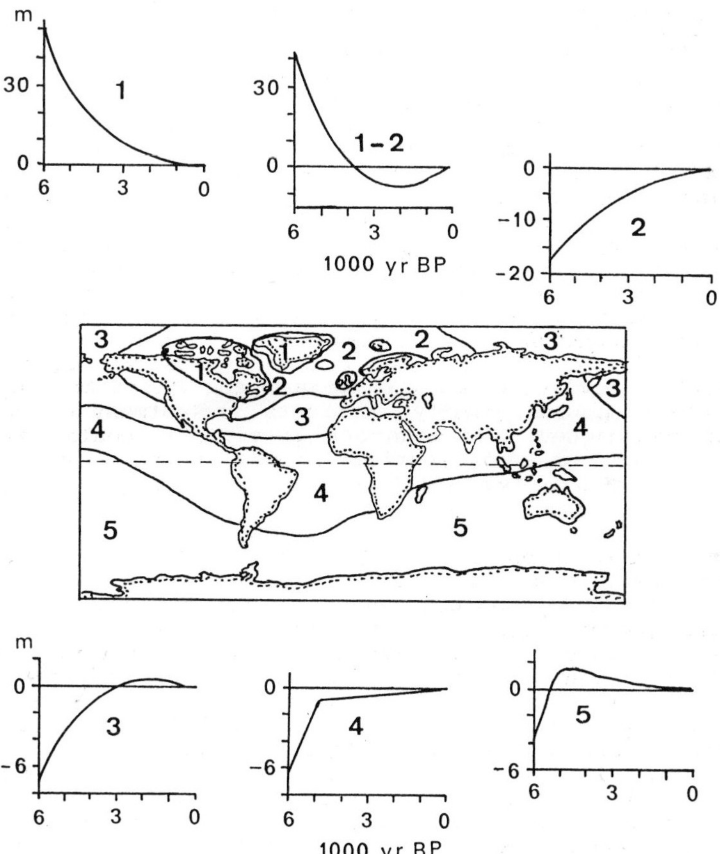

Fig. 1-4. Distribution of the sea-level zones and typical relative sea-level curves predicted for each zone by Clark et al. (1978) under the assumption that no eustatic change has occurred since 5000 yr BP. Zone 6 (not represented) includes all continental margins except those that lie in Zone 2 (from Clark and Lingle, 1979, adapted).

One of the aims of IGCP Project 200 was to collate published sea-level curves, in order to update the "Atlas of Sea-Level Curves" prepared by Bloom (1977), thus continuing the work carried out by the French "sea-level working group" (NIVMER Informations, 1978-1982) who reproduced in newsletters and widely circulated a number of sea-level curves from all regions, and by Ota et al. (1981), who compiled summaries of sea-level work carried out in Japan. However, apart from very accurate contributions from Japan (Ota et al., 1987a,b), only a few participants in IGCP Project 200 sent reprints of new sea-level curves they had published, in spite of several appeals made in newsletters with a wide circulation. The preparation of a new sea-level Atlas by the end of IGCP Project 200 was also prevented from materialising, due to a lack of adequate financial support.

The opportunity to complete the work finally occurred when a call for research projects was made by the Commission of European Communities (DGXII) in 1986. A broader project "Investigation of past and future sea-level changes and their impacts" was proposed jointly by the Centre National de la Recherche Scientifique (Paris and Meudon), the Natural Environment Research Council (Birkenhead), the Rijkswaterstaat ('s Gravenhage) and the Universities of Cork, Durham, Naples and Norwich (East Anglia). Accepted one year later by the CEC, this project enabled several new European sea-level investigations to be carried out and provided, among others, a financial contribution to the preparation of this Atlas, which was undertaken by the C.N.R.S. between late 1987 and early 1991.

POSSIBLE CAUSES OF SEA-LEVEL CHANGES

Sea level is highly variable in both space and time. If tides, differences in water density, currents and atmospheric forcing could be disregarded, the mean sea level (MSL) would adjust to take the form of an equipotential surface of the earth's gravity field (Pugh, 1990). Yet sea level is of course modified by a variety of factors, including all those affecting the gravity field: from outer space (astronomical impacts), from the atmosphere (meteorological influences), from addition/removal of masses to/from the earth's surface (formation or melting of ice sheets, growth of volcanic masses, accumulation of sediment deposits, erosion processes, etc.), from the oceans (density changes, ocean circulation), and finally from processes acting at the earth's interior.

Processes which can cause a change in the global (eustatic) sea level depend therefore on possible variations in: 1) the volume of ocean basins; 2) the mass of ocean water; and 3) the ocean water density (*steric* changes). A variation in the relative sea level at any particular location may also be caused by: 4) local or regional uplift or subsidence of the land; 5) gravitational and rotational variations; 6) changes in atmospheric pressure, winds, ocean currents, etc.

1) The volume of the ocean basins can be modified because of changes in the rate of plate divergence (seafloor spreading), plate convergence (subduction, overthrusting), epeirogenic changes in the elevation of the seafloor (mid-ocean volcanism), marine sedimentation, or isostatic adjustment of the earth's crust

beneath the sea resulting from glaciation or deglaciation on land.

2) Changes in the mass of ocean water can be caused by the melting or accumulation of continental ice (glacio-eustatic) or by increasing or decreasing retention of liquid water underground or on the surface of the continents, or by slow release of juvenile water from the earth's interior. The mass of water at or near the earth's surface is almost constant for time scales of the order of 10^4 yr or less. What is important for eustatic sea-level changes is the way water is distributed between the major hydrological reservoirs and, regarding regional or local sea-level changes, what can be the isostatic consequences of this partition.

3) Exchanges between the oceans and the major continental hydrological reservoirs modify durably the salinity of ocean water; related climatic changes modify the water temperature. As the density of sea water depends on both salinity and temperature, this implies steric changes in the volume of the water without changes in the water mass.

4) Uplift or subsidence movements at the surface of the earth are very common occurrences and may have several causes. Isostatic adjustments can be due: a) to temperature/density changes of materials inside the earth's interior [called *thermo-isostasy* (Pirazzoli and Grant, 1987) when temperature is the predominant factor, e.g. near a hot spot], b) to increased or decreased loads on the earth's surface (*glacio-isostasy* when the load is ice, *volcano-isostasy* when it is extruded magma, *sedimento-isostasy* when it is sediment deposition or erosion, *hydro-isostasy* when it is a layer of water). Important vertical movements of land can be due to tectonic phenomena, at time scales varying from a few seconds (co-seismic displacements) to several million years. Subsidence phenomena can also be due to compaction of sediments, eventually accelerated by human activities. As noted by Revelle et al. (1990), land subsidence, which is now occurring in many coastal areas, will reduce the volume of sedimentary aquifers above sea level, and thus have an effect similar to that of erosion or of decline in the level of the water table, due to decreased precipitation. Local land subsidence may contribute therefore to a global sea-level rise.

5) Gravitational and rotational variations include changes in the earth's rate of rotation and of the tilt of the rotation axis (Munk and Revelle, 1952; Fairbridge, 1961; Barnett 1983; Mörner, 1988; Sabadini et al., 1990; Peltier, 1990).

6) Among various possible causes of short-term relative sea-level changes, atmospheric depressions can cause a local sea-level rise of some decimetres and variations in the runoff of large rivers of as much as 1 m. In relatively shallow water, winds can pile water up against the shore or drive it away; in exceptional circumstances (North Sea, Chinese coast, Bay of Bengal) sea level may rise 5 m or more in a storm surge, under the combined effect of deep atmospheric depression, strong winds and local hydrodynamic phenomena (Revelle et al., 1990).

Though some of the above processes may be insignificant on the time scale of interest, all of them need to be considered. In particular, ephemeral variations in sea level may leave durable marks or deposits on the shore, which may subsequently be misinterpreted as being due to a lasting sea level different from the present one.

Fig. 1-5. Submerged quarries near Cap Couronne (Martigues, France). These quarries were used between the 4th c. B.C. and the 1st c. A.D. to build ancient Massalia (Marseille) and indicate a relative sea-level rise of 0.5 ± 0.2 m since that time (Guéry et al., 1981).

Fig. 1-6. Slightly submerged remains of a Roman fish tank at Astura (Italy), indicating a relative sea-level rise of at least 0.5 m since 27 B.C.- A.D. 14 (Pirazzoli, 1976a).

Fig. 1-7. A double notch appears at low tide at Miyagi-jima, east coast of Okinawa Island. The lower notch corresponds to the present-day tidal range. In the upper part of the notch roof, erosion marks belonging to a previous notch, about 0.75 m higher, correspond to a slightly higher sea-level stand.

Fig. 1-8. Remnants of six stepped Holocene shorelines are visible at Kwambu, Huon Peninsula, Papua New Guinea. The position of one of the former sea levels is clearly indicated by this tidal notch.

HOW SEA-LEVEL CHANGES ARE STUDIED

A great amount of sea-level data is contained in the scientific literature, with contributions from several disciplines: archaeology, geomorphology, geology, geophysics, oceanography, geochemistry, ecology, etc. Obtaining shoreline-related data involves a number of procedures and interpretations and includes several steps where error and uncertainty may enter into the results. The usual procedure consists in identifying former sea-level indicators, locating the position of a former water plane and measuring the elevation of this paleolevel, and determining the time at which water stood at this specific level.

Sea-level indicators

Several sea-level indicators can be used (Groupe NIVMER, 1979-80; Van de Plassche, 1986). Among archaeological remains, the most useful are obviously those implying specific activities and classes of people who were closely related to the sea (sailors, fishermen, pilots, boatbuilders, salt or shell gatherers, etc.) (Flemming, 1979-80; Pirazzoli, 1979-80, Martin et al., 1986) (Fig. 1-5, 1-6), whereas it will usually not be possible to specify for most other artifacts how far from the shoreline they were constructed.

Marine erosion features such as benches, pools or notches (Pirazzoli, 1986c) may provide quite accurate determinations of former sea level positions (Fig. 1-7, 1-8), though they are often inadequate to date the former sea stand in the absence of other related sea-level indicators.

Marine deposits are most common and their shape or stratigraphy may be clearly identifiable as typical coastal features (beach, littoral bar, estuary or lagoon floor, beachrock, coral conglomerate, foreshore delta, etc.) (e.g. Gross et al., 1969; Andrews, 1986; Roep, 1986; Hopley, 1986a). Easily datable by reworked shell or coral fragments (which however can be older than the deposit containing them), these features can imply wide vertical uncertainty ranges when they are used to reconstruct former MSL positions, due to the tidal range and to possible storm events. Increased accuracy can be obtained by studying assemblages of marine molluscs quantitatively (Petersen, 1986), or of microfauna such as foraminifera (Scott and Medioli, 1986) and ostracodes (Van Harten, 1986). However, as discussed by Andrews (1989), it is often difficult to determine how closely datable organic materials are associated with a specific paleolevel. Shells lying within beach gravels can be reworked from older deposits offshore, or material can be carried to the sea after erosion of pre-existing fossiliferous strata. Wood incorporated into beach gravels can be reworked from older units; articulated marine bivalves represent an in situ horizon within the sediment (Fig. 2-33), though they can seldom be unambiguously associated with a specific paleo sea level.

The most reliable indications on the elevation and age of former sea levels are probably given by reefal or encrusting marine organisms collected in growth position. This is the case for coral reefs (Hopley, 1986b) and coralline algae (Adey, 1986), especially

Fig. 1-9. The upper boundary of living *Dendostraea hyotis* marks an almost horizontal line, situated between 20 and 30 cm above MSL in Ago Bay, Mie Prefecture, Japan, and is an excellent sea-level indicator. Mean spring tidal range is 1.35 m, water level is 75 cm below MSL.

Fig. 1-10. The growth of *Porites* coral microatolls is limited upwards by the low tide level, thus giving excellent sea-level indications in the absence of moating phenomena. In the background: displaced reef blocks. Okinoerabu Island, the Ryukyus, Japan. Mean spring tidal range is 1.4 m, water level is 20 cm below MSL.

Fig. 1-11. A sequence of intercalated marine and terrestrial deposits in a sand spit near Calais, France. The "Dunkirk" transgressions represented here by marine sediment layers may correspond to changes in the local tidal range related to breaks in the sand spit rather than to eustatic changes. Mean spring tidal range is over 6 m in this area.

Fig. 1-12. An international field excursion is examining peat cores sampled from De Kolken, near Groningen, the Netherlands. In subsiding areas, where all evidence of former sea levels is submerged, coring is essential to sea-level research.

Fig. 1-13. The lower boundary of mangrove growth is a precise intertidal sea-level indicator. Saloum River estuary, Sénégal.

Fig. 1-14. Pneumatophore roots of *Avicennia* extend slightly below the lower limit of growth of these mangrove trees. Near Fandrina, NW Madagascar.

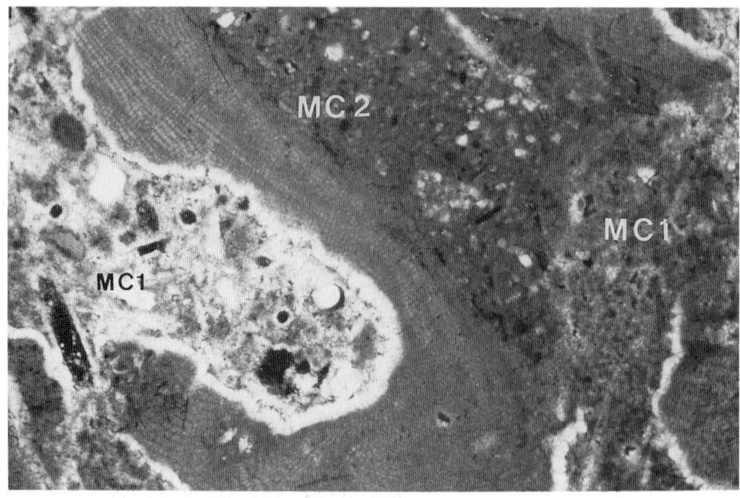

Fig. 1-15. In a marine incrustation now emerged in Rhodes Island (Greece), two different generations of micrite cements (MC1, affected by microsparitization, then MC2, occupying the residual pores and exhibiting little or no microsparitization) occur at an elevation (3.2 m above MSL, see Fig. 2-16, sample 10 in crustal block A) which has experienced two different phases of marine phreatic deposition and cementation, separated by a phase of subaerial exposure. (Photo L.F. Montaggioni).

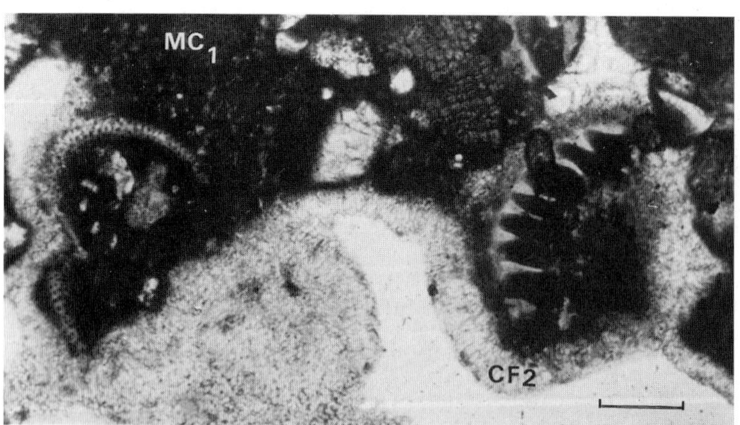

Fig. 1-16. Example of marine phreatic (= subtidal), then marine-freshwater mixing vadose (= intertidal to supratidal) cementation. The earlier generation of cements includes dense, structureless micrite matrices (MC_1), whereas the later generation exhibits isopachous high-magnesian calcite rims of blunted fibres (CF_2). A relative sea-level drop occurred after deposition of the earlier generation of cements. Scale is 0.1 mm. From a sample of coral-rich conglomerate, Tupai Atoll, French Polynesia. (Photo L.F. Montaggioni, from Pirazzoli et al., 1985).

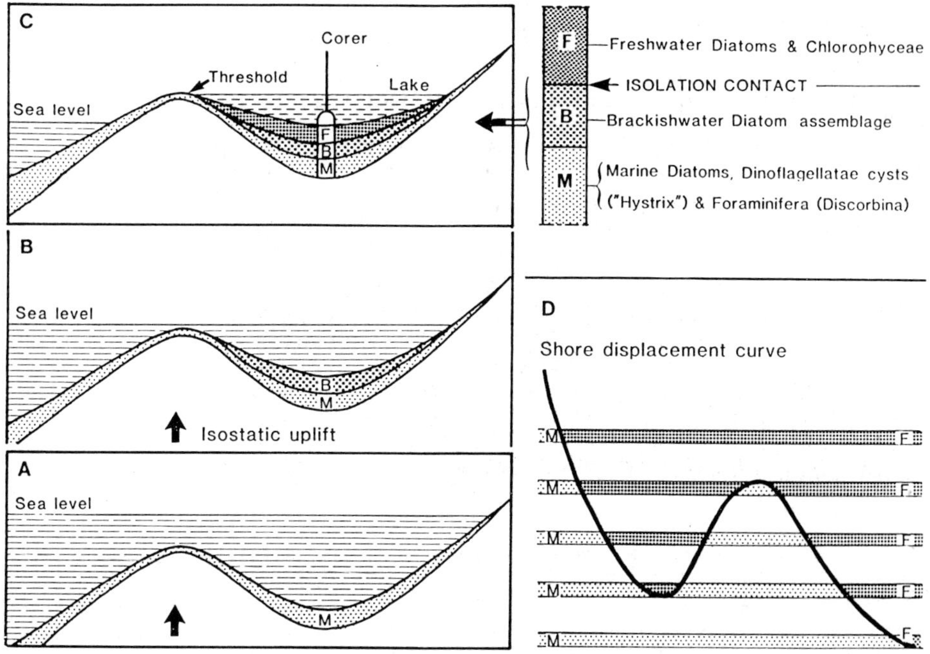

Fig. 1-17. Depositional conditions 1) when a trough-formed part of the sea floor is isolated from the sea during an isostatic uplift (A, B, C) and 2) when a eustatic rise in sea level exceeds the isostatic uplift, causing marine transgression (D) (from Hafsten, 1983).

for species living close to the low-tide level, for vermetid gastropods (Sanlaville,1973; Laborel, 1979-80, 1986), certain marine biogenic constructions (Laborel, 1987), barnacles (Pirazzoli et al., 1985a), oysters, etc. (Fig. 1-9, 1-10).

Coastal plant remains are often employed in sea-level studies (Behre, 1986). The most frequently used material is indeed marshy peat, epecially when it is intercalated with marine deposits (Fig. 1-11, 1-12). In order to minimize errors due to compaction phenomena, which are often important (Kaye and Barghoorn, 1964; Greensmith and Tucker, 1986) and exert the greatest influence at low rates of sea-level rise (Fletcher, 1987), the preferred peat layer for paleo-altitude estimation is the base one, possibly resting on a hard older substratum. In addition, peats can scarcely move round as sea level moves up and down, for they would simply disintegrate if they moved at all. Thus, to find peat resting on former subaerial exposure surfaces is a good indication of the time at which the sea first inundated the former subaerial landscape (Matthews, 1990).

Useful indications can also be given by so-called submerged forests (Heyworth, 1986; Fujii et al., 1986) (Fig. 2-6), mangroves (Woodroffe,1988; Ellison, 1989; Ellison and Stoddart, 1991) (Fig. 1-13, 1-14), diatom assemblages (Palmer and Abbott, 1986), etc. Yet paleoshorelines which are now submerged (submerged forests, saltmarshes, freshwater peats buried beneath marine sediments, submerged littoral facies and landforms) can only be studied by an elaborate process of marine geophysical surveys and coring.

Diagenetic changes occurring in nearshore marine carbonates can be used to deduce diagnostic criteria in terms of the sea level position (Coudray and Montaggioni, 1986) (Fig. 1-15). In elevated coral-rich conglomerates, in particular, the former low-tide level existing at the time of the conglomerate lithification can be determined with good accuracy (Montaggioni and Pirazzoli, 1984) (Fig. 1-16).

In high-latitude areas, sea-level research is often based on *ad hoc* sea-level indicators, like those obtained in uplifting Scandinavia from the study of isolation processes in lake basins (Fig. 1-17), the use of whalebones to date raised marine terraces in Arctic Canada (e.g. Blake, 1975) or even of penguin guano deposits in Antarctica (Baroni and Orombelli, 1991).

Measuring elevation is only possible in relation to a reference level. When a permanent benchmark is not available near the area surveyed, a local datum-level has to be determined. The most reliable local datum-level is usually the present (active/living) counterpart of the fossil sea-level indicator which is being studied (Fig. 1-18, 1-19). When this counterpart is lacking locally, it is not advisable to refer to a similar counterpart at some distance away, since exposure to waves and to meteorological and hydrological conditions may vary along the coast, shifting the indicator considered to a higher or a lower level by an unknown amount. In these cases a "local tidal datum-level" has to be chosen (Jardine, 1986). This can be the level of the sea during calm conditions at a certain moment (to be related to the records of a tide gauge or to tidal predictions), or well defined biological intertidal levels, such as the upper limit of growth of barnacles (Fig. 1-20, 1-21), oysters, seaweed such as *Fucus vesiculosus*, coralline algae, *Patella*, date mussels, vermetids, corals, etc, the vertical range of growth of *Spartina alterniflora* (which usually develops in temperate salt marshes between the marsh surface and just below MSL), or of mangroves in tropical marshes (Fig. 1-13, 1-14), or finally the lower limit of continuous land-based vegetation (Fig. 1-22, 1-23, 2-26). The biological level chosen will have to be referred subsequently to the local tide levels. A non-biological surrogate can be the upper limit of sea-drifted materials, particularly seaweed, the flat surface of the icefoot in the Canadian Arctic, etc. Error margins in altitude determinations will have to be estimated realistically, taking into account the accuracy of each step in the estimation, from the uncertainty range affecting the identification of a datum level, to that relating the present position of the fossil indicator to a former sea level (which may be poorly defined) and lastly to that associated with the surveying technique. The possibility that tidal ranges varied as the geometry of the water body changed in response to oscillations in sea level (e.g. Grant, 1970, Jardine, 1975, Scott and Greenberg, 1983) has also to be considered and the corresponding changes in elevation estimated. The error induced by tidal changes may be especially significant in macrotidal areas and reaches a maximum, as noted by Fletcher (1987) in coastal Delaware, at locations which possess the greatest shelf width. A detailed discussion on assessment of various types of errors was published by Heyworth and Kidson (1982).

Fig. 1-18. Near Cape Flomés, on the south coast of Crete, Greece, remains of a raised rim of *Dendropoma* and calcareous algae (A) show, with great precision, a shoreline at 7.0 m above the presently living counterpart (B) (from Laborel et al., 1979).

Fig. 1-19. A spectacular bioconstructed vermetid-oyster cornice near Çevlik (Hatay, Turkey), about 0.8 m above its active counterpart, was probably raised by an earthquake in A.D. 551 (Photo J. Laborel) (Pirazzoli et al., 1991).

Fig. 1-20. Barnacle development suddenly ends here between MSL and the mean high water level. Below, green algae do not grow much higher than MSL. This type of clear biological zonation is often of great help when measuring former sea levels differing from the present one. Kannoura, Muroto Peninsula, Shikoku, Japan. Mean spring tidal range is 1.45 m, water level is near MLWST.

Fig. 1-21. Fossil barnacles, found during the excavation of the Roman harbour of Marseille (France), indicate a sea level about 0.5 ± 0.1 m lower than the present one (from Pirazzoli and Thommeret, 1973; Pirazzoli, 1976a).

Identification of former shorelines and reconstruction of ancient sea-level positions is therefore a difficult task, mostly based on observation, in which interpretation takes a prominent part.

Dating Holocene shorelines

Since the late 1950s, radiocarbon dating is the current major method for providing a time scale estimation of Holocene sea-level changes. Though well established, this technique presents certain problems and uncertainties however, which may affect sample collection, data evaluation and interpretation. A discussion of these problems was published by Mook and Van de Plassche (1986), to which the reader can refer for details and references up to 1983. More recent developments can be found by consulting the journal *Radiocarbon*. Discussion will be limited here to a few points which can help when a comparison between sea-level curves by different authors has to be made.

^{14}C is continuously produced by the interaction of cosmic ray neutrons with nitrogen atoms in the atmosphere. Living plants and animals fix ^{14}C in their tissue. After their death, ^{14}C disappears gradually at a known rate by radioactive decay. The amount of ^{14}C left that can be measured at the present time in a sample is therefore proportional to the age of this sample.

^{14}C ages are given in yr BP (years Before Present), i.e. before A.D. 1950, with a ± error of a certain number of years. This error is the standard deviation (1σ) and consists solely of the error of counting random events. Other errors involved in the method can be assessed mathematically, but the true error is likely to be larger than the one given, since 1σ implies that 33% of the determinations are liable to fall outside the limits given.

Conventional ^{14}C ages are defined following a number of assumptions and conventions, among which: 1) the ^{14}C activity of carbon-containing material during its formation, defined internationally, has always been the same; 2) there has been no inclusion of new ^{14}C in the sample after its death; 3) the ^{14}C half life used is 5568 years.

In fact, as concerns 1), dendrochronological calibrations have shown that ^{14}C activity has varied in the past, at least until 13,300 yr BP (Stuiver and Kra, 1986), and this implies that corrections are necessary before conventional ages can be expressed in sideral years BP or in calendar years. Bard et al. (1990) extended calibration of the ^{14}C time scale over the past 30,000 years, using mass spectrometric U-series ages from Barbados corals. Bard et al. (1990) showed that radiocarbon ages are systematically younger than Th/U ages, with a maximum difference of about 3500 years at c. 20,000 yr BP.

Datation laboratories may differ in their procedures and calculations. Regarding marine samples in particular, uncertainties due to isotope fractionation may not necessarily compensate uncertainties related to reservoir effects, so that the calibration curves proposed by Stuiver et al. (1986) are more reliable than previously published calibration tables and curves, though obviously they were not used in earlier sea-level publications. Anyway corrections are necessary each time radiocarbon dates of marine and terrestrial samples have to be

Fig. 1-22. The lower limit of terrestrial vegetation on the windward side of San Sebastián Bay, Tierra del Fuego, Argentina, is clearly marked by *Lepidophyllum cupressiforme* shrubs, while *Salicornia* is gradually killed, buried by sediments. Mean spring tidal range is about 8.7 m.

Fig. 1-23. The lower limit of land-based vegetation is clearly marked along tidal channels near Manori, north of Bombay, India. Mean spring tidal range is about 3.5 m.

plotted in the same graph.

Assumption 2) is often unrealistic and various cases of possible contamination, especially in peat samples, are discussed in detail by Mook and Van de Plassche (1986). Regarding convention 3), though the value 5568 years is known to be in error by 3%, to avoid confusion the better value 5730 years should not be used. Some authors however (e.g. in Taiwan) use the value 5730 years currently and this is not always specified in their publications.

Other methods of dating Holocene shorelines include the U-series dating procedures (though they are more adapted to late Pleistocene samples), archaeology (which can give very precise dates, but only at sites where well preserved cultural remains are available), dendrochronological dating (Heyworth, 1986a), varve counting, palynological estimations, etc.

The accuracy of sea-level curves

The advent of radiocarbon dating, with its systematic computation of standard deviations, has led to higher precision when plotting the age of sea-level index points. Since the pioneer publication by Van Straaten (1954), who was among the very first to produce a sea-level graph showing probable altitudinal and temporal uncertainties in error boxes, many sea-level curves have been published. Most of them are however just continuous lines fitted through sea-level index points, with little regard for the errors likely to have affected them. In other cases, probable errors arising from the sampling and dating techniques employed have been represented by error-rectangles, -segments, -ellipses or -bands and plotted with the sea-level curve proposed. It is clear, as stated by Chappell (1987, p. 313), that "data can be placed on a common footing by duly estimating error sources. Only when this is done consistently can the real differences between different sites be perceived and measured".

Most sea-level curves are not deduced mathematically from the data, but interpreted by their author in order to follow more or less closely certain index points considered more reliable and to express trend variations deduced from other information and from interpretation of observations which are not always strictly quantifiable. From this point of view, as in many scientific fields depending mainly on observation, a linear sea-level curve is constructed not only from data, but also from subjective ideas, and in some cases preconceived theories of their author. How can we explain otherwise that some authors find oscillating sea-level curves everywhere, while others only obtain curves showing regular trends?

As discussed by Shennan and Tooley (1987, p. 373-374): "A single line or an age/altitude graph is a poor summary of sea-level changes if the variation in data points is not shown. Furthermore the interpretation of changes in sea level is scale dependent. Whether or not sea level has been characterized by low amplitude fluctuations during the Holocene is only important over the specific period of study (Shennan, 1982b). For time scales of the order of 1000 years the sea-level reconstruction may be generalised and alternations of periods of positive and negative tendencies may become insignificant as the time scale increases to 10,000 years. At this scale, sea-level altitude data are required in more general terms since the other variables that are studied, e.g. regional crustal movements, ice-sheet dimensions and ocean volumes, cannot be correlated precisely with sea level. As the resolution increases to 100 years the local scale variables contributing to the spread of the sea-level index points on the age/altitude graph increase in importance in this study, even though their precision of measurements is unchanged. Generalisations that were valid for the longer periods cannot now be made. Long-term variables may have changed their status from dependent to independent and the shorter term and local spatial scale variables from indeterminable to determinable. Thus the final summary of the sea-level data will vary according to its specified or potential application. Subsequent users of the information should be able to use original sea-level index points rather than the curve which may be an inappropriate summary for their analysis".

This lucid approach should always be in borne in mind by sea-level workers when they interpret their data plotted on a graph. Unfortunately, relatively few sea-level curves found in the literature are provided with error boxes for each sea-level index point. If comparisons had to be limited to these curves only, this Atlas would be restricted to very few areas. In addition, superimposition of various sea-level curves on the same graph, each with all its error boxes, would soon become illegible. In some plates of this Atlas (e.g. Plates 6, 12, 15, 18, 24, 50, 64, 72, 75, 77) error boxes estimated by authors have been reproduced with one or more curves, in order to give an idea of the deviations existing between field data and a linear curve intended to summarize them. In other cases, only error bands have been given.

It seems more advisable that individual sea-level index points be collected, together with their standard error margins and other relevant information, in appropriate sea-level data bases, which would be gradually extended geographically and updated. Comparisons of sea-level error bands between two or more regions, for a given period of time, would then be as accurate and realistic as permitted by the data available. Some sea-level data bases of this kind already exist.

The "Palaeogeodesy" data base assembled by the late W.S. Newman is probably the best known (e.g. Marcus and Newman, 1983; Newman and Baeteman, 1987; Newman et al., 1987; Pardi and Newman, 1987;). It contains more than 4000 radiocarbon-dated sea-level indicators secured from the littoral of many of the world major land masses as well as from oceanic islands. "An elevation vs. time plot for the past 16,000 years demonstrates that the magnitude of eustatic sea-level rise is exceeded by both neotectonic as well as geoidal (usually referred to as isostatic) changes in level (Fig. 1-24). For example, formerly glaciated areas reveal late-glacial sea levels of more than 200 m above the contemporary datum, while zones marginal to these glaciated areas find earlier sea levels as much as 160 m below the present datum. Thus, the sea-level geoid of the last glacial age now exhibits a relief of more than 360 m and therefore exceeds the largest estimated postglacial eustatic rise in sea level by a factor of two or even more." (Newman, 1986).

In the United Kingdom, a database of 904 radiocarbon-dated sea-level index points was assembled by Shennan (1989), of which 429 validated peat samples were plotted in Fig. 1-25. At the

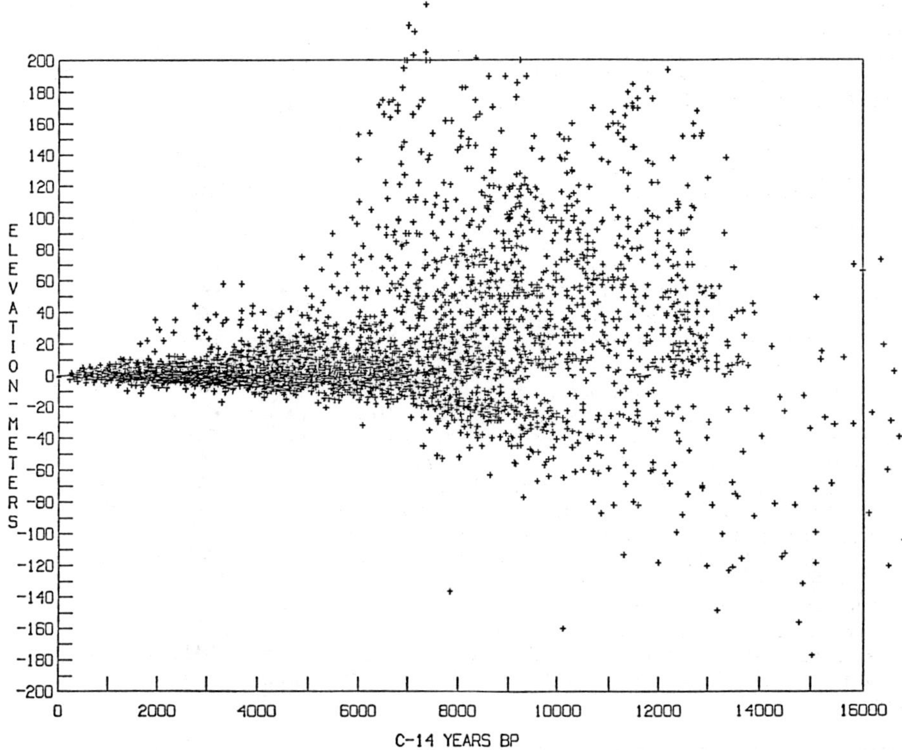

Fig. 1-24. Elevation versus time plot of dated indicators of sea levels for the past 16,000 years (Newman, 1986).

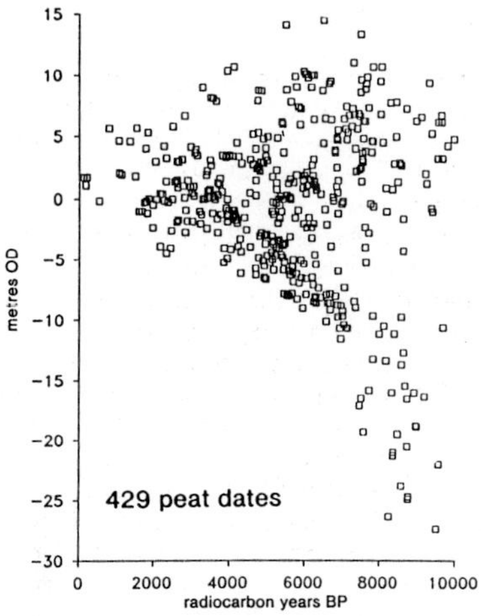

Fig. 1-25. Age-altitude distribution of 429 sea-level index points in Great Britain (from Shennan, 1989).

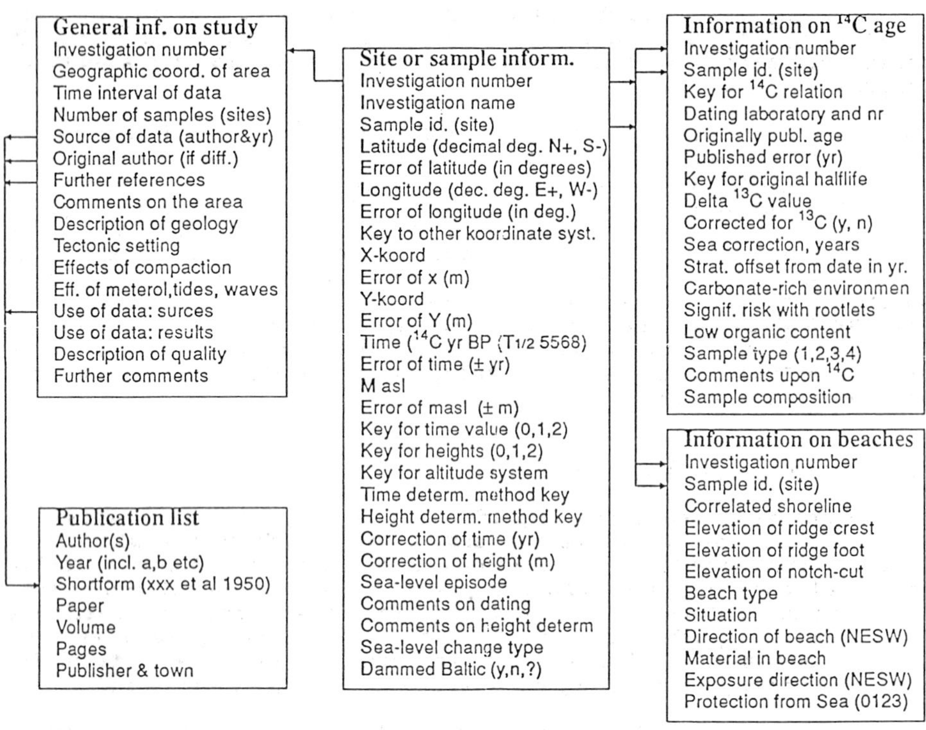

Fig. 1-26. Preliminary structure of the Swedish shoreline database (N.O. Svensson, personal communication, 1990).

University of Toronto, relative sea-level data from almost 400 sites on the earth's surface have been assembled in a global data base (Peltier, 1988).

Shoreline data from Sweden, derived from individual geographical sites where a ^{14}C chronology is available, are being collected at the University of Lund (N.O. Svensson, pers. comm., 1990). They include lake-isolation dates (the lake threshold corresponding to sea level) and elevation data from significant shorelines such as the highest shoreline formed at deglaciation, the Baltic Ice Lake shoreline formed just before the final drainage, the shoreline formed by the Ancylus transgression, and the shoreline(s) of the maximum postglacial transgression(s). The structure of the Swedish database is shown in Fig. 1-26. According to Svensson, a sea level data bank with the same structure is being set up on a worldwide scale at the Geophysical Institute of Kiel, and a national one is in preparation in Finland.

Gornitz (1991) has developed a global coastal hazards data base, intended to provide an overview of the relative vulnerabilities of the world's coastlines. To date however, information on variables associated with potential inundation and erosion hazards (relief, rock type, landform, local relative sea-level trend, rate of present shoreline displacement, tidal range and wave height) have been compiled for the United States, and parts of Canada and

Mexico only. It would certainly be useful to encourage the development of existing and new sea-level data bases and integrate their information in the framework of an international cooperation programme.

For the instrumental period, the PSMSL (Permanent Service for Mean Sea Level, Bidston Observatory, Birkenhead, Mersyside L43 7RA, United Kingdom) has collected monthly and annual mean values of sea level computed at 1243 tide-gauge stations worldwide (Pugh et al., 1987). This data bank is regularly updated (3000 new station-years of data were entered in 1989, over 2600 in 1990) (PSMSL Newsl., Dec. 1990). Other series of tide-gauge mean data which are available for analysis, in addition to those included in the PSMSL databank, have been listed by Spencer et al. (1988). An archive of parallel important data sets (i.e. geological and archaeological data sets or geodynamic models, has also been assembled at the PSMSL.

Sea-level predictions

In regions which have been uplifting at a regular rate, shoreline diagrams are used to represent and correlate shoreline altitudes. Where a shoreline diagram can be constructed, new local relative sea-level curves can be provided using only a few additional sea-level index points. This, and the fact that emerged data are easier to study than submerged ones, explain why more sea-level curves have been published for Fennoscandia and Canada than for other regions. Two main types of shoreline diagrams are used.

The method of *shoreline relation* (GrØnlie, 1940; Donner, 1965; Andrews, 1970; MØller, 1987) is based on the assumption that there is a direct proportionality between the altitude of a certain raised shoreline and the altitude of other raised shorelines, within the region. The postglacial transgression-maximum shoreline (marine limit), though strongly diachronous, or the Younger Dryas level, or in Fennoscandia the Tapes or the Litorina shorelines, are often used as a regional reference. A shoreline relation diagram, in particular, which enables the age of a given shoreline to be estimated, is a graphical device on which synchronous shorelines are plotted as straight, inclined lines (Andrews et al., 1970). However, it is uncertain whether the regional postglacial shoreline displacement in many regions of Fennoscandia and Canada has been proportional through time. In addition local disturbances near ice margins (due to glacio- and hydro-isostatic processes, postglacial faulting, etc.) may have caused regional variations. Marthinussen (1960) pointed out that relation diagrams are not suitable for large regions.

Equidistant shorelines diagrams are cross-sections perpendicular to isobases, i.e. to imaginary contours between sites of equal emergence operating over the same length of time, displaying the elevation and tilt of shorelines of different ages. It is assumed with this method that 1) the directions of the isobases are known, 2) the isobases are parallel within the area, and 3) the isobases remained unchanged during the time for which the shorelines are shown. This method has been preferred by several authors (e.g. Gray, 1983; Kjemperud, 1986; Svendsen and Mangerud, 1987).

Geophysical models to predict relative sea-level changes forced by Pleistocene deglaciation, based on the mathematical analysis of the deformation of a viscoelastic earth produced by surface mass loads, were initiated in the early 1970s by the pioneering work of Walcott (1972a,b), Peltier (1974, 1976) and Cathles (1975). They were first used by Peltier and Andrews (1976), on the assumption that meltwater from ice sheets was distributed uniformly over the global ocean (Peltier, 1990). In these models the earth's interior is assumed to be radially stratified having an elastic structure with a rheology linearly viscoelastic of the "Maxwell" type in which the initial response to an applied shear stress is elastic, but the final response is viscous. Farrell and Clark (1976) constrained the meltwater distribution within and among the ocean basins mathematically in such a way that the instantaneous geoid remained an equipotential surface at all times. This enabled Clark et al. (1978) and Peltier et al. (1978) to delineate six broad zones of differential movement (Fig. 1-4), demonstrating that the rate, direction and magnitude of crustal movement must have varied from place to place and therefore that no region could be considered as vertically stable.

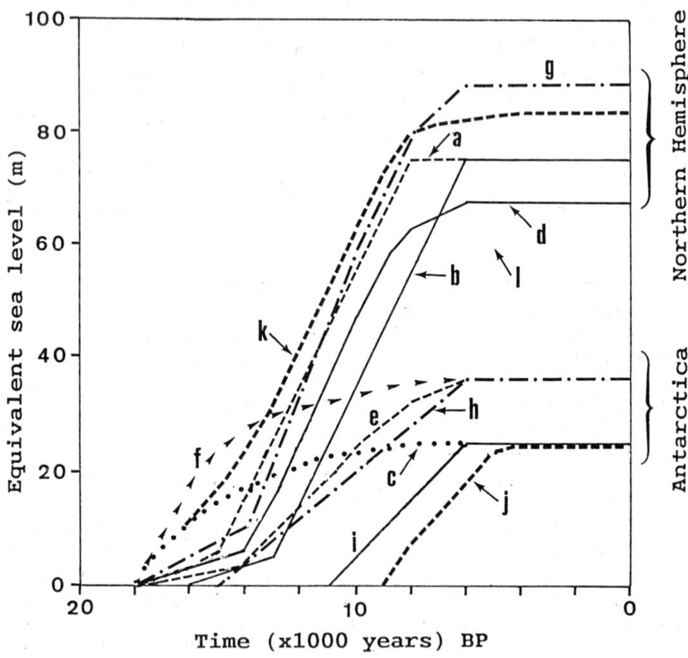

Fig. 1-27. Estimations of the global eustatic rise of sea level due to the melting of ice sheets employed by geophysical modellers. **a**: ICE-1 model (Peltier and Andrews, 1976) according to Clark et al. (1978); **b**: Clark et al. (1978); **c**: Clark and Lingle (1979), Lingle and Clark (1979); **d**: Arctic component of the ICE-2 model (Wu and Peltier, 1983) according to Nakada and Lambeck (1987) = ARC 2 model (Nakada and Lambeck, 1987); **e**: ANT 1 model (Nakada and Lambeck, 1987); **f**: ANT 2 model (Nakada and Lambeck, 1987); **g**: ARC 3 model (Nakada and Lambeck, 1989); **h**: ANT 3 model (Nakada and Lambeck, 1989); **i**: delayed Antarctic component of an ICE-2 model variant (Peltier, 1988). Contributions of the individual geographic regions to the eustatic curve for ICE-3G (Peltier, 1991) are **j** for Antarctica and **k** for Eurasia and North America.

In predictions of sea-level variations, a melting history has to be assumed for all the continental ice loads that existed at the time of the last glacial maximum. Various melting histories used in models have been summarized in Fig. 1-27. Peltier and Andrews (1976) used a first approximation model, called ICE-1, considering only deglaciation of Northern Hemisphere ice sheets (curve **a**). Clark et al. (1978) adopted model ICE-1, but the ice melting history was shifted 2000 years in time (**b**). Clark and Lingle (1979) and Lingle and Clark (1979) also took into account a partial melting of the Antarctic ice sheet (**c**), which was later tabulated by Wu and Peltier (1983). The Arctic component was adjusted to make the predicted rebound consistent with some observations of sea-level change, and tabulated by Wu and Peltier (1983) (**d**). The ICE-2 model (Wu and Peltier, 1983) consists of the combination of curves **c** and **d**. Nakada and Lambeck (1987) chose the Arctic part of ICE-2 (which they called ARC 2) and considered two possible Antarctic contributions: ANT 1 (**e**) and ANT 2 (**f**). Nakada and Lambeck (1989) added to the ICE-1 model an extra 12 m in equivalent sea-level rise from the Barents-Kara ice sheet, obtaining model ARC 3 (**g**), that they combined with a slightly modified model (ANT 3) for the Antarctic ice sheet (**h**). Peltier (1988) employed model ICE-2 (**c** + **d**) but made comparisons also using curve **i**, which differs from **c** for a 7000-yr delayed Antarctic melting. More recently, Peltier (1991) introduced model ICE-3G, in which two components are distinguished: **j** for Antarctica, and **k** for Eurasia and North America. There is considerably more ice in ICE-3G than in ICE-2, due primarily to the addition of a presumedly thick ice cover over the Barents and Kara Seas.

In Part 2, predictions by models have been compared with field sea-level data 109 times: 3 from Clark (1977), 2 from Clark (1977b), 1 from Clark (1980), 4 from Clark and Bloom (1979), 2 from Clark and Lingle (1979), 1 from Lambeck (1990), 8 from Lingle and Clark (1979), 3 from Nakada and Lambeck (1987), 12 from Nakada and Lambeck (1989), 2 from Nakiboglu et al.(1983), 40 from Peltier (1988), 7 from Peltier (1991), 13 from Quinlan (1985), 1 from Quinlan and Beaumont (1981), and 10 from Quinlan and Beaumont (1982). These examples mainly concern the east coasts of the Americas (56 sea-level predictions), Oceania (25) and Arctic Canada (23). The predictions selected in this Atlas usually correspond to sites for which a relative sea-level curve deduced from field data was also available. When various successive variants were predicted by the same author(s) at the same site, the most recent prediction was preferred. When several variants corresponding to different models were given in the same paper, the variant preferred was the one considered most reliable by the author. An exception was made for the study by Peltier (1991), which was not communicated to the present author until after all the plates in this Atlas had already been completed: among the 18 new relative sea-level curves predicted by Peltier (1991), only seven could be inserted in the corresponding plates. Most of his predictions correspond therefore to those published in a previous study (1988).

Peltier (1988) provided sea-level predictions at several locations which are far removed from the major ice sheets. Two kinds of examples have been considerd. The sea-level bands predicted in the East Carolines (**Plate 42**, band **G'**), Huon Peninsula and New Caledonia (**Plate 43, B** and **G**), Oahu (**Plate 44, R**), Wairau (**Plate 46, I**), Christchurch (**Plate 47, D**), Moruya (**Plate 49, H**),

Townsville (**Plate 50, B**), Surinam, Panama and Belize (**Plate 58, B, E** and **G**) were obtained by employing the ice models **c + d** (Fig. 1-27) (i.e. the ICE-2 model) for the curve delimiting the upper part of the band, the ice models **c + i** (i.e. with a 7000-yr delayed Antarctic melting) for the lower part and assuming that the lithospheric thickness is 196 km, the upper mantle viscosity 10^{21} Pa s and the lower mantle viscosity 2 x 10^{21} Pa s. In each of the above mentioned sites, the upper curve of the sea-level band is characterized by emergence earlier than observed and the 7000-yr delay in the melting of Antarctic ice produces a marked improvement in the fit to field data. In other locations, however, this 7000-yr delay has disastrous effects on the fit to the relative sea-level observations. In order to rectify part of the misfit induced at such sites, Peltier (1988) reduced the lithosphere thickness from 196 km to 71 km, obtaining sea-level bands which he considered would "restore the good fit to the data". Examples of this "restored fit", where the upper curve of the predicted sea-level band corresponds again to the ICE-2 melting chronology and the lower curve to a 7000-yr delay of Antarctic melting, have been included here for five sites: Cap Blanc (**Plate 28**, band **S**), Bahia Blanca, Rio de la Plata, Iguape and Recife (**Plate 56, E, F, I** and **N**).

For North American sites located more or less close to former ice sheet locations, bands of relative sea-level changes predicted by Peltier (1988) are based again on a lithospheric thickness of 196 km, using the ICE-2 model and a 7000-yr Antarctic delay. Comparisons with field data are given for Bermuda (**Plate 57**, band **H**), the NW Gulf of Mexico and the "continental shelf of the United States of America" (**Plate 59, J** and **Q**), Daytona Beach, Port Royal and South Port (**Plate 60, E, F** and **J**) Cape Charles and Bowers (**Plate 61, C** and **M**), Brigantine and Clinton (**Plate 62, E** and **N**), Boston (**Plate 63, K**), Isles of Shoals and Addison (**Plate 64, B** and **I**), St. John, N.B. (**Plate 65, D**), Halifax (**Plate 66, E**), Tignish, P.E.I. (**Plate 67, H**), Rivière du Loup et Ottawa (**Plate 68, D** and **G**), NW Newfoundland and St. George's Bay (**Plate 69, D** and **L**), Goose Bay (**Plate 70, P**), Churchill and Southampton (**Plate 72, I** and **M**).

Sea-level predictions by Nakada and Lambeck (1987) for Brigantine (**Plate 62, D**) were closer to field data using the ARC 2 ice model and a lithosphere thickness of 220 km, whereas at Boston (**Plate 63, J**) and in Newfoundland (**Plate 69, N**), again using the ARC 2 ice model, results were improved with a lithospheric thickness of 50 km.

Predictions reported by Nakada and Lambeck (1989), concern several far-field sites in Oceania, which lie at considerable distance from both the Arctic and the Antarctic ice sheets (see **Plates 44-50**). These predictions are based on the ice models ARC 3 + ANT 3 and on a mantle model characterized by a lithospheric thickness of 50 km and a lower mantle viscosity of 10^{22} Pa s. The predicted sea-level bands obtained correspond to variations in the upper mantle viscosity between 10^{20} Pa s (lower curve of the bands) and 2 x 10^{20} Pa s (upper curve). When the band obtained was very narrow, only one sea-level curve has been reproduced.

Sea-level predictions by Peltier (1991), using the ICE-3G model (i.e. curves **j + k + l**, Fig. 1-27) have been included for four North American sites [Victoria BC. (**Plate 52**), Boston (**Plate 63**), Churchill Man. (**Plate 72**) and Alert Ell. (**Plate 76**)], one site in

North Western Europe [(Stockholm (**Plate 9**)] and two far field sites [Reunion (**Plate 30**) and Wairau Valley NZ. (**Plate 46**)]. Predictions of shoreline positions in Antarctica by Lingle and Clark (1979) and in the Laurentide Ice Sheet area by Quinlan and Beaumont (1981, 1982) and Quinlan (1985) are commented in Part 2.

AIMS OF THIS ATLAS

This study has been undertaken to assess geological trends in sea level during the Holocene and to summarize the state of the art in the field, with a view to producing a reference compilation of Holocene sea-level studies. In order to compare different trends in various regions of the earth and confront them with present-day trends, over 900 curves of relative sea-level change already published for all areas have been assembled and redrawn at a uniform scale in 77 regional plates. Whenever possible, an explanatory text analyses very briefly for each curve the type of data and main assumptions adopted; in some cases, comments contain criticism.

As there is a great number of publications dealing, at least in part, with sea-level changes during the Holocene, some selection has been necessary. Isolated observations, though in some cases useful when deducing local trends in vertical movements, have generally been disregarded; it was simply not possible to collect all these data, in all languages, from all regions and to localize and plot them in a reasonably limited number of graphs. Such a collection would indeed be very useful, but more appropriate to a data bank than to an atlas and would require the international cooperation of many motivated regional specialists, far beyond the scope of a European research project. Several studies including various sea-level data had also to be discarded when the data were not represented graphically, or were just plotted in a graph as index points without any indication of error bars, or of an uncertainty band specifying the expected former sea-level positions in relation to the samples and the inferred trends.

Local relative sea-level curves found in the literature were assembled systematically according to geographical regions. Although index points with error bars are required to assess the reliability of a sea-level curve, a simple superimposition of several sea-level curves at the same scale in the same graph, as in this Atlas, is a necessary first step in the analysis and a very effective way of detecting regional differences directly. These may also appear clearly without error bars, if one considers each curve as the centre of an uncertainty band of unknown but not unlimited width. The next step, often outlined in the annotated text, is to describe and interpret the differences observed, taking into account possible causes of deviation, and in some cases of misinterpretation. An extensive bibliography (over 750 references) and two indexes are intended to help the reader in further related studies, e.g. the preparation of more detailed regional surveys, palaeogeographic maps, the testing of new geophysical models, the comparison of recent geological trends with present trends deduced from tide-gauge records and with near-future trends predicted by climate models, the inference of geological, climatic, and other correlations, etc.

The present author apologizes if, during the preparation of this Atlas, some important publications with reliable sea-level curves have unintentionally been neglected or some ideas distorted. His excuse is that he could visit only a very small number of the localities mentioned, that his linguistic knowledge is limited and that it has been materially impossible to consult all important studies made on all the coasts throughout the world.

Sea-level curves have been reduced to the same scale using a special laser photocopy machine, in order to minimize graphic distortions. This does not preclude the possibility of subsequent drawing errors for which the author accepts full responsibility. It is hoped that this compilation will be useful.

To make a preliminary comparison easier between long-term Holocene trends and recent sea-level tendencies shown by tide-gauge records, eight regional graphs, taken from an earlier publication now out of print (Pirazzoli 1986b), have been reproduced in this Atlas (Fig. 2-3, 2-9, 2-17, 2-23, 2-25, 2-31, 2-32, 2-41).

HOW TO CONSULT THE ATLAS

Part 2 consists of 77 regional plates and an accompanying text of comments. The order in which the plates are arranged follows the shoreline, from Northern Europe to the Mediterranean, Africa, the Indian Ocean, East Asia, Oceania, the west coasts of America, Antarctica, the east coasts of America, Arctic Canada, and Greenland. Rapid access to the plates of interest and the corresponding annotated text is possible by following the above geographical order, or by consulting the locality or author indexes. Regarding geographical names, local spellings have usually been preferred (e.g. fjord in Norway, but fiord in Canada; Chebogue Harbour, but New York Harbor).

Each plate contains 4 to 20 sea-level curves or bands (on average a dozen curves or bands), most of them drawn to the same scale. The X-axis represents ages expressed in years Before Present (BP), i.e. in uncalibrated radiocarbon years. Only in a few cases (e.g. in the middle graph of Plate 15, or in the lower graph of Plate 17) is the time scale expressed in calendar years (B.C. and A.D.). In exceptional cases (indicated in the text), sea-level curves represented in calendar years had to be plotted together with curves presented in radiocarbon years, in the same graph. In these (fortunately rare) cases, visual comparisons between curves must be done cautiously, a graphic deviation being introduced by the unhomogeneous time-scale representation. The Y-axis represents elevations in metres above and/or below sea level.

In the plates, each local sea-level history is summarized by curves or by uncertainty bands - sometimes by error boxes of samples - and identified by a capital letter. The same letter is used to specify the area of study in a location map and to refer to the author name(s) and year of publication corresponding to the full bibliographic reference, which is given in Part 4. When a sea-level curve is related to a local datum different from MSL, this is specified in the reference list, where the following abbreviations have been used (in Plates 15-23, 30-31, 34, 43, 50, 52 and 61-67) (for a more complete list of definitions of tide levels, see Van de Plassche, 1986a):

HAT: highest astronomical tide;
MHWST: mean high water spring tide;
MHW: mean high water;
MLW: mean low water;
MLWST: mean low water spring tide.

The field datum HHW [higher high water, often used in Atlantic Canada, defined by Brookes et al. (1985) as "the lowest occurrence of *Iris versicolor*"] and the French "le niveau des plus hautes mers" are probably very close to HAT. Relative sea-level histories predicted by models are marked "predict." in the reference list.

A comparison between sea-level curves referring to different tidal data in the same graph may introduce inaccurate visual appreciations, depending on the possible occurrence of palaeotidal changes, the amount of which is generally unknown. Another deviation in altitudinal comparisons may occur when a local sea-level curve is referrred to a datum other than MSL, but the local elevation of this datum in relation to present MSL is not given by the authors. In addition, certain authors neglect to specify to which "sea level" they are referring and this is especially regrettable in macrotidal areas. In order to help the reader to estimate the range of these kinds of uncertainties, significant values of the local spring tidal range are given (in metres) in the location maps of the plates.

PART 2: REGIONAL SEA-LEVEL CHANGES

SVALBARD - FRANZ JOSEPH LAND (Plate 1)

The Svalbard archipelago lies at the northwestern margin of the shallow Barents Shelf, to the north of Norway and northwest Russia. Franz Joseph Land lies on the same shelf to the northeast. The Quaternary evolution of the Svalbard area was summarized by Héquette (1988).

Hughes et al. (1977) postulated a Barents ice sheet at 18,000 yr BP, confluent with the Scandinavian ice sheet. This view, supported by several authors, was suggested by the 6500 yr BP isobase, which rises to the east over Spitsbergen and to the west over Franz Joseph Land. As shown by Boulton (1979) however, an analysis of uplift curves from the Svalbard archipelago indicates that the strongest early Holocene uplift occurred over northern Spitsbergen and eastern Nordhaustlandet, falling eastwards and westwards, and that the centre of the uplift migrated to the southeast during the Holocene. Stratigraphic evidence suggests the absence of an important glacial event in 18,000-20,000 yr BP. Furthermore, no evidence for a Barents Sea ice sheet has been found on the continental shelf of the Arctic Sea (Biryukov et al., 1988). Direct evidence of glacier fluctuation on the other hand indicates an important stage of glaciation at about 11,000 to 10,000 yr BP, much later than in Scandinavia, with dominant ice masses centred over the eastern part of the archipelago (Boulton, 1979). The decay of this ice mass could be primarily responsible for the pattern of the Holocene uplift which can be observed in the field.

The sea-level indicators used in this Arctic area are often whale bones, which are assumed to correspond to high-tide or storm deposits. This explains why some sea-level curves of **Plate 1** (**I**, **H**, **D**)reach their present level above zero.

The largest rates of a relative sea-level drop were reported by Salvingsen (1981) from the Kongsøya area, in the eastern part of the archipelago (curve **I**), where the isostatic recovery totally obliterates the eustatic sea-level rise and the land is still rising today at the rate of 3 mm/yr. The rates decrease rapidly however towards the west. Near Svartknausflya (curve **H**), a striking reduction in the rate of emergence appears between 7500 and 5500 yr BP, corresponding to the Tapes transgression in Scandinavia, followed by a remarkable increase between 5500 and 3500 yr BP (Salvingsen, 1977). If a similar pattern is not evident in curve **C**, proposed by Feyling-Hanssen and Olsson (1959) for the central part of Spitsbergen, this is probably because no radiocarbon date midway between 7400 and 4000 yr BP was available to these authors. On the north coast of Spitsbergen, a relative balance between eustasy and isostasy has been reached since about 6000 yr BP, as Salvingsen and Österholm (1982) in curves **F** and **G**. On the east cost, curves **B** (Landvik et al., 1987, 1988), **D** (Forman et al., 1987) and **E** (Héquette and Ruz, 1989) suggest the occurrence of one or two submergence episodes; curves **B** and **D** place an emergence phase around 6000 or 5000 yr BP, on the basis of whale bones of this age found behind the modern storm beaches; according to Héquette (1988) and Héquette and Ruz (1989), on the

Plate 1: SVALBARD -FRANZ JOSEPH LAND

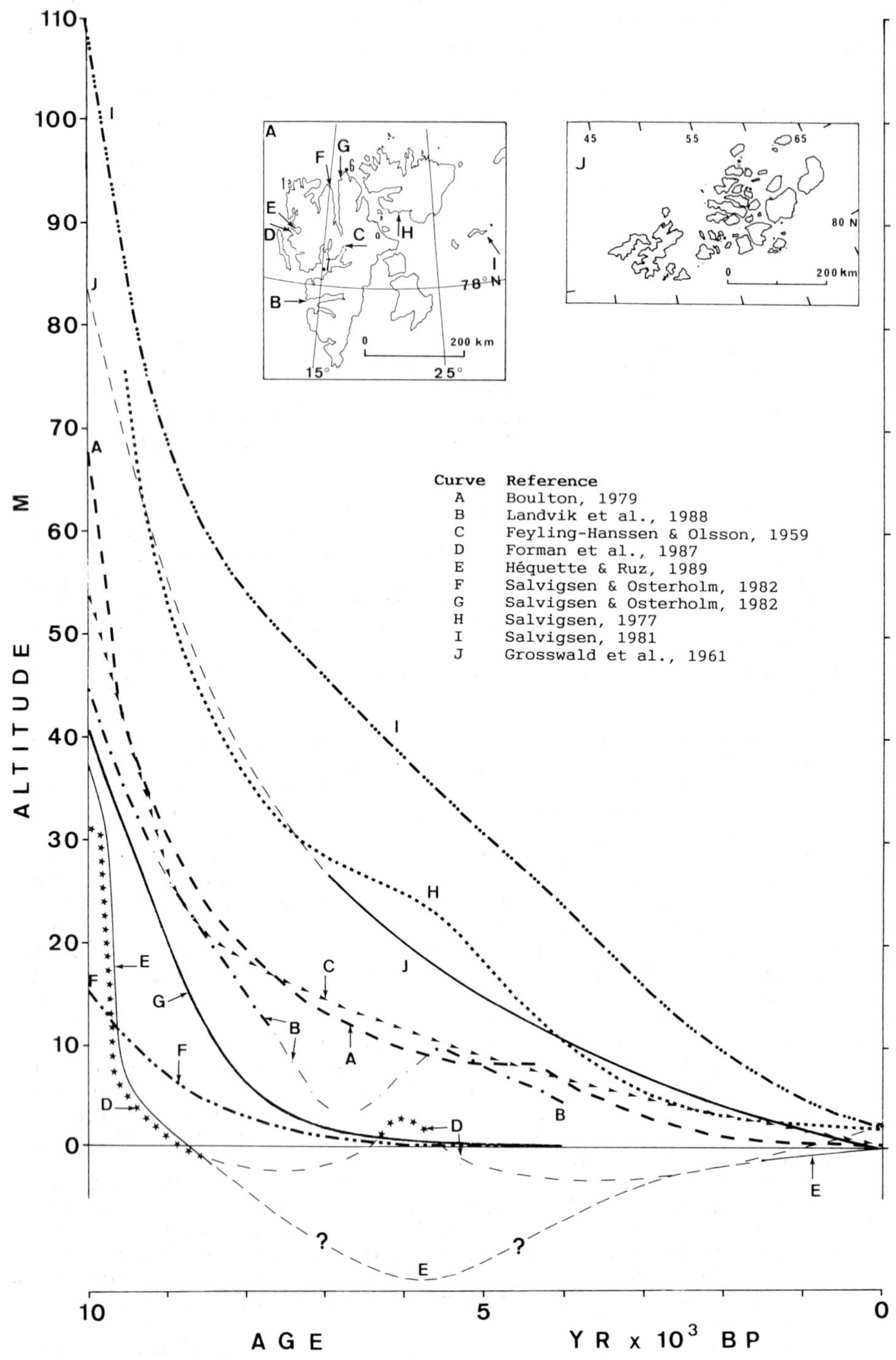

other hand, the sea level of 6000 yr BP could have been at -15 m, as suggested by the occurrence of various submerged features, including beach pebbles forming a bar at -15 m, which are however not dated. Nevertheless evidence of recent ongoing transgression is frequently reported from the east coast of Spitsbergen.

Finally curve **A** is considered by Boulton (1979) to represent a general pattern for the archipelago. For the sake of comparison, relative sea-level curve **J**, by Grosswald et al. (1961), shows a similar pattern; curve **J** was reproduced from Badyukov and Kaplin (1979), who however did not specify the locations investigated, the type of data used nor the complete bibliographic reference.

FENNOSCANDIA: INTRODUCTION

During the Holocene period, the geologic history of Fennoscandia and of the south Baltic areas has been controlled mainly by the effects of a deglaciation phase. During the last glaciation (locally called the Weichselian stage) a thick ice sheet, probably extending across the North Sea as far as Scotland, covered the whole area.

At the beginning of the Holocene, in spite of the short-lived interruption of the Younger Dryas, the deglaciation processes were very rapid. They were to lead to the total disappearence of the ice sheet around 7500 yr BP, when the last parts of northern Sweden were uncovered near the centre of the former ice sheet. The main effects of this melting ice on the relative sea level have been a contribution to the global eustatic rise, with an amount of the order of 40 m, and locally, of much greater vertical isostatic movements.

Since the last glaciation period four main stages have been recognized in the Baltic area (Hyvärinen and Eronen, 1979; Fredén, 1979). The stage of the Baltic Ice Lake and that of the Yoldia Sea are dated about 10,500-10,200 yr BP and 10,000 yr BP respectively (Fig. 2-1). In the early Holocene, the Ancylus Lake stage (named after the mollusc *Ancylus fluviatilis*), corresponds to a damned up freshwater basin, at a level that may have been higher than in the ocean. The transition between the Yoldia and Ancylus stages occurred when the saline water intrusion into the Baltic basin was stopped, near Degerfors, by land uplift and its development became dependent on the evolution of the southern part of the Baltic: due to land uplift there was a regression phase in the north and a transgression in the south. Dates for the Ancylus transgression range between 9500 and 8800 BP and the stage ended when the bathymetric conditions in the Danish straits and Öresund allowed a new connection to be established with ocean water entering the Baltic basin once again. This was the Litorina stage, which is still ongoing, so called because of the gastropod *Littorina littorea* found there. The Tapes transgression described in some areas of Norway corresponds to the Litorina transgression in the Baltic. Transitional brackish conditions between the freshwater Ancylus stage and the warmer and saltier Litorina stage, called the Mastogloia phase owing to the presence of the brackish-water diatom *Mastogloia smithii*, have been recognized by several authors. This phase is metachronous to some extent, depending on the hydrographical conditions in relation to the land uplift and the bathymetry. North of Stockholm, the Mastogloia phase

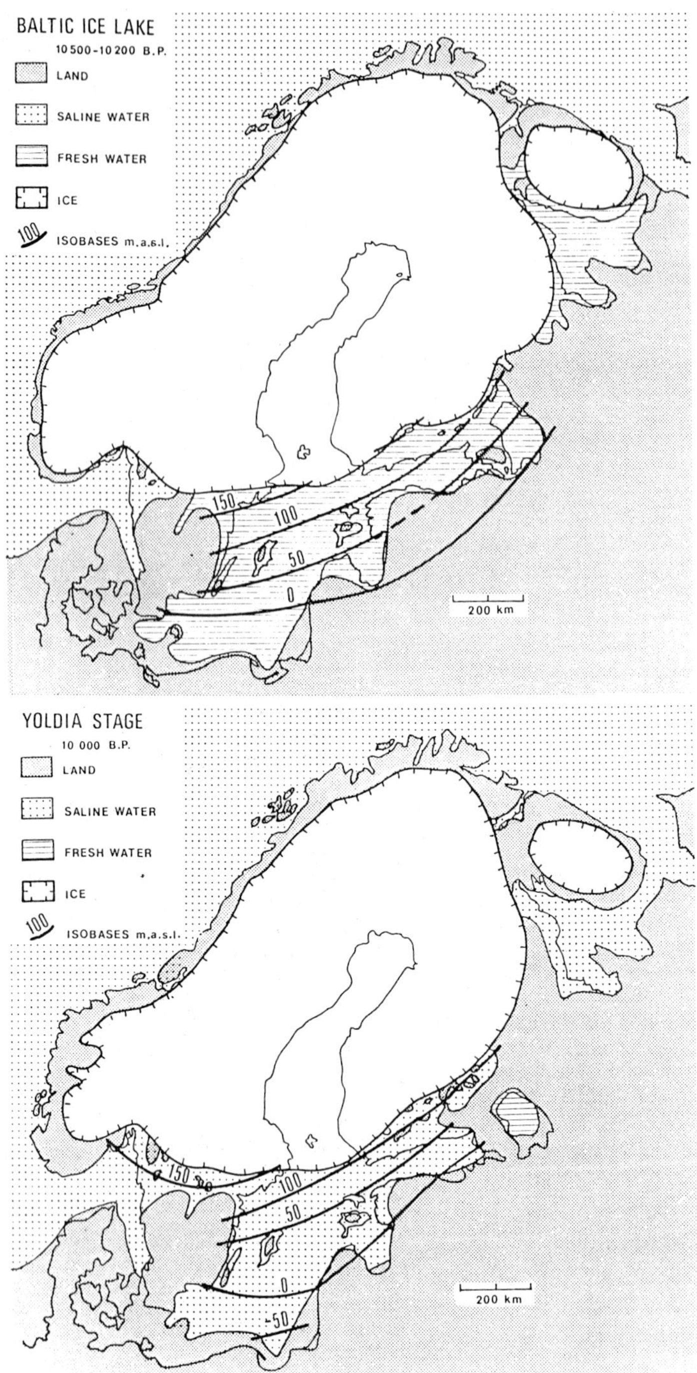

Fig. 2-1. The Baltic Ice Lake and the Yoldia Sea stages, from Eronen (1983).

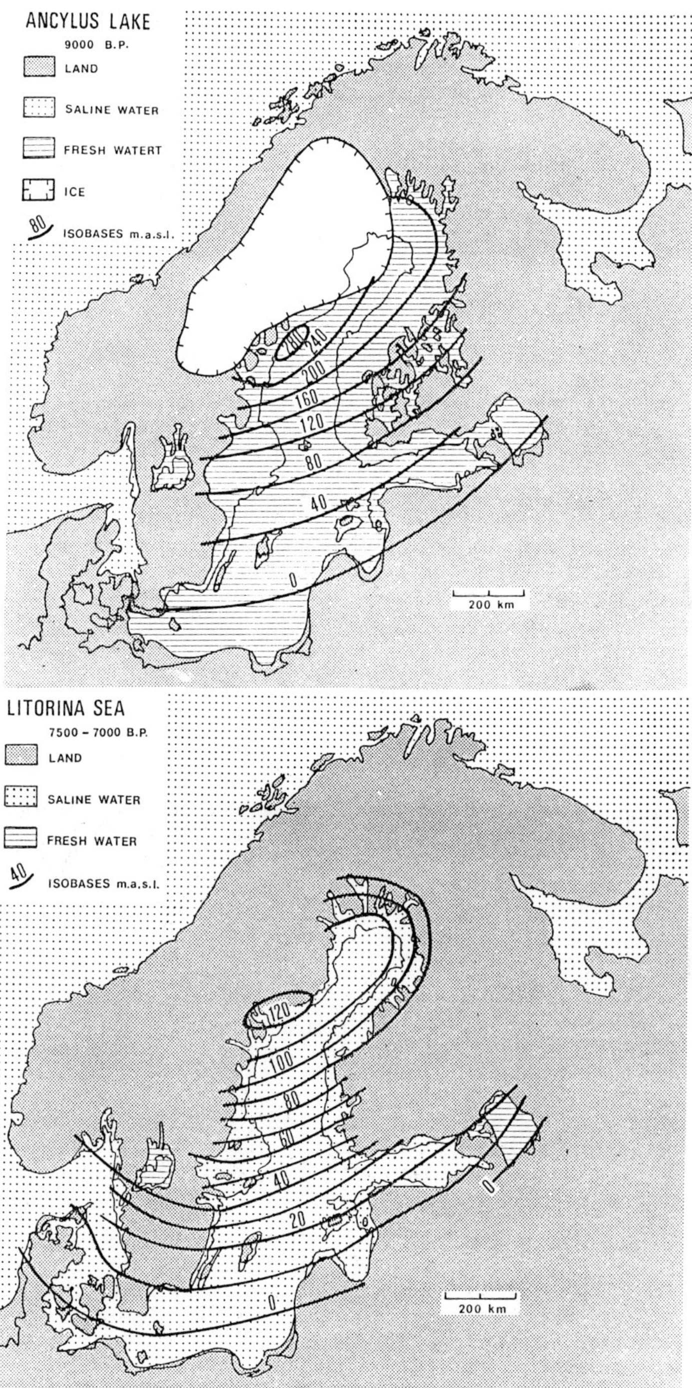

Fig. 2-2. The Ancylus Lake and the Litorina Sea stages, from Eronen (1983).

corresponds to the salt water limit, also called Clypeus limit (after the brackish-water diatom *Campylodiscus clypeus*), which denotes the beginning of the Litorina stage, ca. 7400-7300 yr BP. This limit reaches a height of slightly over 120 m above present sea level on the western side of the Gulf of Bothnia.

A number of transgressions of the Litorina Sea have been recognized by several authors: e.g. Mörner (1969) at Halland, Berglund (1971) in southern Sweden, Digerfeldt (1972) at Öresund. In other areas, however, only one or two transgressions are recorded and Eronen (1974) reaches the conclusion that only one major transgression is identifiable in the shoreline displacement pattern for the whole Baltic basin during the Litorina period.

Due to the impressive shoreline displacements experienced during historical times, Nordic countries have a long tradition and specific methods for the study of relative sea-level changes. The development of the main ideas and results on postglacial uplift in Fennoscandia have been summarized by Mörner (1979b).

The principal method used to identify former sea level positions is related to the study of isolation processes in lake basins, which have been abandoned by the sea as a consequence of the isostatic uplift of the land. Thick sequences of organic deposits frequently occur in these lakes. Diatom analyses of such sequences have been made since the 1920's and have been supplemented by pollen analyses and later by radiocarbon datings.

Uplifted marginal or sandur deltas are related to the position of the water level beyond the retreating ice margin at the time they were active. In favourable areas, the retreat of the sea may be tied to the counting of annual varves.

Other methods of study consist in morphological observations of raised beaches and fossil erosion scarps, in the determination of the highest limit of wave washing on till slopes, in the archaeological excavation of dwelling sites of Stone Age fishing populations, etc.

When presenting the results of shoreline displacement studies, diagrams have been used to show the effect of the tilting of shoreline surfaces away from an uplift centre. In "relation diagrams", points are measured from a baseline which runs parallel to an uplift isobase, which is usually the highest Litorina shoreline. In "distance diagrams" (also called "equidistant diagrams"), the distance between the observations is calculated at right angles to the trend of the isobases. When a relation or a distance diagram is used for a large area, it is assumed that the relationship between the altitudes of the shorelines is constant in the whole area.

With such techniques, a few new sea-level data may be sufficient to construct new shore displacement curves, on the assumptions however that: (1) the shorelines are straight lines, (2) the earth's crust has risen without any marked flexures, and (3) the rate of crustal upwarping has decreased continuously with regard to time. These simplifying assumptions thus introduce methodological inaccuracies which are increased by the fact that former shorelines are metachronous in nature in an area of dome-shaped uplift. Fortunately the vertical displacements to be measured are in most cases greater than the methodological ranges in uncertainty.

NORWAY (Plates 2-6)

Plate 2 summarizes the main results obtained by two important papers, published by Hafsten (1983), for the southern part of Norway, and by MØller (1989) for the Nordland and Finnmark regions. Hafsten (1983) has compared 15 relative sea-level curves, 14 of which concern Holocene periods of time, for the area extending from Oslo to 65° N in latitude. They are numbered **7 to 20** in **Plate 2.** The review by Hafsten is based on results obtained primarily from bio-lithostratigraphical investigations of bog and lake deposits, to provide transgression-regression sequences, and from chronological pollen data adjusted with the help of radiometric dates. In diagrams **7 to 20**, the periods during which the regions were covered either by ice or by sea are indicated by shading.

The highest recorded relative sea level appears to be located at much greater heights in the eastern areas (curves **9** and **15-20**, Plate 2) than in the western and extreme southern areas. According to Hafsten (1983), the highest marine level in Norway is recorded in the Oslo area, at over 220 m in altitude. In the same area, the maximum altitude of the 10,000 yr BP shoreline is about 140 m, slightly less than in the Trondheim Fjord (e.g. curve **D** in **Plate 5**).

Shore-level displacement curves appear distinctly different in the extreme western coast regions and in the eastern fjord districts. In the eastern fjords (Oslo, Trondheim), the postglacial isostatic uplift brought about an uninterrupted recession of the shore level, although the rate of recession slowed down gradually (the present-day rates at Oslo, S. Østfold and Frosta, are of the order of 3-4 mm/yr). Here the strong postglacial isostatic recovery has prevented a eustatic submergence. In the western coastal area of South Norway, on the other hand, where glacial depression and the subsequent isostatic uplift were quite moderate, the eustatic sea-level rise exceeded the isostatic recovery for a long time. The transgression peak reached a maximum of 11-12 m near BØmlo and Sotra.

MØller (1989) combined sea-level data from sixty localities, from three local relative sea-level curves and seven equidistant shoreline diagrams, to obtain the best three-dimensional trend surface for fitting Northern Norway. This model proposes a traditional shore-level displacement curve at any particular geographic location, e.g. the curves **1-6** reproduced in **Plate 2.**

NORTH NORWAY - KOLA PENINSULA (Plate 3)

Donner et al. (1977) have proposed four curves for the Finnmark region: **A**, **B** and **C**, for the Varanger Fjord area, and **D**, for the outer coast of the Varanger Peninsula and the Porsanger Fjord; the association of inner fjord and outer coast data in the same curve (**D**) may however result in an unreliable mixture of different trends in isostatic recovery. The dates of curves **A** and **D** have

Plate 2: NORWAY (General)

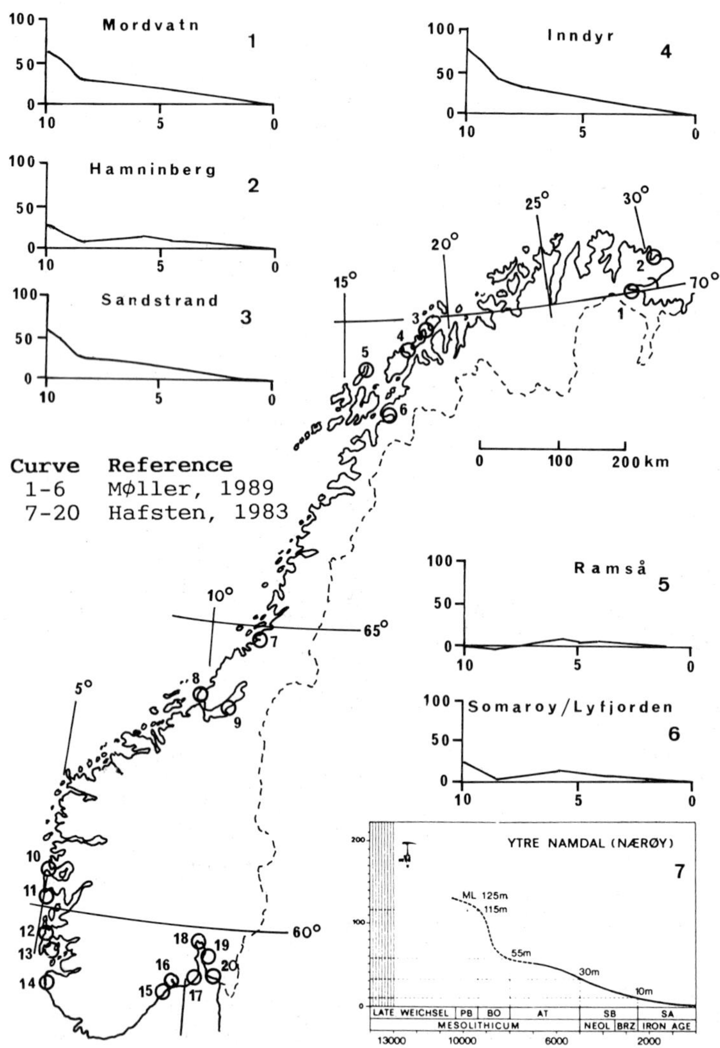

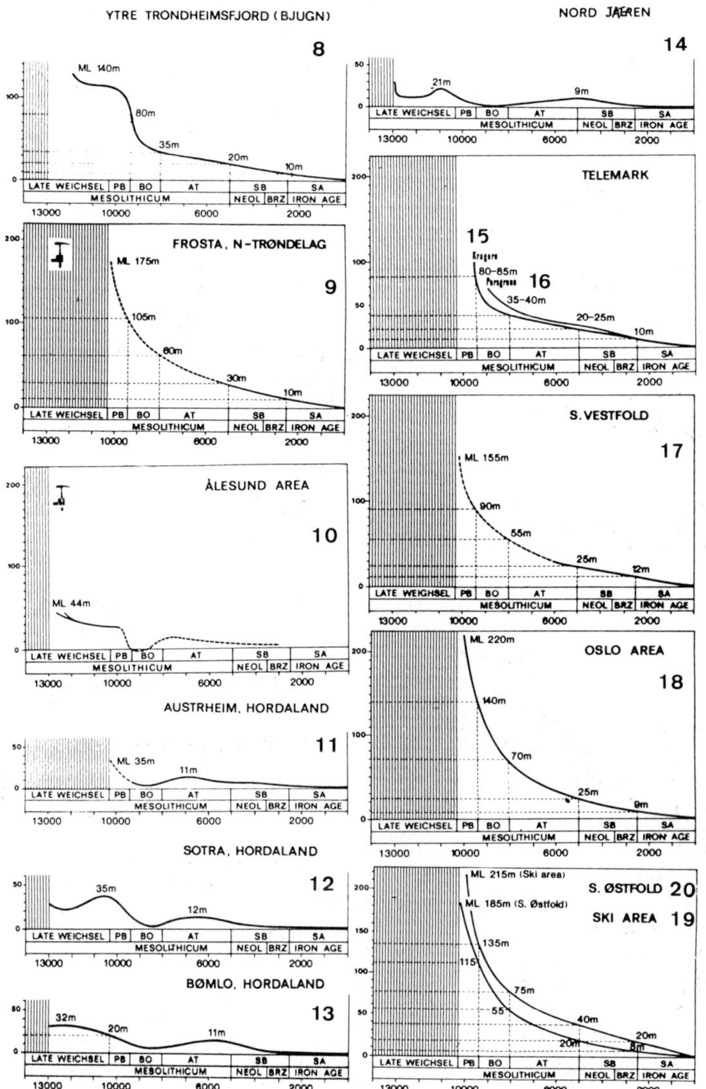
YTRE TRONDHEIMSFJORD (BJUGN)
8
ML 140m
80m
35m
20m
10m
FROSTA, N-TRØNDELAG
9
ML 175m
105m
60m
30m
10m
ÅLESUND AREA
10
ML 44m
AUSTRHEIM, HORDALAND
11
ML 35m
11m
SOTRA, HORDALAND
12
35m
12m
BØMLO, HORDALAND
13
32m
20m
11m
NORD JÆREN
14
21m
9m
TELEMARK
15
80–85m
16
35–40m
20–25m
10m
S. VESTFOLD
17
ML 155m
90m
55m
25m
12m
OSLO AREA
18
ML 220m
140m
70m
25m
9m
S. ØSTFOLD 20
SKI AREA 19
ML 215m (Ski area)
ML 185m (S. Østfold)
135m
115
75m
55
40m
20m
20m
LATE WEICHSEL
PB
BO
AT
SB
SA
MESOLITHICUM
NEOL
BRZ
IRON AGE
13000
10000
6000
2000

Plate 3: NORWAY I (North) - KOLA PENINSULA

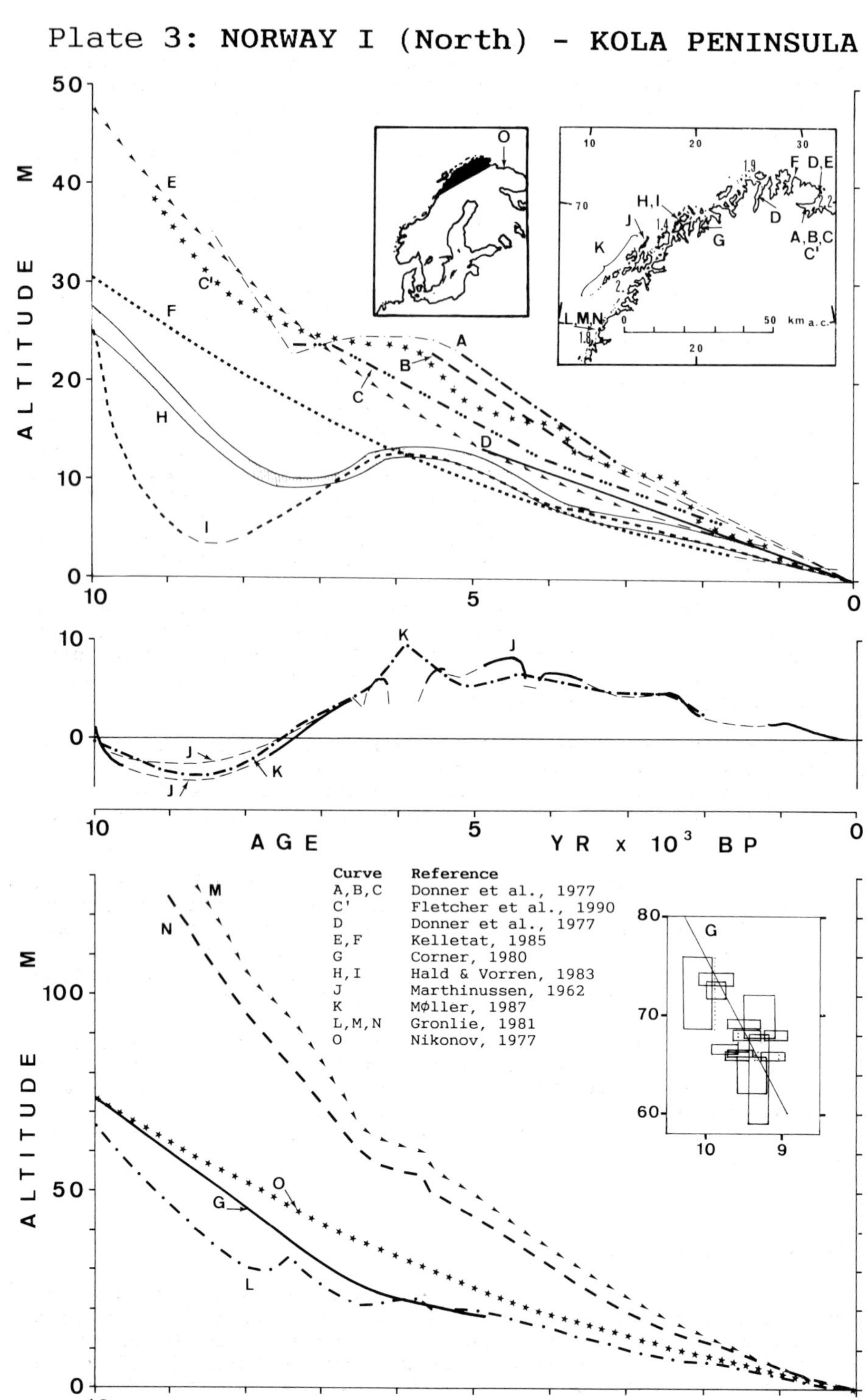

been deduced from raised beach shells, those of **B** from charcoal collected from archaeological sites. Elevations have been measured in relation to the *Balanus balanoides* line, which coincides approximately with the upper limit of *Fucus vesiculosus*, situated between the half-tide and the mean high-tide levels, which is often used as an altitudinal datum on the Norvegian coasts. Curve **C** is based on the same data as curve **B**, but after calibration of the radiocarbon ages. The difference between curves **B** and **C** (or **C'**)illustrates clearly the need to know in detail the data used to construct a sea-level curve before attempting its interpretation.

For the sake of comparison, curve **O** for the coasts of Kola Bay has been reproduced from Badyukov and Kaplin (1979) who quote (without giving the bibliographic reference) Nikonov (1977).

Curves **E** and **F** were established by Kelletat (1985) for the outer parts of the Varanger Peninsula, after compilation of data already published by other authors. Curve **G** (Corner, 1980) concerns the Tromsø region, northeast of the Lofoten Islands; it is based on morphological marks of the marine limit and on an equidistant shoreline diagram proposed for the region by Marthinussen (1960). The vertical scale has been reduced in the lowest graph of **Plate 3**. Details on the uncertainty limits of the data used to construct curve **G** are given in the inset for the period 10,000-9000 yr BP.

Curves **H** and **I** (Hald and Vorren, 1983) have been constructed for the Lyfjord region, NW of Tromsø. **H** has been obtained by compiling already published data; **I** is a variant based on new local data which suggest the occurrence of a wide fluctuation in the relative sea level during the early Holocene.

Curve **J** (Marthinussen, 1962) is composite, with data from Andøy Island and from Finnmark and Tromsø sites. A marked relative sea-level oscillation, deduced from alternations of peat layers and marine deposits, may have been emphasized by peat compaction phenomena.

In the Lofoten and Vesterålen, Møller (1987) obtained a shoreline displacement curve (**K**) related to the present MTL for the period 10,000 to 2000 yr BP. This curve is also composite, the main assumption used by Møller being that within one region there is direct proportionality between the altitude of a certain raised shoreline and the altitude of other raised shorelines.

Grønlie (1981) studied the postglacial isostatic rebound across a geotraverse going from the Baltic to the Arctic Ocean shores. In Norway, near the Arctic Circle, three shore-level displacement curves have been provided: at Messingslett (**M**), Mo i Rana (**N**) and Traena Island (**L**) respectively. They show a decreasing isostatic rebound towards the Ocean.

CENTRAL NORWAY (Plate 4)

Thirteen relative sea-level curves from central Norway are compared at a reduced scale in **Plate 4.** The curve **A** from Ytre Namdal, published by Hafsten (1983), is the same as in Plate 2. Several curves were provided by Svendsen and Mangerud (1987) in a regional review. Other major contributions are due to Kjemperud (1981, 1986), especially in the Trondheimsfjord area. Most

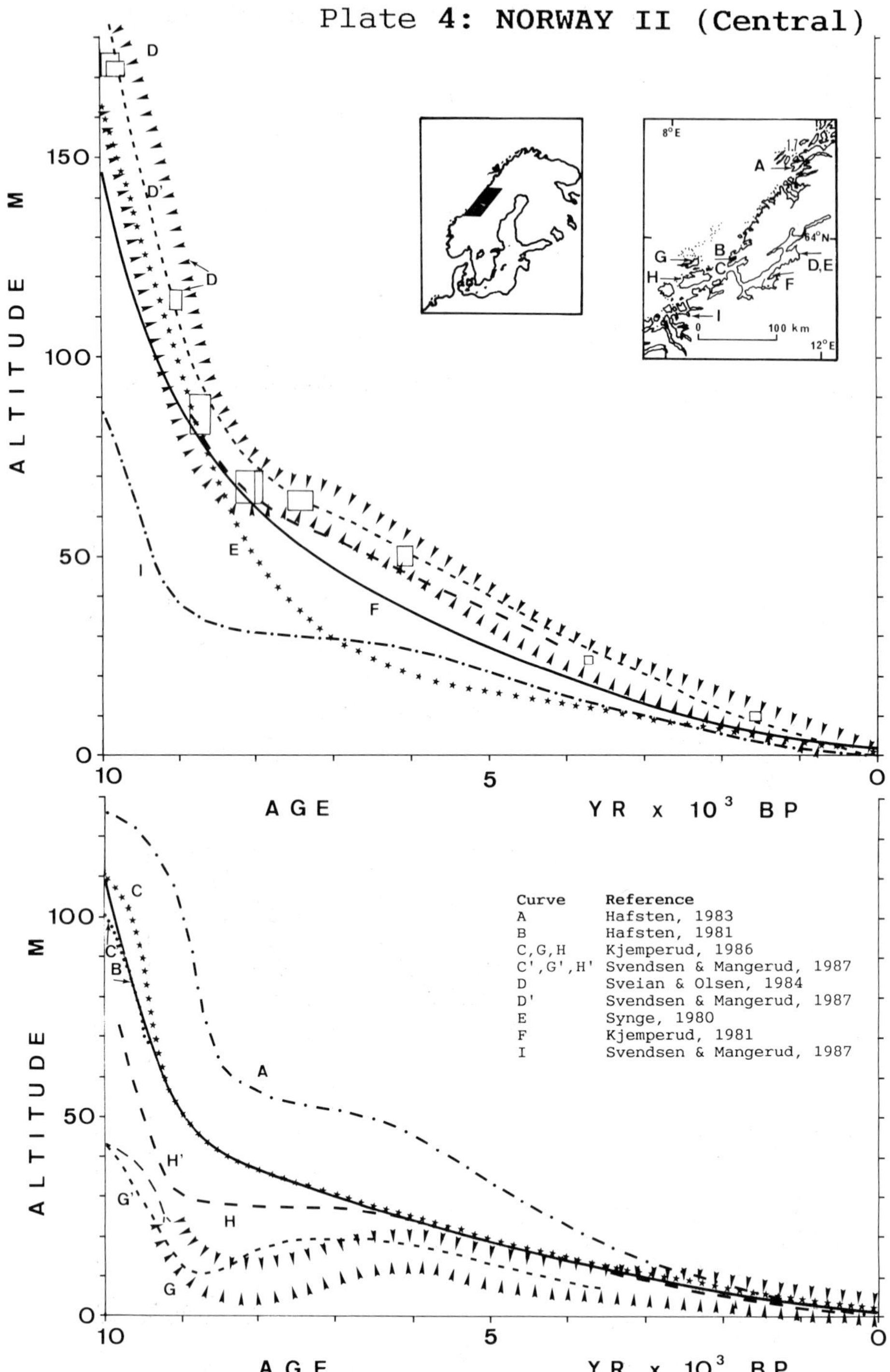
Plate 4: NORWAY II (Central)
ALTITUDE M
150
100
50
0
D
D'
D
E
F
I
10
5
0
AGE
YR x 10³ BP
8°E
A
G
B
C
H
64°N
D,E
F
I
0
100 km
12°E
ALTITUDE M
100
50
0
C
C'
B
A
H'
G'
H
G
10
5
0
AGE
YR x 10³ BP
Curve Reference
A Hafsten, 1983
B Hafsten, 1981
C,G,H Kjemperud, 1986
C',G',H' Svendsen & Mangerud, 1987
D Sveian & Olsen, 1984
D' Svendsen & Mangerud, 1987
E Synge, 1980
F Kjemperud, 1981
I Svendsen & Mangerud, 1987

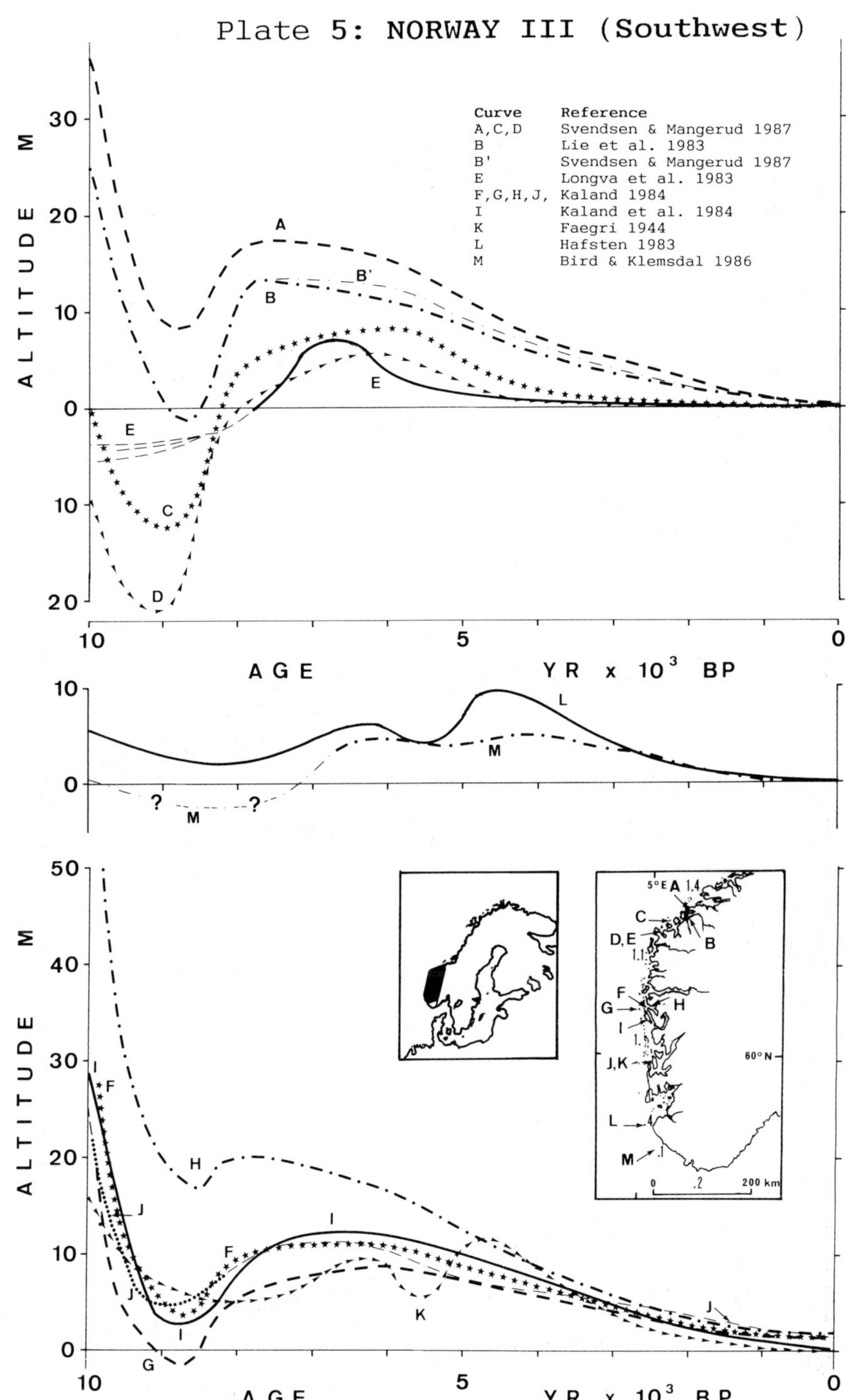
Plate 5: NORWAY III (Southwest)
Curve Reference
A,C,D Svendsen & Mangerud 1987
B Lie et al. 1983
B' Svendsen & Mangerud 1987
E Longva et al. 1983
F,G,H,J, Kaland 1984
I Kaland et al. 1984
K Faegri 1944
L Hafsten 1983
M Bird & Klemsdal 1986
ALTITUDE M
AGE
YR x 10³ BP
5°E
60°N
200 km

striking in Plate 4 is the difference between the inner fjord curves (e.g. from Verdalsøra (**D**, **D'**, **E**) or Frosta (**F**), where the shore-level displacements have been extremely rapid, expecially in the early Holocene, and the relatively moderate uplift movements observed in some outer coast localities, where a long relative sea-level standstill can be observed in Hitra (curves **H**, **H'**) or Tingvoll (**I**), or even a transgressive episode in Frøya (curves **G**, **G'**). The curves from Bjugn (**B**, **C**, **C'**), show obviously intermediate trends between those of the preceding areas.

SOUTHWEST NORWAY (Plate 5)

In the Sunnmøre region, Svendsen and Mangerud (1987) published relative sea-level curves for Stette (**A**), Sula (**B**), Leinøy (**C**) and Borgundvåg (**D**). Variants for Sula (**B'**, from Lie et al., 1983) and Borgundvåg (**E**, from Longva et al., 1983) have also been reproduced in **Plate 5**. Svendsen and Mangerud (1987) consider the Sula curve (**B**) as "the only Holocene curve from the Sunnmøre that is reasonably well established", though it is based on a single date after 7000 yr BP.

In the Bergen district Kaland (1984) provided shore displacement curves at Fonnes (**F**) and Bømlo (**J**) and proposed a model for constructing an equidistant shoreline diagram, which has been used to obtain curves calculated at Ostereidet (**H**) and Feidje (**G**). The Bømlo curve **J**, also reproduced as graph **13** in **Plate 2**, is a revision of curve **K** by Faegri (1944), but its double oscillation was not confirmed by Kaland.

The shore displacement curve for the island of Sotra (**I**) reproduced from Kaland et al. (1984), and that for Nord Jæren (**L**) from Hafsten (1983), correspond to the graphs 12 and 14 in **Plate 2**. Lastly, curve **M** was deduced for the coastal zone of Brusand by Bird and Klemsdal (1986) from local stratigraphical information and comparisons with relative sea-level changes in neighbouring localities.

In conclusion, the southwest coast of Norway shows, during the Holocene, a similar relative sea-level history, but with rates of change and oscillations amplitudes decreasing from North to South: first a rapid sea-level drop until about 9000 yr BP, then a transgression between 9000 and 8000 yr BP and finally a gradual regression at decreasing rates since 6000 yr BP, which is still underway.

OSLOFJORD (Plate 6)

In the outer part of Oslofjord, no evidence has been found of the early Holocene transgression sequence observed along the southwest coast of Norway, not even on the outer coast of the Trondheimsfjord. In the Telemark county, Stabell (1980) used diatom analysis to construct two radiocarbon dated relative sea-level curves for the Kragerø (**A**) and Porsgrunn (**B**) areas, which are reproduced at reduced vertical scale in **Plate 6**. The first part of curve **A**, which appears too steep to be continued back without imposing a stage of transgression or syngression, may however contain some errors in age determination. Indeed the

Plate 6: NORWAY IV (Oslofjord)

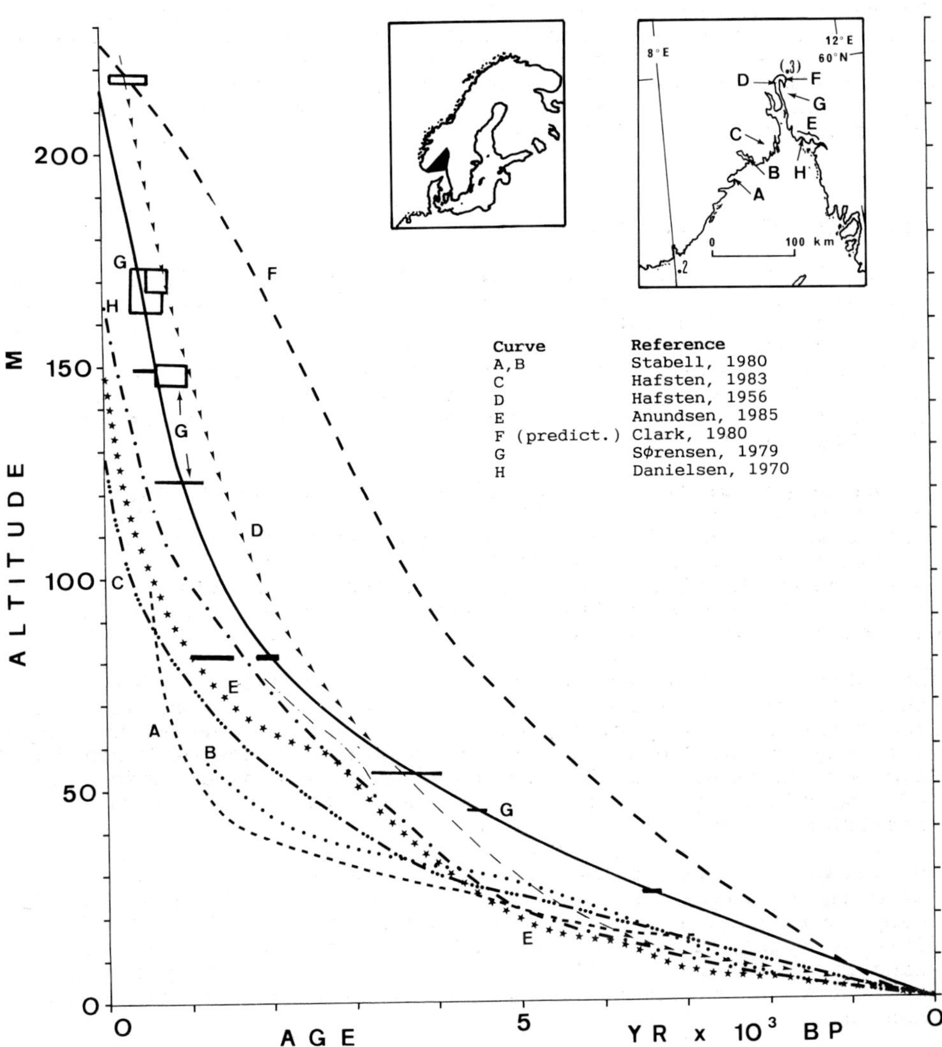

nearest marks of the Tapes transgression along the south coast of Norway are reported near Kristiansand, over 100 km SW of Kragerø. Curve **C** reported by Hafsten (1983) for South Vestfold is the same as graph **17** in **Plate 2.**

For the inner fjord area, curve **D** by Hafsten (1956) is based mostly on diatom investigations and pollen dating. The main conclusions of this study are still valid, although no radiodarbon dates were provided. Curve **F** shows the emergence predicted by Clark (1980) in the same area, using a numerical model. A comparison between **D** and **F** shows that even if steep regression trends appear in both the curves, the differences in elevation are of the order of 35-65 m, and this is certainly not negligible when the model has to be used for predictions of relative sea-level changes in localities where sufficient field data are not available.

In the Ski area (**G**), Sørensen (1979) constructed a detailed sea-level curve, which is reproduced in **Plate 6**, together with the uncertainty limits of former sea level reconstruction for the samples dated: gyttja or wood (filled bars) and marine shells or snails (open bars).

Lastly, in the south Østfold district, Danielsen (1970) obtained a shore-level displacement curve (**H**) from pollen analysis, which was later corrected by Anundsen (1985) to take into account the fact that the uplift gradient has been decreasing with time (curve **E**). Although the curves for south Østfold have still not been radiocarbon dated, it is interesting to note the flat part of curve **E** during the period when the Tapes transgression was found in other areas too.

DENMARK - THE KATTEGAT (Plate 7)

In Denmark, Mikkelsen (1949) published a shore displacement curve for Præstø (**C**), which was later adapted to the radiometric chronology by Krog (1979), corrected with a change in origin (curve **D**) and compared with the tentative curve **B** obtained by pollen analysis from the Store Belt area. Both these areas of South Denmark, which are situated on the +1 m isobase of the maximum level reached by the Litorina Sea deposits, were dry continental land in Late Weichselian times and the relative sea-level rise since 10,000 yr BP has been about 35 m here. Connections between the Kattegat and the Baltic were established about 8000 yr BP, above a threshold situated at -26/-27 m in the Store Belt, and above the Öresund threshold at -7 m more than one thousand years later.

In northern Jutland a very steep curve of local sea-level change, showing a rise of 28 m in 880 ^{14}C years around 8000 yr BP (curve **A**), was deduced by Petersen (1981) from pollen analysis of basal freshwater sediments and radiocarbon dates of marine shells collected from a borehole. These and several other data from Denmark were compiled by Petersen (1985) in a three dimensional computer plot (**E**), in which Denmark is viewed from the northeast at an angle of 30° and the time the transgressive movement occurred over the country (as a linear function) appears as the third dimension.

On the glacio-isostatically tilted coasts of the Kattegat, especially in SW Sweden, morphological marks of several shorelines, corresponding to postglacial transgression maxima, have been identified by Mörner (1969, 1976), followed for some 300 km in the direction of tilting and dated by over 100 radiocarbon dates. The amplitudes of the interjacent regressions have been determined from the stratigraphy, with an accuracy estimated at ±0.1 m. In the lowest graph of **Plate 7**, curve **L** is considered by Mörner to correspond to the tilting axis, and therefore to the regional eustatic factor. The other curves are expected to give the shorelevel displacements in the northeastward direction of tilting, at distances from the axis of -50 km (curve **M**), +50 km (**K**), +100 km (**J**), +150 km (**I**), +200 km (**H**), +230 km (**G**) and +250 km (**F**) respectively.

Plate 7: DENMARK - SWEDEN I (Kattegat)

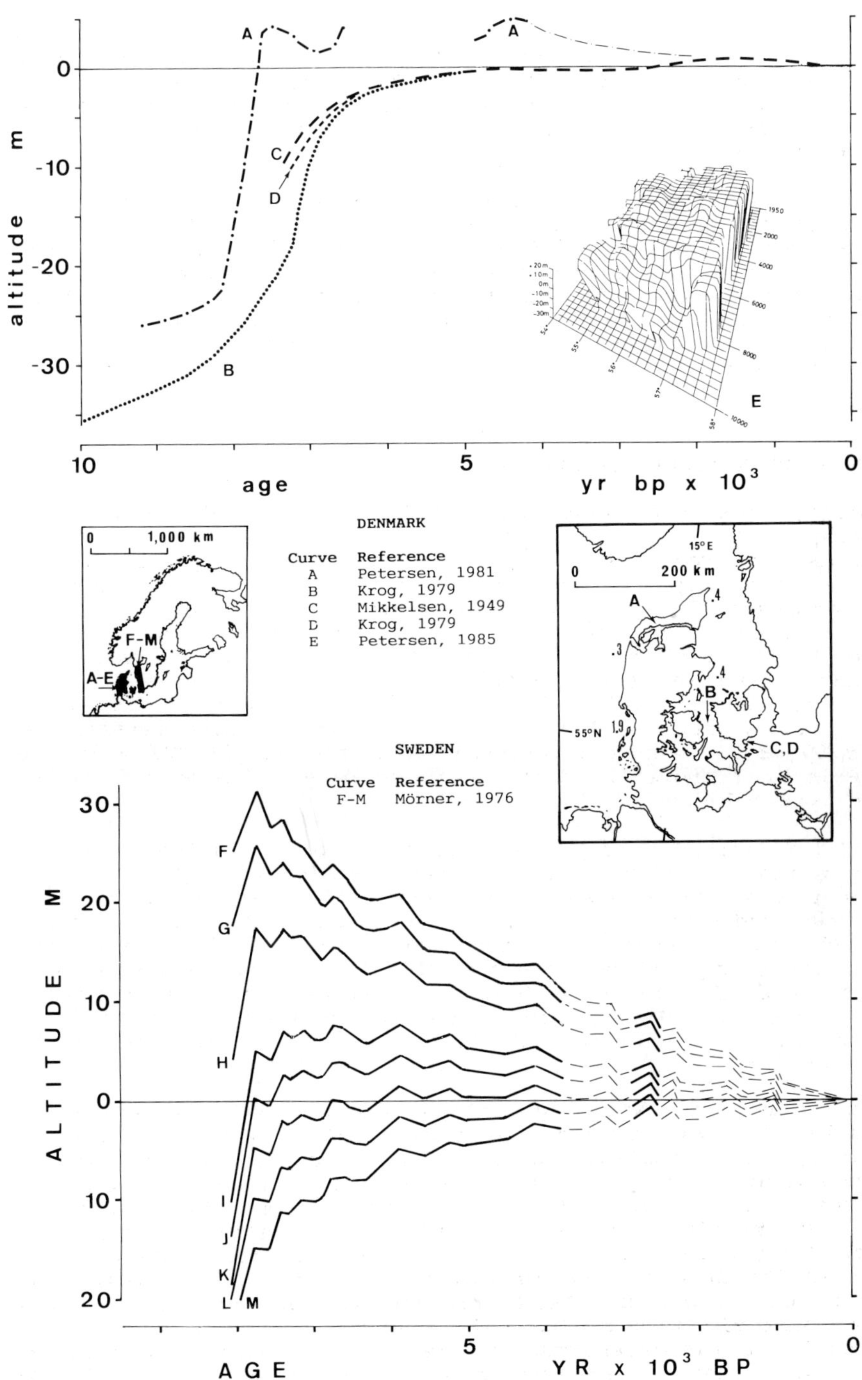

SOUTHWEST & SOUTH SWEDEN (Plate 8)

On the SW coast of Sweden, the early shore displacement curves **A** and **C**, published by Sandegren and Johansson (1931) and Sandegren (1952), already suggest a fairly credible general pattern, which may be compared with similar trends recognized in SW Norway.

More recently, Påsse (1987) studying the time of isolation from the sea of several basins, confirmed with pollen analysis and biostratigraphical correlations that a rapid regression occurred in the northern Halland region during the Late Weichselian, until about 9200 yr BP (curve **B**), followed by a transgressive trend characterized by a series of oscillations, which could be identified as having lasted until about 7200 yr BP, and by a final regression trend. The uplift rate has been estimated here by Påsse at 2-3 mm/yr during the last 7000 years, and recently at 1 mm/yr.

In the Båstad-Torekov area (**D**), Mörner (1980) established by geomorphology and stratigraphy means, with the help of 37 radiocarbon dates, a very detailed relative sea-level curve for the last 8000 years. Several low amplitude fluctuations (i.e. not exceeding 1 m) appear, which, according to Mörner, have a "regional eustatic" significance in the northeast Atlantic region; as a matter of fact, similar, though not identical, fluctuations have been found by several other authors in northwest Europe. In western Skåne, Digerfeldt (1975) recognized five transgressions between about 7000 and 4000 yr BP (curve **E**), using diatom and pollen analyses and several radiocarbon dates.

Regarding the coastal Blekinge area in South Sweden, following the early studies by Sandegren (1940) (curve **F**) and Nilsson (1953) (curve **J**), Berglund (1964) proposed a detailed shore displacement curve (**H**) for the Karlskrona Archipelago, based on geomorphological observations, pollen and diatom analyses and radiocarbon dates. Curve **H** shows a rapid sea-level drop around 9000 yr BP, which has been ascribed to the drainage of the Baltic Ice Lake (caused by the opening of a connection with the ocean through the sounds of central Sweden), followed by a Litorina Sea transgression and by a series of small oscillations after 7000 yr BP. Some of these oscillations differ however from those published by Berglund and Welinder (1972) for Siretorp (curve **G**). A slightly different shore displacement curve for the Litorina period, based on 50 radiocarbon dates, has even been postulated as being valid for the whole coast of Blekinge by Berglund (1971) and was later adjusted by Aaby (1974) (curve **L**), who claimed it gave evidence of a 520-yr periodicity. Lastly in the same region, the shore displacement curves **K** by Björk (1979) and **I** by Liljegren (1982) provided two additional variants, during the Litorina Sea period, which nevertheless remain very close to the most reliable ones among the preceding curves.

SOUTHEAST SWEDEN (Plate 9)

In the Oskarhamn area, pollen and sediment analyses from two lakes, 16 m and 26 m above sea level, were interpreted by Svensson (1985) as showing a regression, shortly after 10,000 yr BP, followed by a transgression, dated c. 9500 yr BP, and a new regression around 9000 yr BP (curve **M**). Similar sequences were found in an early study by Granlund (1932) at a higher altitude

Plate 8: SWEDEN II (Southwest & South)

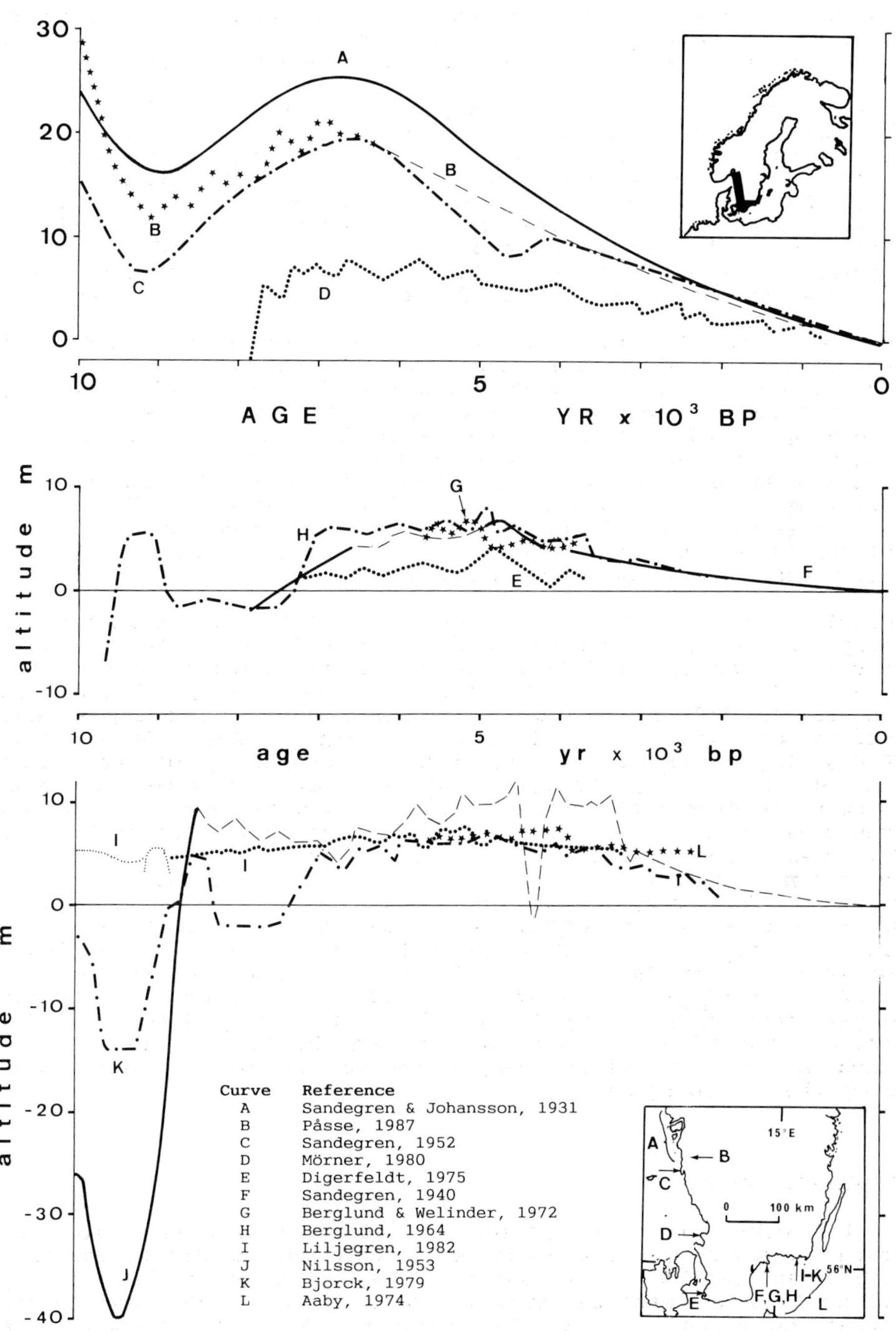

Plate 9: **SWEDEN III (Southeast)**

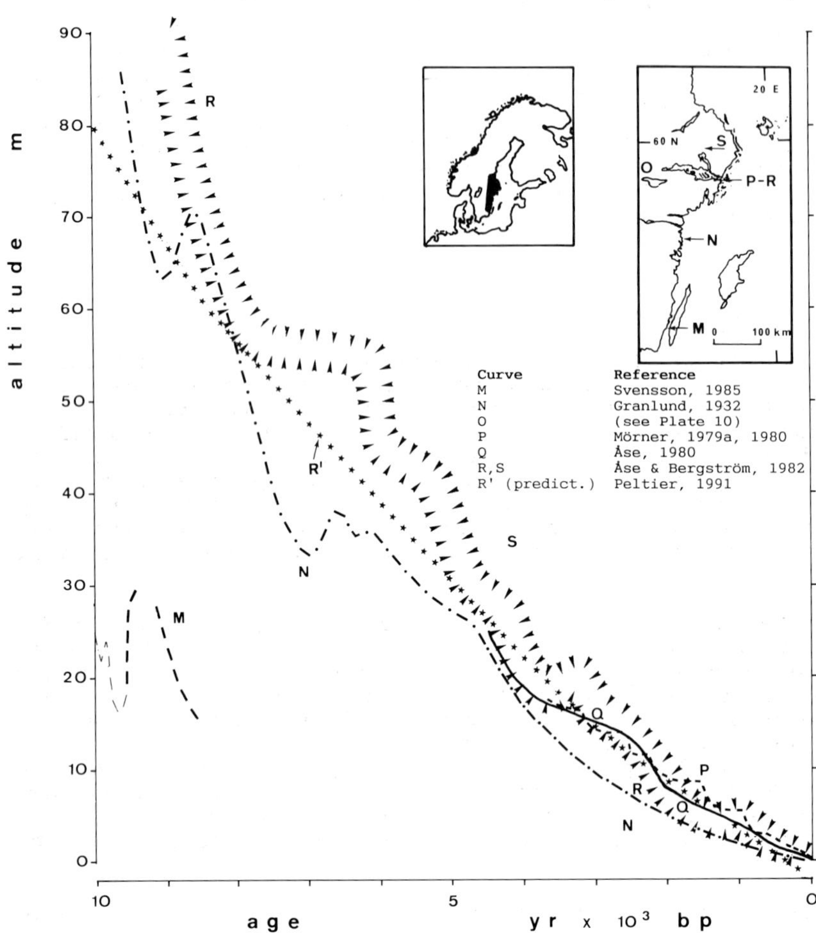

near Gusum (**N**), around 9000-8000 yr BP, and again around 7000-6000 yr BP.

In the Stockholm area, where the tide gauges indicate a rate of present relative sea-level drop of over 4 mm/yr, a large number of historical and archaeological data make possible the construction of a detailed sea-level graph for the last millennium, with a reasonably detailed continuation backwards via geological data (curves **P** and **Q**). According to Mörner (1980), a rapid regression took place at about 950 yr BP and there are indications of a small transgression maximum at 650-600 yr BP. Curve **Q** by Åse (1980), which shows the shore displacement during the last 4000 years, is similar though not identical to curve **P** by Mörner (1980), since differences in elevation between the two curves may be greater than 1-2 m even for the historical period.

A more complete, even if preliminary, shore displacement band (**R**), covering the last 9000 years was provided by Åse and Bergstrom (1982) for the Stockholm area, on the basis of stratigraphical data of bogs or lakes isolated from the Ancylus Lake or the Litorina Sea and lake thresholds, dated palynologically or by radiocarbon. Band **R** indicates four periods characterized by stagnation of the sea level. Slightly to the north, a similar preliminary band (**S**) was proposed by the same authors for the Uppsala area simply "by adding 20% to the altitude" of band **R**, under the assumption that the land uplift rate has been 20% faster near Uppsala than near Stockholm. Lastly curve **R'** corresponds to the predictions by the Peltier (1991) model at Stockholm.

NORTHEAST SWEDEN (Plate 10)

Magnusson (1970) published for the Örebro area (loc. O in Plates 9 and 10) a simplified shore displacement curve starting about 7900 B.C., when the ice sheet retreated to the north and the Yoldia Sea started forming beaches on hill slopes. The highest shoreline, determined as the line up to which the till was wave-washed, has been found at about 170 m in altitude. Although some regression contacts indicated by the diatom flora have been dated by radiocarbon, the curve **O** remains uncertain in the older part. All the curves in **Plate 10** have been represented at a reduced vertical scale.

South of Gavle (T) and north of Umeå (V), the shorelevel displacement curves used by GrØnlie (1981) for his Scandinavian geotraverse (see Plate 3) cover the last 9500 and 9000 years respectively. It can be noted that in both curves **T** and **V** the part older than 6000 yr BP is duplicated by a dotted, lower curve. The dotted variants indicate the corresponding ocean level positions, whereas the dashed, upper variants correspond to the surface of the Baltic basin which was higher during the Ancylus Lake stage than the ocean level. Since scarsely less than 6000 yr BP, the Baltic is considered to have been approximately level with the ocean.

In the Västernorrland (U), Lidén (1938) obtained a remarkable shorelevel displacement curve, very correctly dated by sequences of varves in deltas ranging from the upper shore down to the present coast. This curve is reproduced in Plate 10 in the slightly modified version (**U**) proposed by Mörner (1979b).

The Västerbotten coastal region lies within the area of present maximum land uplift in Scandinavia, the greatest present value being reached at Furuögrund, with about 9 mm/yr. Here several relative sea-level curves are available. Sea-level band **Y** by Renberg and Segerström (1981) was obtained in the southern part of the province by making varve counts of sediment cores from lakes situated at different altitudes above present sea level; it shows a smaller estimate of the rate of land uplift than other curves previously available, e.g. curve **Z** by Granlund (1943) or **X** by Bergsten (1954), which, however, concern areas from the northern part of the province. Curve **Z** was based by Granlund (1943) upon the present known rates of landrise, the so-called "Salt Water

Plate 10: **SWEDEN IV (Northeast)**

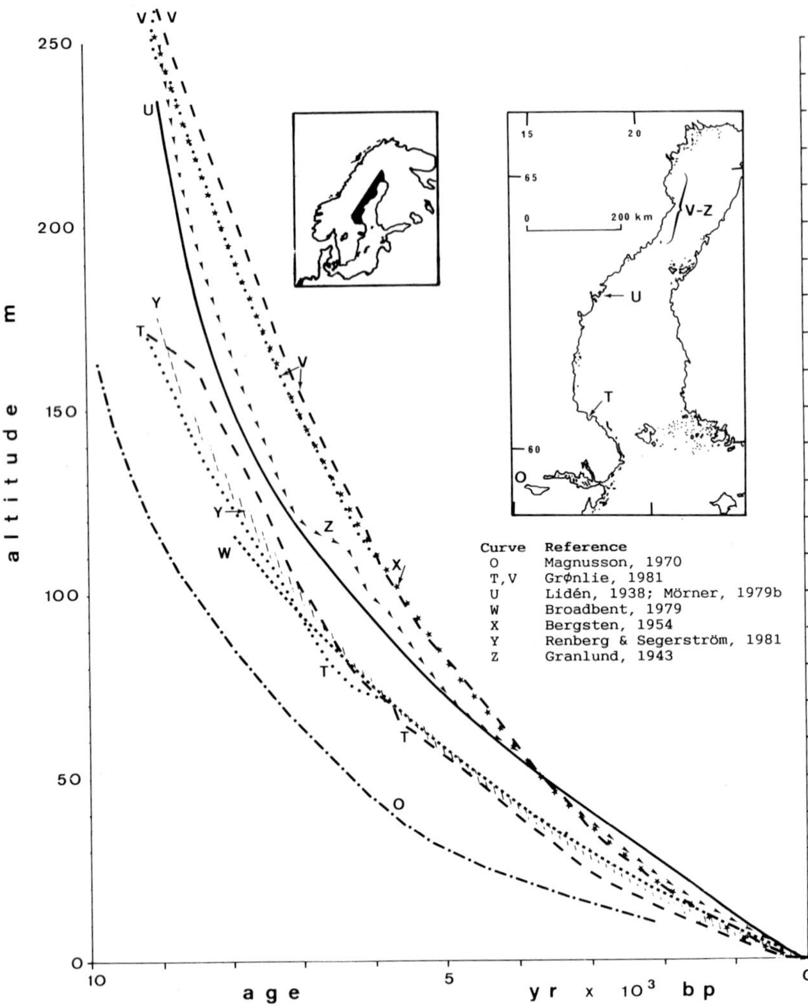

Limit", at an altitude of about 120 m, and the curve's form previously proposed by Lidén. Curve **X** results from extrapolation of exponential and polynomial equations deduced by Bergsten (1954) from watermark data. Later Broadbent (1979) obtained for northern Västerbotten a shore displacement curve almost coincident with the Lidén curve **U**. Finally the **W** curve reproduced in **Plate 10** is a calibrated version of the **U** curve and should be read in calendric years instead of years BP.

FINLAND (Plates 11 and 12)

In the area north of the Gulf of Bothnia, the highest shoreline following deglaciation developed over 200 m above present sea level about 9000 yr BP, during the Ancylus Lake phase. Here, by dating the isolation of successively lower small lakes from the waters of the Baltic, determined from sediment and stratigraphical changes, Saarnisto (1981) obtained a local shoreline displacement curve (**A, Plate 11**, at a reduced vertical scale). About half of the total emergence, i.e. 100 m, occurred within approximately a

Plate **11: FINLAND I (West)**

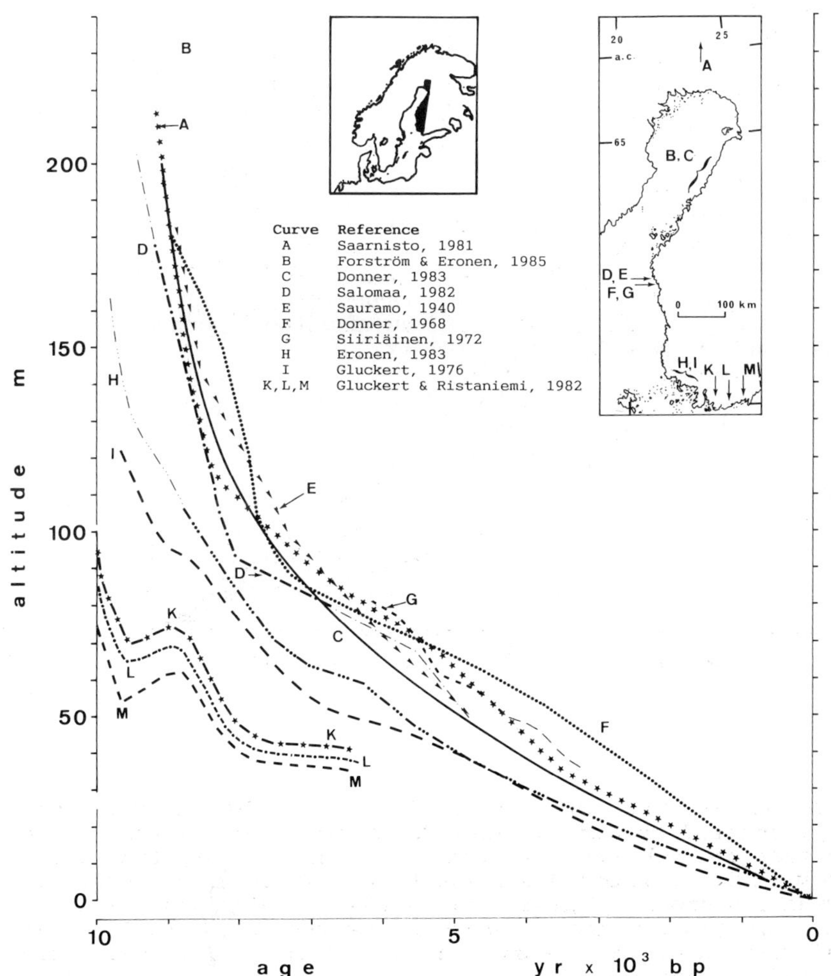

1000-year period prior to 8000 BP, whereupon a gradual retardation took place.

East of the Gulf of Bothnia, a schematic shore displacement band (**B**) was proposed by Forsström and Eronen (1985), which is very similar to the curve **C** given by Donner (1983) for the Pohjanmaa (Ostrobothnia) region.

Less gradual, the shoreline displacement curve **D** deduced by Salomaa (1982) from the study of sediment cores collected from six lakes in the Lauhanvuori area, suggests that a distinct retardation in the regression trend, which was not apparent in the earlier curve **E** by Sauramo (1940) in the same area, took place around 8000 yr BP. A similar retardation can be noted in other Finnish sea-level curves in Plate 11, although at slightly changing dates, e.g. at Kristiinankaupunki (curve **F**), where it coincides with the appearance of an increase in salinity in the diatom flora (Donner, 1968), or along the southern coast (curves **H**, **K**, **L**, **M**). A variant (**G**) for Kristiinankaupunki was proposed by Siiriäinen (1972) for the period 6000-3000 yr BP.

In the Turku area, the shoreline displacement curves **H** by Eronen (1983) and **I** by Gluckert (1976) show a roughly similar trend, but with differences in elevation between the two curves exceeding one dozen metres, which probably result from the fact that the sea-level index points used are not situated exactly on the same isobase of land uplift.

West of Helsinki, Gluckert and Ristaniemi (1982) constructed three shore displacement curves for the period 10,000-6000 yr BP at Karjalohja (**K**), Lohja (**L**) and Espoo (**M**), which show slightly metachronous dates for the beginning and the end of the Ancylus Lake transgression; the end of the transgression, in particular, is slightly older in the NW part of the area, as a consequence of differences in the amount of land uplift.

In the upper graph of **Plate 12**, the differences in isostatic crustal uplift along the south coast of Finland are illustrated by a comparison between the curves **K**, **S** and **T**. The **K** curve, which is an enlargement of the Karjalohja **K** curve in Plate 11, indicates, according to Ristaniemi (1984) that the Ancylus transgression began at 9600 yr BP, culminated at 9300 yr BP at an altitude of 74 to 77 m, and was followed by a fast regression which slowed down one thousand years later. The **S** curve, constructed by Eronen (1983) for the 30-m Litorina Sea isobase using biostratigraphical data from the Porvoo-Askola area, shows a trend almost parallel to that of **K**, but with a certain time lag and with differences in level slightly decreasing with time. Lastly the **T** curve, was obtained by Eronen (1976) using samples cored from a raised bog near Anjalankoski (formerly Sippola). It shows during the Yoldia stage a relative sea-level drop at the rate of almost 6 cm/yr until 9600 yr BP, followed by the Ancylus Lake transgression, from 9500-9300 yr BP to about 8900 yr BP, to a height not exceeding 55 m (probably higher than the contemporaneous level in the ocean) and by a regression until about 8300 yr BP.

In the lower part of **Plate 12** all the curves represent shorelevel displacements in the Helsinki area according to various authors. Curve **N** by Eronen and Haila (1982) summarizes several shore-displacement data projected on to the 34 m isobase for the Litorina Sea. Curves **O** (Donner, 1980) and **Q** (Alhonen, 1979) show a

Plate 12: FINLAND II (South)

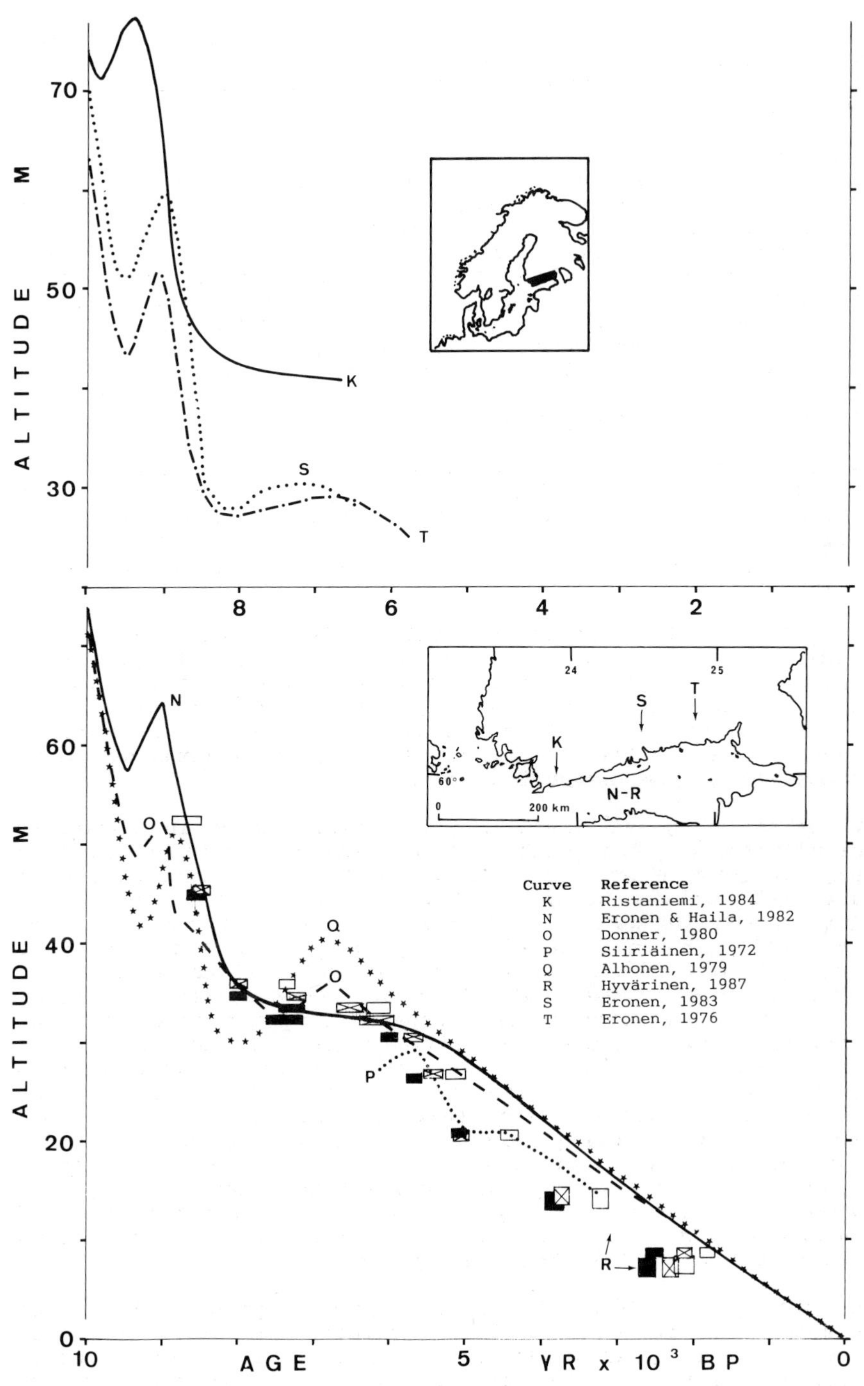

Plate 13: EAST BALTIC

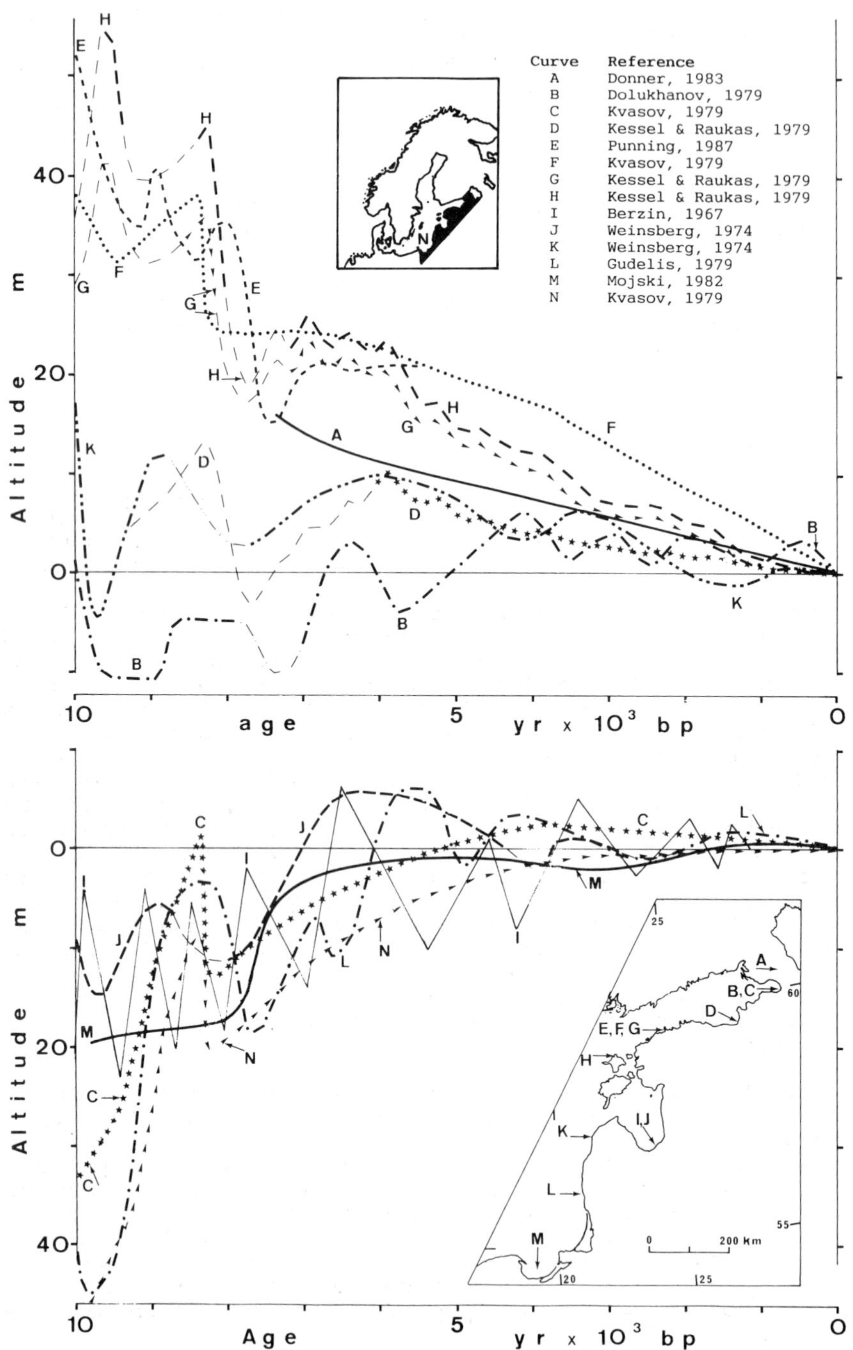

culmination of the Ancylus stage much lower than in **N** and a marked transgression peak around 7000 yr BP. The **P** curve by Siiriäinen (1972), originally expressed in sideral years and limited to the period 4200 to 1250 B.C., reproduced in **Plate 12**, is expressed in radiocarbon years BP without previous calibration and this has probably caused some deformation. Lastly, the **R** curve by Hyvärinen (1987) is based on the radiocarbon dating of sediments showing the isolation contact of 14 small lakes from the Baltic; the uncertainties in threshold altitude and in dating for curve **R** are expressed by crossed boxes for the dates from isolation contact layers, with full boxes for the dates from below the isolation contact and with open boxes for dates from above the contact. The main conclusion by Hyvärinen (1987) is that during the last 8500 years none of the basins studied showed evidence of renewed connection with the Baltic, since they had become isolated, and that both isostatic uplift and the eustatic movement must have been even in the Helsinki area.

EAST BALTIC COASTS (Plate 13)

In a paper devoted to a comparison between Eemian and Weichselian deposits in Finland, Donner (1983) provided a rather gradual "land/sea curve, beginning at 7000 yr BP, for the central part of the Karelian Isthmus" (**A**), for which, however, no details were given. Other curves proposed for the same area are much more fluctuating. In the Leningrad area, according to Kvasov (1979), the local shorelevel history since about 11,000 yr BP (**C**) would have been marked: (1) by a regression, bringing about the drying-up of the strait in the northern part of the Karelia Isthmus, when Lake Ladoga became isolated; (2) at a date that Kvasov situates slightly before 8000 yr BP, the peak of the Ancylus transgression would have been for a short time higher than the runoff threshold of Lake Ladoga; subsequently (3) a transgression on the southern shores of Lake Ladoga would have been caused by a rise of the runoff threshold; and lastly (4) the spilling across the watershed in the southern part of the Karelia Isthmus resulted in the development of the Neva River, whose deep erosion led to a fall in the level of Lake Ladoga. The main differences between curves **B** (Dolukhanov, 1979) and **C**, which concern about the same area, seem to be due to a time lag, since both the regression (1) and the peak (2) are situated by Dolukhanov earlier than 10,000 yr BP, and to various fluctuations in **B**, two of which have been deduced from the stratigraphy and some others, since 4000 yr BP, from systems of raised bars.

On the Estonian coasts, isostatic uplift was more intensive in the NW part of the Republic than in the NE (and SW), as shown by a comparison between the curves of "water-level changes of the Baltic Sea" obtained by Kessel and Raukas (1979) at Köpu (**H**), Tallin (**G**) and Narva (**D**), which also indicate metachronous dates for the highest level of the Litorina Sea. Two other curves for the Tallin region, (**E**, for the period 10,000-5000 yr BP by Punning, 1987; **F**, for all the Holocene, by Kvasov, 1979) show patterns roughly similar to that of the curve **G**, though with some altimetric shifts, which may depend on the location of the specific areas studied.

Plate 14: GERMANY I (Baltic)

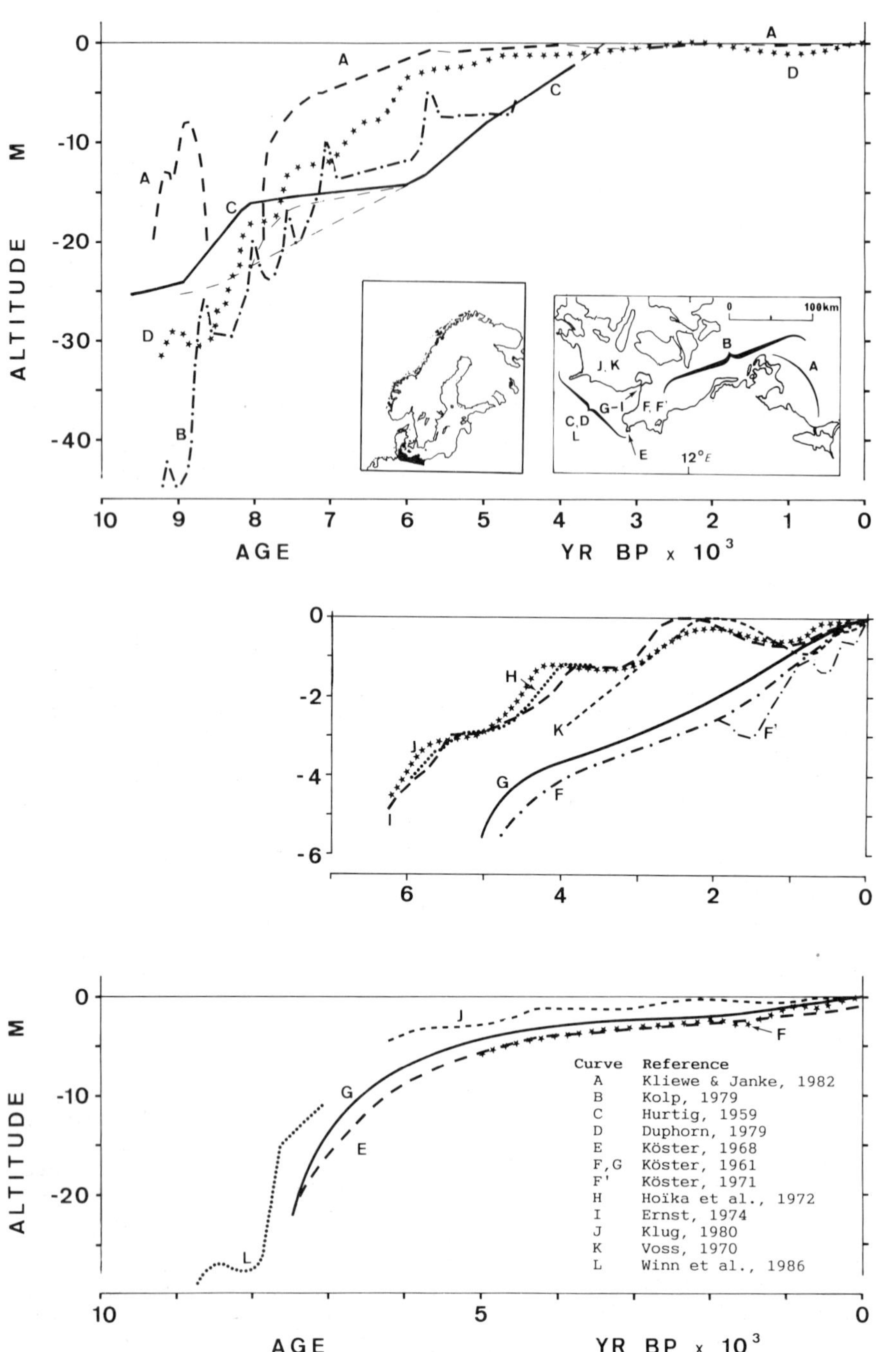

In the "upper part of the Gulf of Riga", Berzin (1967) obtained a "curve of sea level oscillations during the whole Holocene" (**I**), showing a toothed pattern, which however seems rather unlikely.

Relative sea-level change curves for the Latvian coasts have also been published by Weinsberg et al. (1974) for the Riga (curve **J**) and the Ventpils (**K**) areas; nevertheless, only a little, second-hand information is available to the present author on these two curves and it is not certain that the time scale used to reproduce them in **Plate 13** is the correct one.

Along the Lithuanian coasts all shorelines are tilted southwards and emerged shores can be found only in the northern coastal areas; southwards from a line joining Nida with Rusnè, they are all situated below present sea level. Curve **L** corresponds to the shore displacements proposed by Gudelis (1979) for the Polanga area (**L**), in which some emergence is still noticeable.

In Poland, the evolution of marine transgression on the southern shores of both Gdansk Bay and the Vistula lagoon, according to Mojski (1982), is summarized by curve **M**. Lastly, Curve **N** corresponds to the trend of the Baltic Sea according to Kvasov (1979).

GERMANY: BALTIC (Plate 14)

On the German coasts of the Baltic, curve **A** by Kliewe and Janke (1982) summarizes the relative sea-level changes for the northeast coast; following a Ancylus Lake peak around 9000 yr BP and a very steep rising slope of the Litorina Sea around 8000 yr BP, sea level has remained almost stable during the late Holocene, with fluctuations not exceeding ±1 m since 6000 yr BP. Much steeper than **A**, at least from 6000 to 4000 yr BP, curve **C** by Hurtig (1959) is based on pollen dating of transgression contacts (the transitions of peat into the overlying marine sediments) in several places along the Baltic coasts of Germany. According to Jelgersma (1961) however, this method (which has been followed by several other authors, e. g., in the same area, by Köster, 1961, see below) is not very reliable, for the surface of the peat may have been eroded by the invading sea, so that the transgression date is too old and the underlying sediments are subjected to compaction.

Kolp (1979) has investigated submarine terraces in the southern part of the Baltic Sea and proposed a step-like sea-level curve (**B**), based on the interpretation that each terrace results from a period of stagnation of the Holocene sea level and on some other assumptions and speculations. In Kiel and Lübeck Bays, data from transgression and regression contacts, buried submarine terraces and archaeological findings have been used by Duphorn (1979) to construct a sea-level curve (**D**) which suggests the occurrence of two regressive phases (around 9000 yr BP and between 2000 and 1000 yr BP) and of at least seven retardation phases. For the inner part of the Bay of Lübeck, relative sea-level curve **E** (**Plate 14**, lower graph) proposed by Köster (1968), partly based on pollen data obtained by Schmitz (1953), shows a much more gradual rise than that of the stepped curves **B** and **D**.

Again for the Bay of Lübeck, a gradually rising curve **F** was published by Köster (1961); this curve was subsequently modified ten years later (Köster, 1971), in order to add three sea-level fluctuations for the last 3000 years (curve **F'**, in the inset of Plate 14). For the Heilingenhafen/Oldenburger Graben area, the relative sea-level curve **G** proposed by Köster (1961) lies slightly above the curve **F**, suggesting a decrease in the subsidence rate. Curve **J**, proposed by Klug (1980) for the Bay of Kiel suggests an even less rapid subsidence rate. The differences between the curves **F**, **G** and **J** can be appreciated more easily in the inset (on a larger scale) in the centre of Plate 14, where they have been compared with three additional curves for the nearby regions: **I** (Ernst, 1974) for Hohwachter Bay, **H** (Hoika et al., 1972) in the Oldenburger Graben/Süssau area, and **K** (Voss, 1970) for the East Schleswig coasts.

Lastly curve **L**, proposed by Winn et al. (1986) for the western Baltic, is based on radiocarbon-dated sediments either overlying or underlying the Litorina Transgression and is expressed in sideral years. Curve **L** shows that the rise in sea level was very rapid between -26 m (when the Kattegat Sea flowed over the thresholds separating it from the Ancylus Lake) and -14 m and slowed down thereafter. The data provided by Winn et al. (1986) do not show any significant deviations between the Kiel, Mecklenburg and Lübeck Bays, and hence do not support the differences in the average regional subsidence rates previously reported by Köster (1961) and Kolp (1979).

Fig. 2-3 summarizes the present-day relative sea-level trends indicated by tide-gauge records in North Europe. In the Baltic area uplift movements increase towards the Gulf of Bothnia, where they reach a maximum, consistently with Holocene data. On the south Baltic coasts, some trends show relative stability, others that a slight subsidence is occurring.

GERMANY: NORTH SEA (Plate 15)

Müller (1962) proposed two curves for the MHW changes along the southern North Sea coast, on the basis of 57 radiocarbon datings and extensive fieldwork: curve **B** corresponds to minimum positions deduced from layers expected to have undergone little subsidence, whereas curve **C** shows possible reconstructed positions of the MHW level. The lower part of curve **B** was connected by Müller to curve **A**, constructed for early Holocene MSL changes by Graul (1959) with data from Holland and East Friesland.

For the Jade-Weser region, Schütte (1939) published a curve of Holocene sea-level rise (**D**) based on pollen analysis of peat layers, suggesting three sea-level oscillations, which however were no longer accepted in later studies. The MSL band **E** has been deduced from a computer time-depth diagram published by Preuss (1980), using corrected sea-level indicators from Nordholz.

In the Eider-Miele area the relative MHW curve **F** by Menke (1976) shows, between 9000 and 7000 yr BP, a steep rise (c. 20 mm/yr) established from dates from basal peat layers. Since 6500 yr BP, the slightly undulating pattern of curve **F** has been deduced from the highest tops of dated sediment beds of marine or brackish origin and from estimates of the former MHW level inferred from

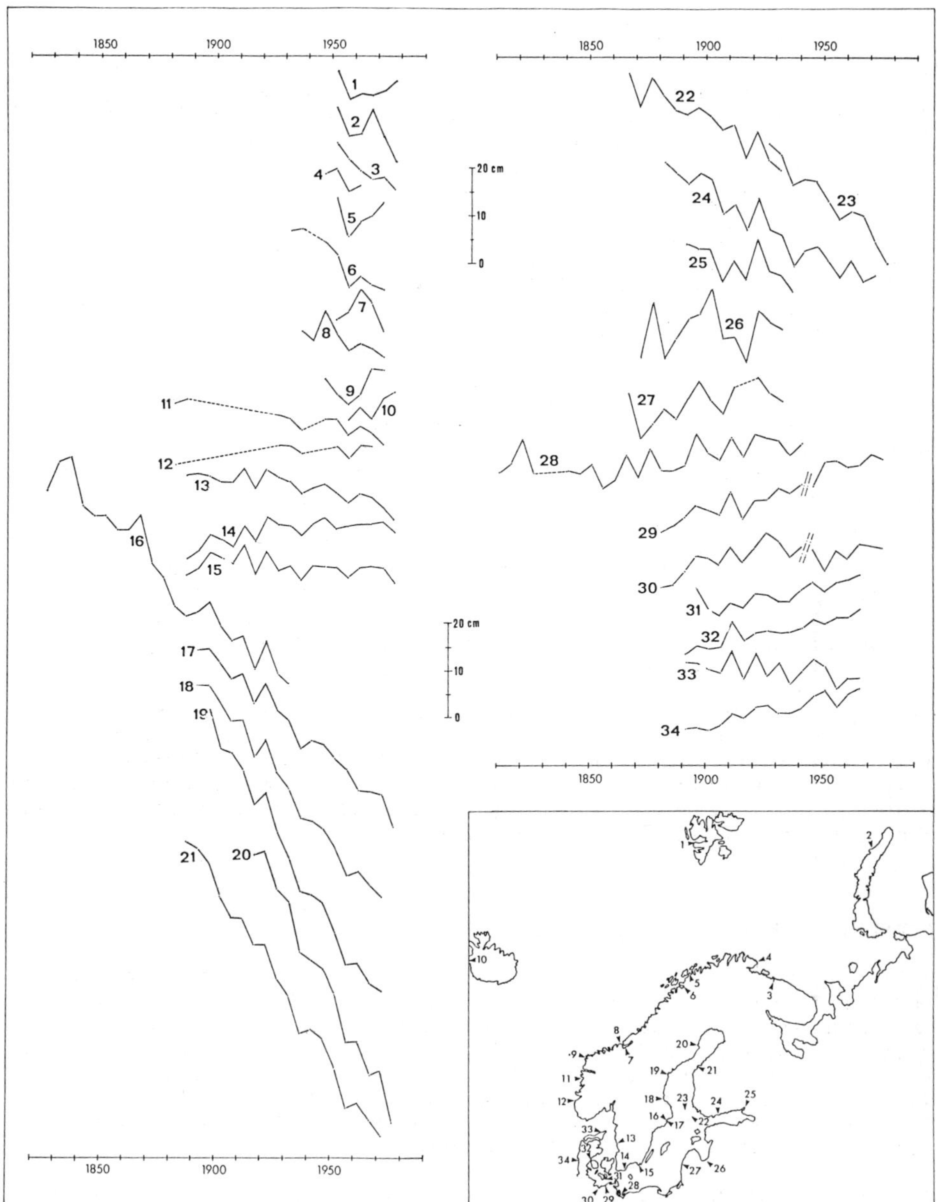

Fig. 2-3. Average 5-yr MSL variations in northern Europe as recorded by tide-gauge stations. 1) Barentsburg; 2) Russgaya; 3) Murmansk; 4) Vardo; 5) Tromso; 6) Narvik; 7) Trondheim; 8) Heimsjo; 9) Maloy; 10) Reykjavik; 11) Bergen; 12) Stavanger; 13) Varberg; 14) Ystad; 15) Kungholmsfort; 16) Stockholm; 17) Nedre Stockholm; 18) Bjorn; 19) Draghallan; 20) Furuogrund; 21) Vasa; 22) Uto; 23) Degerby; 24 Helsinki; 25) Vyborg; 26) Riga; 27) Liepaja; 28) Swinoujscie; 29) Warnemunde; 30) Wismar; 31 Gedser; 32) Fredericia; 33) Hirtshals; 34) Esbjerg. (From Pirazzoli, 1986b).

Plate 15: GERMANY II (North Sea)

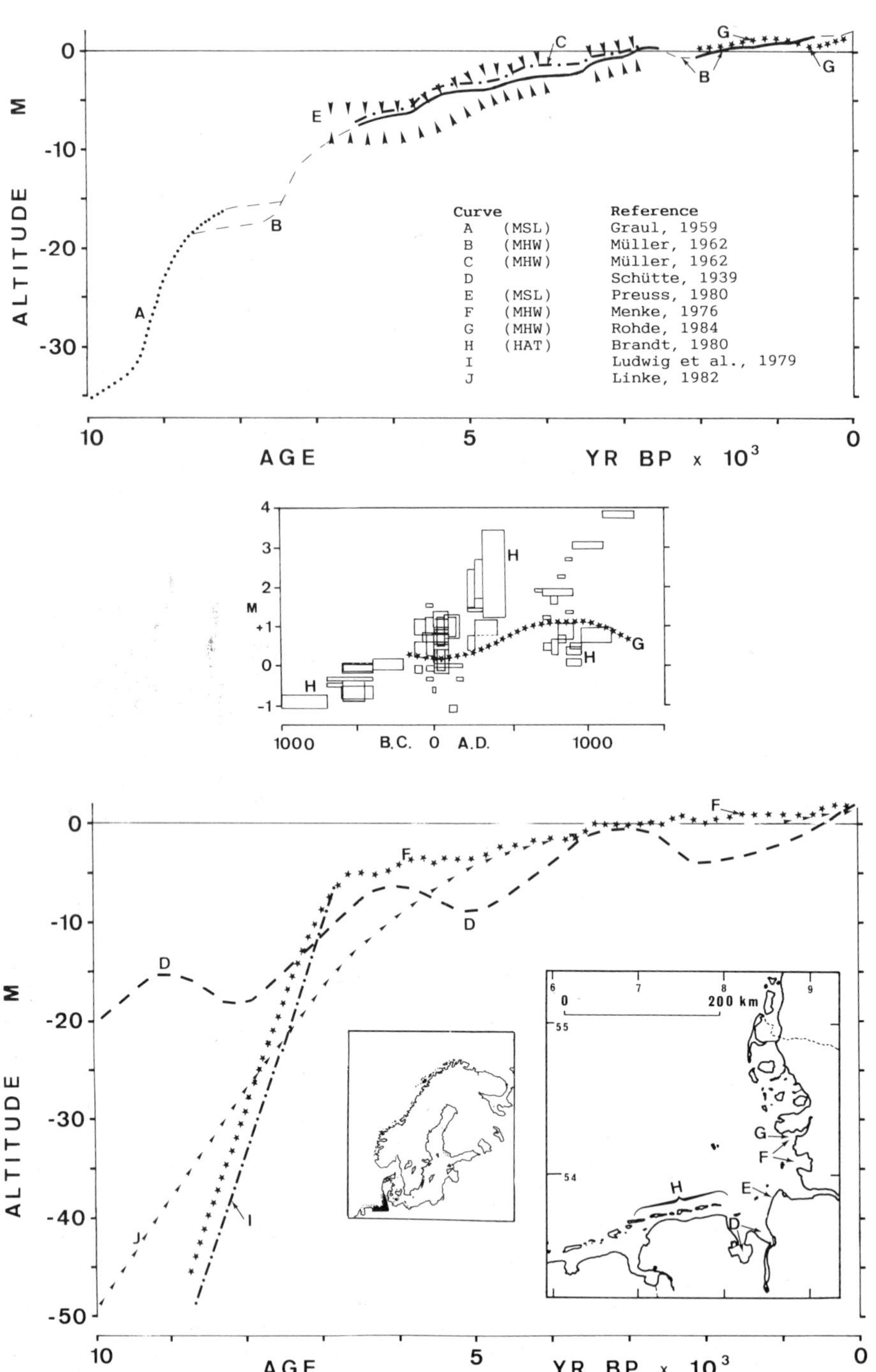

vegetational and sediment changes; for this period curves **B** and **F** are very similar, with differences rarely exceeding ±1 m, i.e. much less than the local tidal range.

MHT curve **G** by Rohde (1984) is based on geological and archaeological data from the German North Sea coast up till A.D. 1500, and on storm surge data and tide-gauge measurements after that date. In the graph on an enlarged scale in the middle of **Plate 15**, curve **G** has been compared with the elevation of prehistoric and early historic habitation sites in the marshlands between the estuaries of the Ems and the Elbe Rivers reported by Brandt (1980). Habitation levels, which mark the beginning of a settlement on flat marshy ground, or on a new layer of a dwelling mound, may be regarded as index marks of the highest water level and have been represented, with their error margins, as boxes (**H**). Although habitation levels of the same age often have different elevations, because in undiked marshlands the tidal range varies according to the distance from the sea, the fluctuations proposed by Rohde (1984) clearly do not apply exactly to the area studied by Brandt (1980) and reveal therefore a local significance.

Finally, for the "German Bight", relative sea-level curves **I** (Ludwig et al., 1979) and **J** (Linke, 1982), reproduced from Winn et al. (1986), are expressed in sideral years.

THE NETHERLANDS (Plates 16 and 17)

The Netherlands, especially Holland, are one of the best studied areas in the world for Holocene sea-level research. From the post-glacial sea-level rise, however, only the uppermost 46 m can be reconstructed on the basis of reliable rediocarbon dates. These were obtained from basal peats resting on Pleistocene deposits, which are overlain in turn by Holocene brackish or marine sediments (Long et al., 1988) (Fig. 1-12). Some parts of this chapter have been inspired by a recent review by Van de Plassche and Roep (1989).

Early sea-level curves are compared in **Plate 16** (above). Most of them are composite, including data collected from various areas. The MSL curve **A** proposed by Umbgrove (1947) and curve **C** by Zwart (1951), which were deduced from coastal data, suggest 2 to 3 m wide fluctuations in sea level that have not been confirmed by subsequent authors. The MHW curve **B** by Nilsson (1948) is based on material dredged from the floor of the southern North Sea region at various places and depths, and on submerged peat layers in Holland and northwest Germany. According to Jelgersma (1961), the samples used by Nilsson (1948) to show a sea-level peak about 4000 yr BP followed by a relative drop, consist of peat, which may be reworked, and of questionable pollen-analysis datings of clays, and are not reliable. The MSL curve **D** by Bennema (1954) is also composite and suggests a gradual and progressive sea-level rise relatively close to that shown by the most accurate curves published subsequently.

The first curve based on a series of radiocarbon dates at one locality is that by Van Straaten (1954), who analysed tidal deposits in a deep tunnel pit at Velsen (**E**). The indicative value of the sedimentary structures used by Van Straaten is considered

Plate 16: NETHERLANDS I

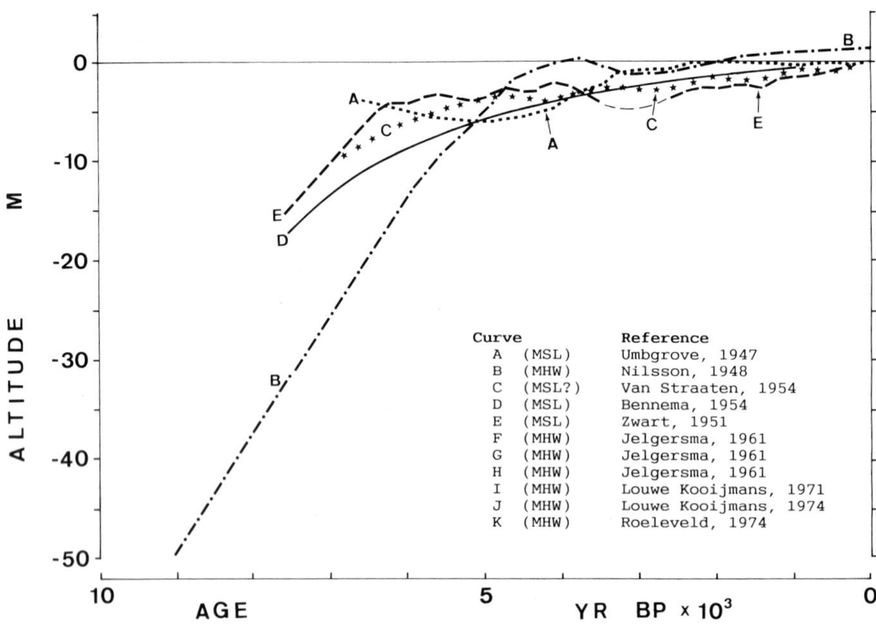

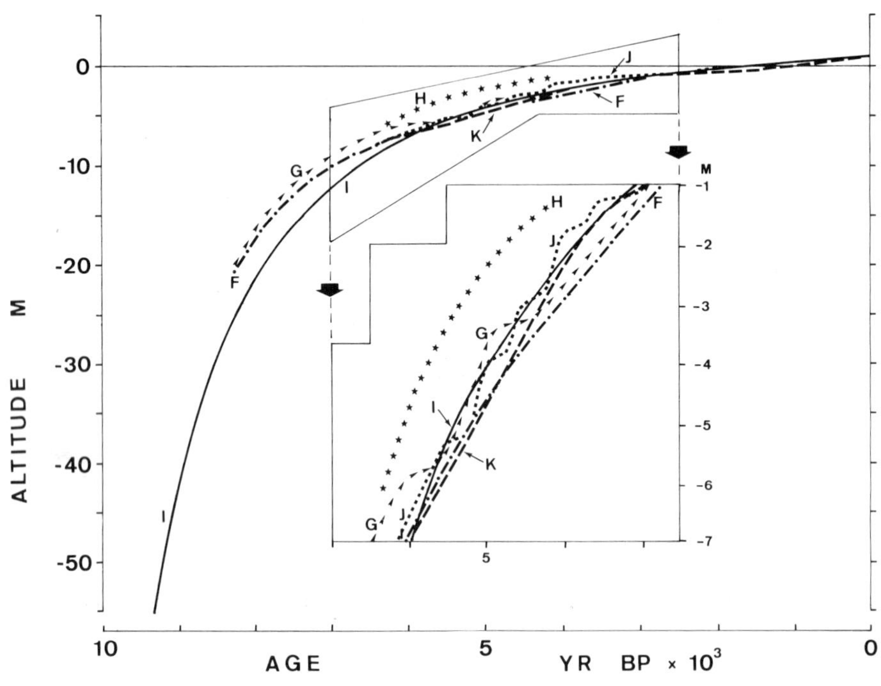

Plate 17: NETHERLANDS II & BELGIUM

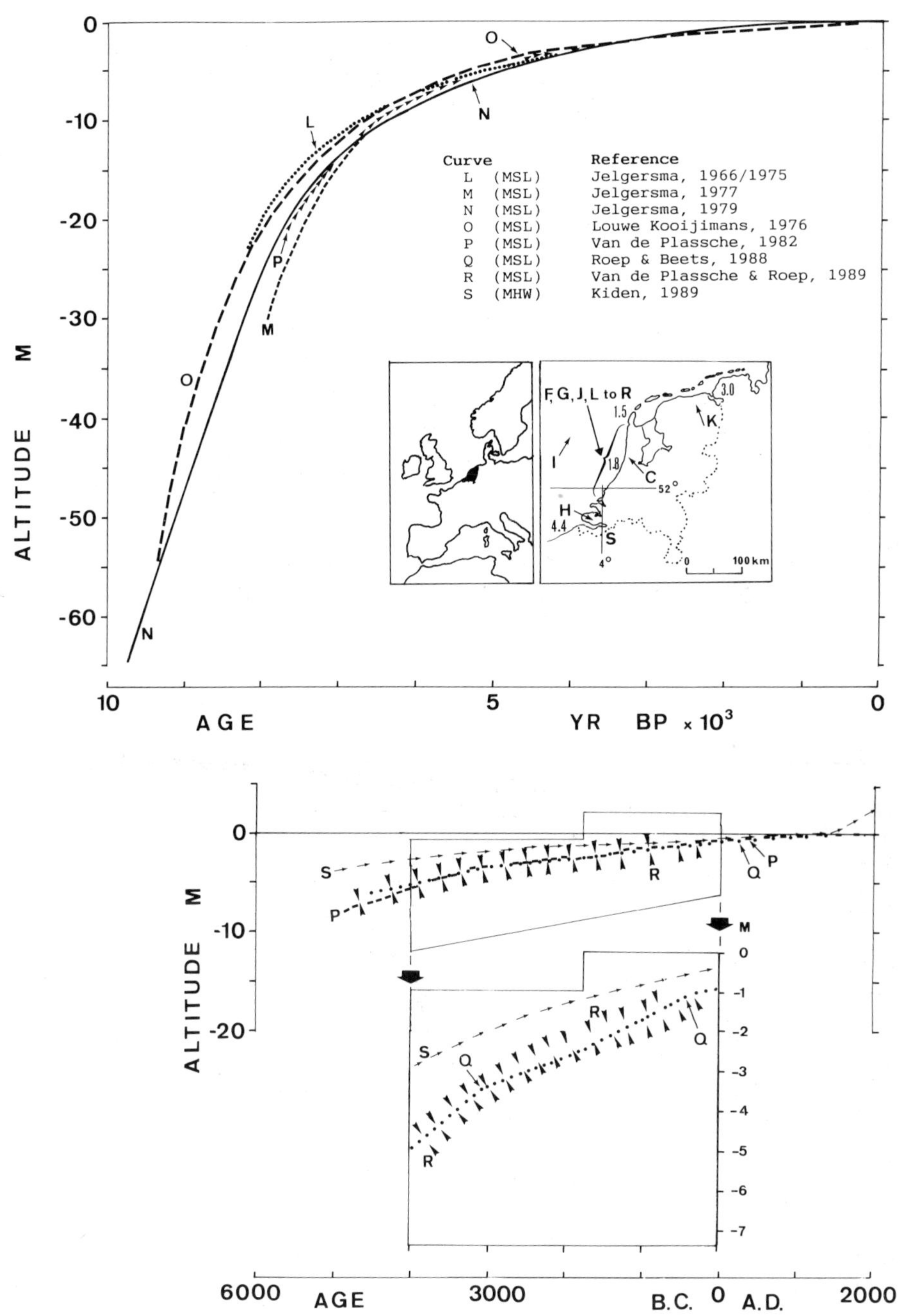

clear and relatively accurate (Van de Plassche and Roep, 1989), but compaction resulted in wide error margins.

The curves reproduced in the lower graph of **Plate 16** are all referred to the MHW level. The best known curve from the Netherlands is the composite one proposed by Jelgersma (1961) for the western and northern Netherlands (**F**), which was published together with variant curves for the Rhine/Meuse delta area (**G**) and for the province of Zeeland (**H**). This well-documented curve, founded on datings of basal peat samples in order to avoid compaction effects, shows, for the last 8000 years, a gradually rising sea-level pattern completely different from the widely oscillating pattern proposed the same year by Fairbridge (1961). Indeed the Jelgersma (1961) paper has been one of the finest contributions of the last three decades to Holocene sea-level research. Its methodological approach is still valid, whereas that used by Fairbridge (1961) or by Shepard (1964), who mixed data from various parts of the world, can be considered obsolete and misleading.

Other MHW curves are **I** and **J**, published by Louwe Kooijmans (1971, 1974), and **K**, by Roeleveld (1974). The **I** curve, rather smooth, is based on bone implements dredged up from the Brown Bank region in the North Sea, and on bored peat samples for the more ancient part; on MHW levels established at prehistoric sites in the western Netherlands for the more recent part; and on other data already published by Jelgersma (1961). A little later however Louwe Kooijmans (1974) published a fluctuating MHW curve (**J**) for the western part of the country, based almost entirely on estimates of former local MHW or of groundwater levels at archaeological sites. The same year, using a completely different method, Roeleveld (1974) obtained a smooth MHW curve for the Groningen coastal area (**K**) showing a trend similar to that proposed by Lowe Kooijmans (1974), but slightly different from the Jelgersma (1961) curve (see **Plate 16**, inset of the lower graph).

Meanwhile (**Plate 17**, all the curves represent MSL) the Jelgersma (1961) curve had been modified slightly, to take into account the Suess effect (Jelgersma, 1966) and the fact that a study on paleotide levels on the coastal barrier deposits had shown that paleo-MHW levels were 0.8 to 1.0 m above the 1961 curve. This led Jelgersma et al. (1975) to conclude that earlier curves by Jelgersma indicate the rise of MSL rather than of MHW. Finally, Jelgersma (1977, 1979) made the earlier part of her curve steeper on the basis of new data from the North Sea and North Holland.

Louwe Kooijmans (1976) reacted to discussions evoked by his (1974) interpretative oscillating curve by producing a methodologically important paper, in which he gave a new smooth MSL curve (**O**), complete with error margins, based on a selection of his most appropriate data. The fact that part of the **O** and **K** curves occurs above the Jelgersma (1979) curve is attributed by Van de Plassche and Roep (1989) to an underestimation of the river gradient effects for the older part of curve **O** and to compaction phenomena for the younger part of **K**.

The most recently published curves (**Plate 17**, lower graph) are a good test of the accuracy of Dutch sea-level data. For the period between 8000 and 2000 years BP, Van de Plassche (1982) provided a very detailed MSL curve (**P**), showing a few small fluctuations derived from c. 60 new basal peat samples collected in western

Holland. Between 4000 and 2000 yr BP, curves **N**, **O** and **P** are almost coincident. The MSL curve **Q** by Roep and Beets (1988) has been obtained from paleotide levels recorded in the coastal barrier area along the open coast of the western Netherlands. Lastly the error margins of band **R** correspond to upper and lower limits of MSL based on peat data and on a variety of radiocarbon-calibrated or archaeologically dated sea-level indicators used by several authors (Van de Plassche and Roep, 1989). It appears that, in spite of the narrowness of band **R**, defining MSL positions derived from basal peat data with accuracies ranging from ±0.75 m to less than ±0.1 m, there is excellent agreement between the sea-level data from near the open coast and from areas farther inland. This confirms the fine accuracy of the sea-level data reported recently from Holland. The corresponding geological trend is a relative sea-level rise of slightly over one metre during the last 2500 years, i.e. an average rate of about 0.4 mm/yr, which however does not include the compaction effects of Holocene sediments.

Excavations in the Belgian part of the lower Scheldt River enabled Kiden (1989) to suggest the occurrence of changes in the local MHW level differing from those taking place along the coast. The gradient of the lower course of the river would have decreased after 5500 yr BP, due to the development of a floodbasin effect, which depressed the local high tide level (in relation to that on the outer coast) between 3000 B.C. and about A.D. 1700. After A.D. 500 the upstream penetration of tidal action increased, however, causing a rapid rise of the local MHW level, the acceleration of which during the last centuries is fairly well documented by historical evidence of storm floods and embankments.

BRITISH ISLES: INTRODUCTION

Knowledge on Holocene sea-level changes in the U.K. improved appreciably in the 1970s and 1980s, essentially due to the creation of an important national working group contributing to the IGCP Projects 61 and 200 and to the careful compilation and screening of sea-level data undertaken by M.J. Tooley and I. Shennan. This has led to regional comparisons and reviews and interesting methodological advances (e.g.: Shennan et al., 1983).

At first sight, there appears to be a great dispersion in the data (Fig. 1-25), and their interpretation will require appropriate models of isostatic and tectonic movements. These include essentially glacio-isostatic movements related to the late-Pleistocene melting of ice caps in Scotland, northern England, the Irish Sea basin and Ireland, which may reveal quite complex local sea-level histories during the Holocene, and to long-term subsidence trends occurring in the North Sea basin. Accurate determinations of former MSL positions are however complicated in many areas by wide tidal ranges and by the possibility of compaction of Holocene sediments, especially of peat, which is the most frequently used sea-level indicator.

SOUTHEAST BRITAIN (Plate 18)

In the Thames estuary, Devoy (1977) provided relative MHWST curves for Tilbury (**B**) and for a number of sites along the south side of

Plate 18: ENGLAND I (Southeast)

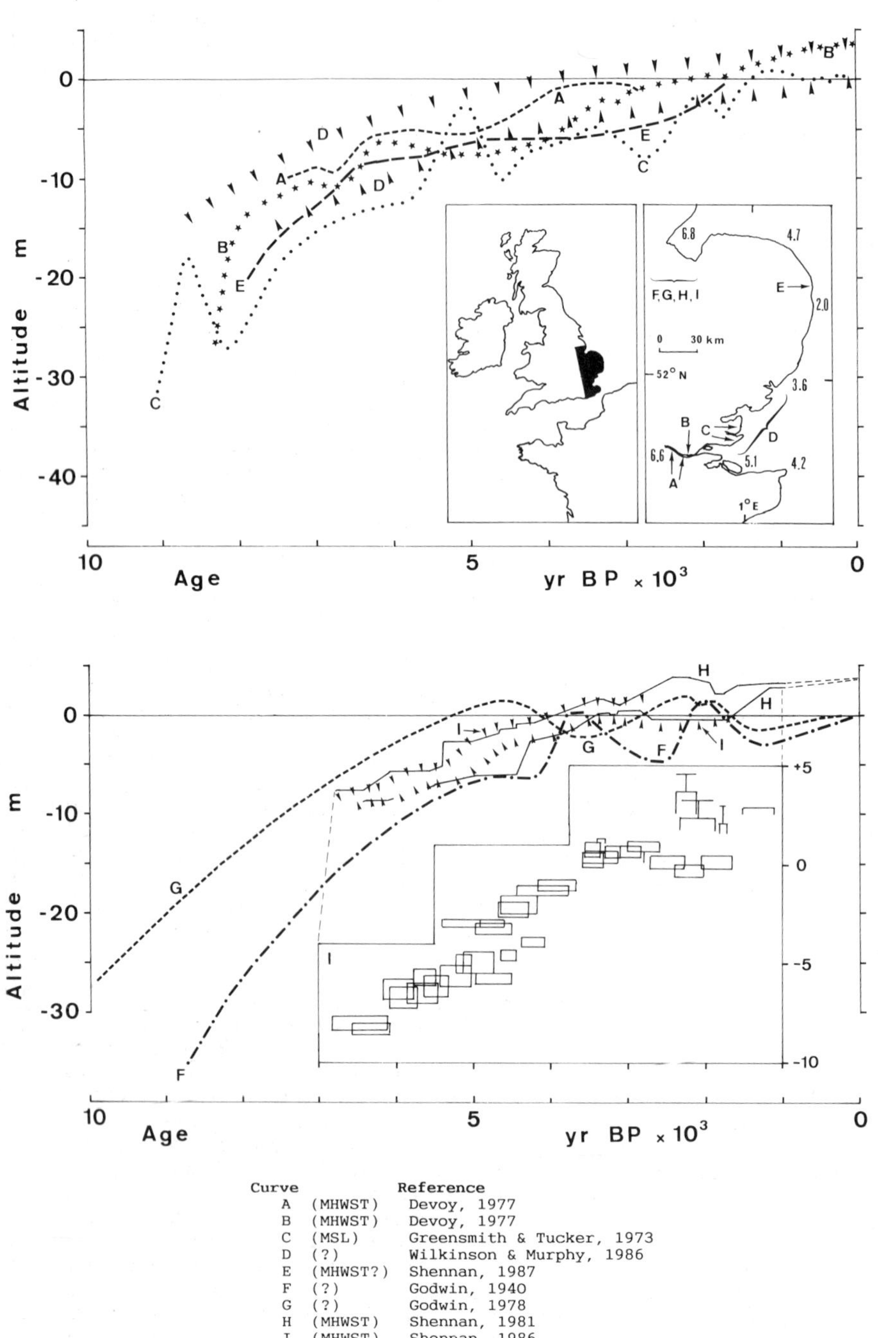

Curve		Reference
A	(MHWST)	Devoy, 1977
B	(MHWST)	Devoy, 1977
C	(MSL)	Greensmith & Tucker, 1973
D	(?)	Wilkinson & Murphy, 1986
E	(MHWST?)	Shennan, 1987
F	(?)	Godwin, 1940
G	(?)	Godwin, 1978
H	(MHWST)	Shennan, 1981
I	(MHWST)	Shennan, 1986

the river (**A**). Five marine transgression sequences were recognized on the basis of pollen, diatom and other micro-fossil analyses. The radiocarbon dating of samples taken from regression-transgression contacts of the deposits permitted the construction of the two curves **A** and **B** in which however "the amplitude of the regressions or negative movements of sea level is not known precisely" (Devoy, 1979, p. 388). The differences in height between the two curves (which may be as much as 5 m) are ascribed by Devoy (1979) to the joint influence of relative land subsidence (compaction and consolidation of sediments, for which no correction has been used in curves **A** and **B**, and downwarping phenomena) and to the effect of varying tidal amplitude and river discharge.

On the Essex chenier plain (**C**) the tentative MSL curve proposed by Greensmith and Tucker (1973) suggests various wide fluctuations which conflict, however, with evidence reported by other authors in nearby areas. The 8-m fluctuation for instance, showing a peak about 5000 yr BP, is based on the interpretation of a single peat sample and it seems very unlikely that in the Essex chenier plain MSL could at that time have been higher than the spring high tide level in the Thames estuary. Other fluctuations of **C** curve are seldom justified by appropriate well dated sea-level indicators and seem to reflect, in some cases, attempts at correlation with the sea-level curves by Fairbridge (1961) and Mörner (1971) rather than local data.

The "sea-level" band **D** proposed by Wilkinson and Murphy (1986), the precise meaning of which is not specified, is based on data reported from the Thames estuary by Devoy (1977), from the Essex coast by Greensmith and Tucker (1973) and on archaeological and biogenic data studied in the Crouch estuary by the authors. Lastly, curve **E** was obtained by enlarging a small graph published by Shennan (1987) for the Norfolk Broads, using sea-level data from sites up to 30 km from the present coast. This may imply changes in the tidal range. For the period after 5000 yr BP, the data points used are mostly from intercalated peats overlying thick sequences of unconsolidated silts, clays and peats, and consolidation effects are more likely than in earlier data from basal peat transgressive overlaps. Although Shennan (1987) did not specify whether curve **E** represents changes in MSL or in the high tide level, it may be assumed from the context to be spring high tide level.

In the fenland of East Anglia, many sea-level data are available. Godwin's pioneer curve **F** (1940) was based on assumptions on the levels of fen clay deposition and peat formation and on datings deduced from pollen assemblages. Although Godwin was very familiar with radiocarbon dating, his more recent curve **G** (Godwin, 1978) seems based on almost the same criteria used 38 years earlier. Two major episodes of marine transgression were recognized by Godwin: the brackish "Fen Clay" transgression in the Neolithic and a quite saline "Upper Silts" transgression in the Roman-British period. In both curves **F** and **G** it was not specified however which "sea level" is being considered. Large tidal changes have certainly occurred in the fenland between the time a given site was located near the shoreline (the present-day average spring tide range is 6.8 m at Boston) and the time when the same site became upstream and isolated from tidal movements by sediment deposition, while the shoreline was displaced horizontally seawards by several tens of kilometres. These tidal changes and effects of sediment compaction make the interpretation of the data difficult, causing uncertainty ranges of the same order, or even larger, than certain sea-level oscillations suggested by Godwin.

Well aware of these difficulties, Shennan (1981) did not draw a single sea-level curve for the Fens area, but a MHWST band (**H**) constructed from a number of data points, each considered with adequate uncertainty ranges. An updated MHWST band (**I**), based on 112 ^{14}C data points, was proposed by Shennan (1986) for the same area. An example of an age-altitude plot of the most reliable indicators used to obtain sea-level band **I** is shown on an enlarged scale in the same plate.

According to Shennan (1982), the sea-level bands should be treated with caution though, since they still include errors resulting from consolidation, palaeo-tidal changes and an inner fenland variation of parameters presently considered as constants. Shennan prefers to determine periods dominated by positive or negative sea-level tendencies which may not have been uniform throughout the Fens since, due to variable local conditions, the maximum extension of marine sedimentation does not seem to have been synchronous.

SCOTLAND (Plate 19)

During the post-glacial period, the melting of the ice cap overlying the Scottish region contributed to the global eustatic sea-level rise and resulted in a local dome-shaped isostatic recovery of the earth's crust (Fig. 2-4). In contrast to areas which had a greater isostatic adjustment, such as the central parts of Fennoscandia where there has been continuous emergence since the time the ice melted, the isostatic rise of the earth's crust has been relatively slow and slight in Scotland. Consequently the glacio-eustatic rise in sea level overtook the isostatic rise at periods varying from place to place, depending on the distance from the uplift centre and on the contemporary rate of eustatic rise. This resulted in a variety of relative sea-level histories, with evidence at some places of several transgression-regression sequences.

The fluctuating trend of relative sea-level curve **A**, reported by Sissons and Brooks (1971) from the Forth area near the central part of the uplift dome, is based on detailed borehole information. Here on either side of the Menteith moraine, the raised mud flats, locally called "carse", which mark the upper limit of the Holocene marine transgression, overlie three regressive buried raised beaches (referred to as the High, Main and Low buried beaches).

The High buried beach, at +12.2 m, dated 10,300 to 10,100 yr BP, is not shown by curve **A**. The Main buried beach, at +9.9 m, is about 9600 years old. The Low buried beach, at +7.2 m, was dated about 8700 yr BP. A subsequent transgression, starting about 8300 yr BP, brought the sea level to a peak at +14.4 m about 6500 yr BP and was followed by a final regression period, probably still underway.

It should be noted that the stratigraphical changes represented by curve **A** (as well as by most other sea-level curves from Scotland)

Plate 19: SCOTLAND - NORTH SEA TRENDS

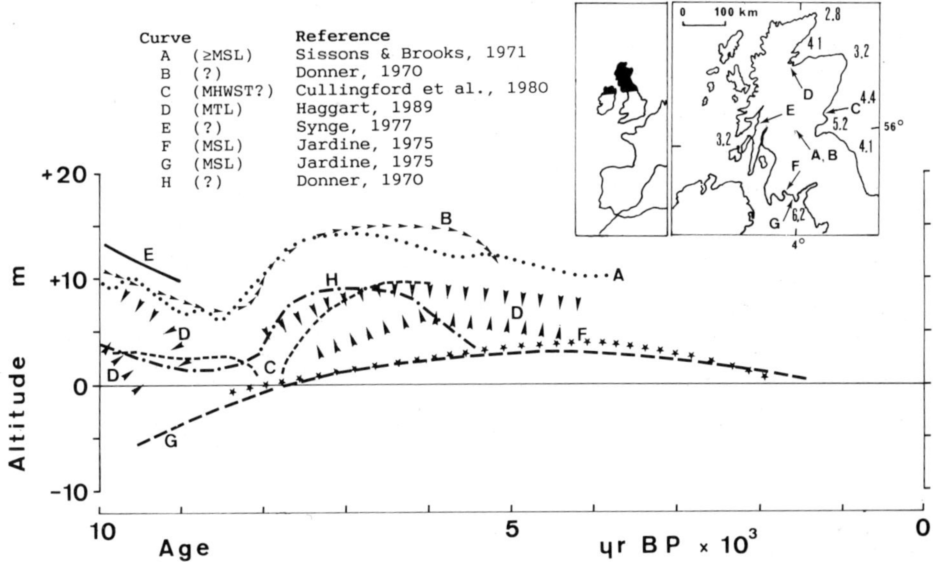

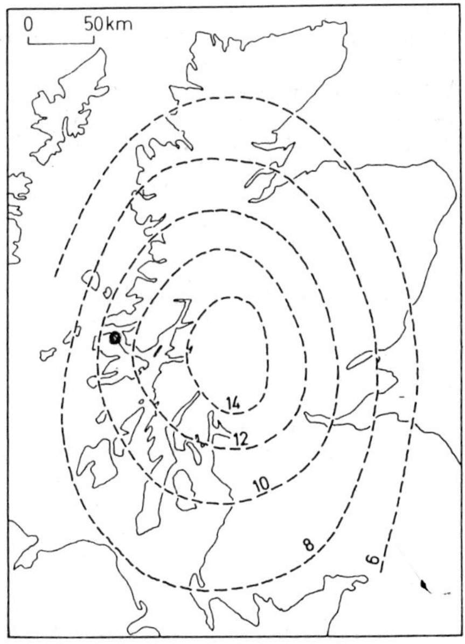

Fig. 2-4. Isobases showing the altitude in m of the Main Postglacial Shoreline, which marks the culmination of the transgression in Scotland (from Sissons, 1983).

are related to the MSL but correspond to phenomena taking place near the high tide level. They may therefore lead to an overestimation by some metres of the altitude at which MSL stood at a certain time. The average spring tidal range varies, along the coasts of Scotland, from 1.6 to 6.2 m.

Curve **B**, proposed by Donner (1970) for "central Scotland", was obtained from five sites located in a relatively small area and shows a shape similar to that of curve **A**, though with less details on the buried beaches. In the Firth of Tay, Cullingford et al. (1980) produced a relative "high water mark" (MHWST?) curve (**C**) showing at a lower altitude a pattern similar to that of curves **A** and **B**, with an additional major regression around 8000 yr BP, which would need however a confirmation and an explanation before being accepted.

In the inner Moray Firth area (**D**), Firth and Haggart (1989) identified former gravel and beach layers and estuarine flats. With the help of seven radiocarbon dates, these relic shorelines were used to construct error bands of local MHWST (Firth and Haggart, 1989) and MTL (Haggart, 1989) variations. The latter band (**D**) is reproduced in **Plate 19.**

In the Oban area (**E**), according to Synge (1977), the relative sea level was gradually dropped during the first thousand years of the Holocene. In southwestern Scotland, Holocene emergence is still of a few metres. Here stratigraphical and geomorphological evidence provided by Jardine (1975) made possible the construction of two conjectural MSL curves, for the eastern part of the northern seaboard of Solway Firth (**G**) and in the area of Wigtown Bay (**F**) respectively. Finally curve **H** was proposed by Donner (1970) for "the marginal parts of Scotland". This curve is probably less reliable, for it is expected to represent a common pattern for sites situated in different geographical positions along the west coast of Scotland, at varying distances from the isobases of the "Main Postglacial Shoreline". Although very few sea-level data from Scotland cover the last two thousand years, continuation up to the present time of a slight relative sea-level drop trend seems very likely in most places.

A comparison between uplift and subsidence trends in coastal areas around the North Sea is now possible. The rates shown in Fig. 2-5 were obtained by Shennan (1987), assuming that the regional eustatic curve proposed by Mörner (1980) could be used as a benchmark for comparison of sea-level records from different locations. The rates obtained for the most recent period vary gradually all around the North Sea, from an uplift between 1.5 and 2.0 mm/yr in the north to subsidence often exceeding 1 mm/yr in the south. It should be noted however that theese rates do not include the compaction of Holocene sediments in areas such as the Netherlands, where the sea-level data have been deduced from basal peats, whereas some compaction is probably included in other localities.

IRELAND (Plate 20, above)

During the last glacial period, the northern two-thirds of the island seem to have been ice-covered, mainly from lowland ice basins in central and north central Ireland, whereas a thicker ice

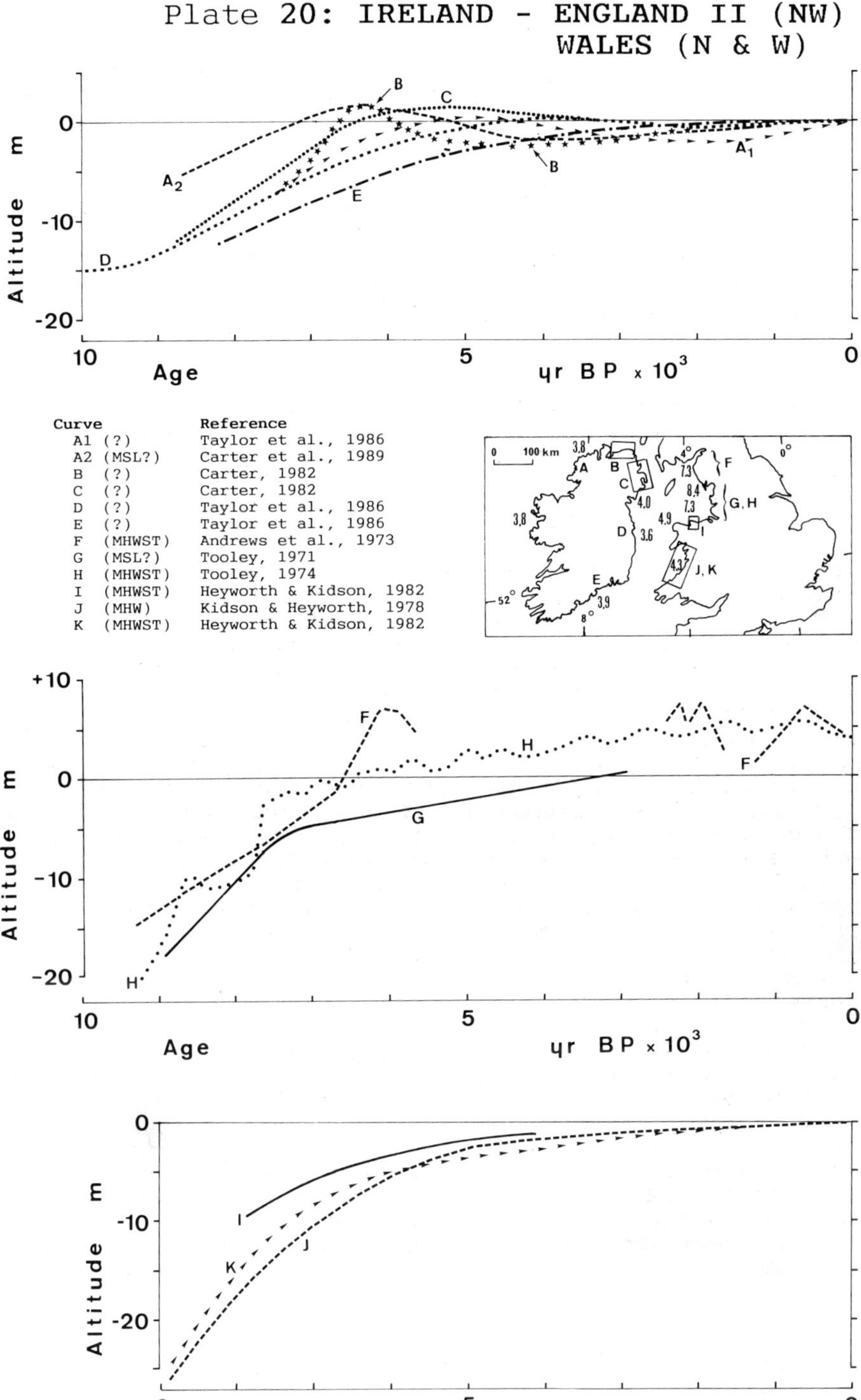
Plate 20: IRELAND - ENGLAND II (NW)
WALES (N & W)
Altitude m
Age
yr BP x 10³
Curve Reference
A1 (?) Taylor et al., 1986
A2 (MSL?) Carter et al., 1989
B (?) Carter, 1982
C (?) Carter, 1982
D (?) Taylor et al., 1986
E (?) Taylor et al., 1986
F (MHWST) Andrews et al., 1973
G (MSL?) Tooley, 1971
H (MHWST) Tooley, 1974
I (MHWST) Heyworth & Kidson, 1982
J (MHW) Kidson & Heyworth, 1978
K (MHWST) Heyworth & Kidson, 1982
100 km

sheet of Scottish origin occupied the Irish Sea and the North Channel. Ice is thought to have been largely clear of Ireland by 13,000 yr BP, although cirque glaciers may have persisted later in some areas. The interaction of isostatic and eustatic factors has produced a variety of sea-level histories in different parts of the island, which were summarized by Carter (1982) in Northern Ireland and by Taylor et al. (1986) and Carter et al. (1987) in a few other areas. The spring tidal range varies along the Irish coasts from 1 to 5 m. The highly indented west coast is subjected to high energy Atlantic swells, whereas the east coast borders the relatively low energy Irish Sea and the north and the south coasts are ones of transition. This frequency of relatively exposed coastal sectors makes the interpretation of many sea-level indicators within the reach of storm waves difficult.

Carter (1982) observed in Northern Ireland a difference between the north coast, where a sea-level peak, slightly before 6000 yr BP, seems to have been followed almost immediately by a sea-level drop (curve **B**), and the east coast where the peak seems to have lasted at almost the same elevation for a longer time (curve **C**).

In the Donegal area, the reasons for the difference between curves $\mathbf{A}_1$ (by Taylor et al., 1986) and $\mathbf{A}_2$ (by Carter et al., 1989), which are presumably based on almost the same data, were not discussed by their authors. In the whole of the north part of Ireland, though the present sea level was reached during the middle Holocene or even earlier, clear emergence marks are lacking and subsidence has been dominant for the last few thousand years, especially on the northwest coast.

On the southeast coast (curve **E**), the present sea level was reached only very recently. Lastly on the southwest coast where only a "conjectural" curve (not reproduced here) was proposed by Taylor et al. (1986), more recent data (Carter et al., 1989) suggest a gradual rise in the relative sea level since about 5000 yr BP.

It should be noted, however, that the sea-level indicators used to obtain the above graphs often seem variable and sparse and, according to Carter (1982, p. 21), many of them "would be inadmissible if more rigorous rejection criteria were to be applied". More generally, according to Devoy (1983,p. 250), "a lack of relevant data of all types" inhibits the development of detailed knowledge on late-Quaternary sea-level events in most parts of Ireland. Fortunately, the most recent work (e.g. Carter et al., 1989) shows that very promising progress is being made.

NORTHWEST ENGLAND (Plate 20, centre)

From the southern Cumberland coast, which became probably ice free about 14,000 years ago, Andrews et al. (1973) reported a discontinuous MHWST graph (**F**), suggesting that sea level could have been slightly higher several times during the Holocene than at present. The authors agreed however that fluctuations of 1 to 3 m are within the range of the chance of possible storm surges and other non-periodic events, which may have deposited above the contemporaneous MHWST level the shell samples considered.

The sea-level change curve proposed for northwest England by M.J. Tooley is well known. In an earlier version of this curve (**G**), "sea level" (MSL ?) was shown to have attained the present MSL about 3000 yr BP. Although only uniform trends of sea-level change are indicated in curve **G**, Tooley (1971) suggested that the sea-level rise was neither linear nor continuous. These ideas were developed further in subsequent papers, leading to a slightly oscillating MHWST curve (**H**) which has been reproduced in a number of papers. During the last 9000 years eleven marine transgression sequences were deduced by Tooley (1974) from stratigraphic and pollen analyses. The amplitude of the oscillations was related by Tooley (1976) "not only to eustatic changes of sea level, but also local and other factors, such as tectonic, material consolidation, tidal inequalities and changes in the geoid".

According to Shennan (1987), the northeast coast of England would be characterized by a zero uplift rate for the past 5000 or 6000 years, with an exponential decrease in the rate of uplift during the preceding period.

NORTH AND WEST WALES (Plate 20, below)

In North Wales, a MHWST curve (**I**) was published by Heyworth and Kidson (1982), on the basis of only two radiocarbon dates. In Cardigan Bay, changes in the MHW level were represented by Kidson and Heyworth (1978) by a smooth curve (**J**), although the authors agreed that the transgression was in fact largely accomplished by discrete events. Four years later, Heyworth and Kidson (1982) published a new MHWST curve for Cardigan Bay (**K**) based on two dozen radiocarbon-dated samples. The differences in elevation between the two curves **J** and **K**, which is systematic before and after 6000 yr BP, is not explained by Heyworth and Kidson (1982). These authors considered however that the possible range of errors in the sample points used were estimated more adequately in the 1982 publication than in previous studies.

BRISTOL CHANNEL (Plate 21, above)

In the Bristol Channel area, whence the largest tidal ranges in the British Islands are reported, a MHWST curve (**A**) was proposed by Devoy (1979), using 18 radiocarbon-dated index points selected from the published literature. In this study, seven marine transgressions are recognized.

The MHWST curve (**B**) by Heyworth and Kidson (1982), updating previous work in the Bristol Channel by these authors, is based on a large number of radiocarbon dates and does not show any fluctuations. Nevertheless, the possibility of occurrences of sea-level oscillations was not excluded by Heyworth and Kidson (1982); they simply cannot be identified reliably when they are smaller than the sum of other uncertainties (a detailed discussion on the various possible errors in this region is included in the above-mentioned paper).

More composite, the HWST curve (**C**) by Hawkins (1971) refers to a wide area with variable tidal ranges. Finally the MHWST curve **D** was deduced by Heyworth and Kidson (1982) from prehistoric wooden

Plate 21: BRISTOL & ENGLISH CHANNELS

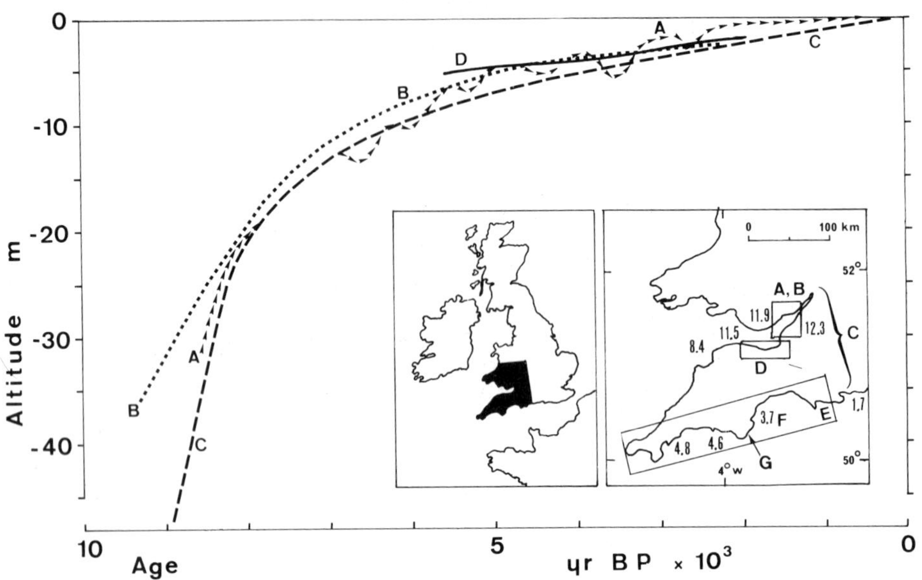

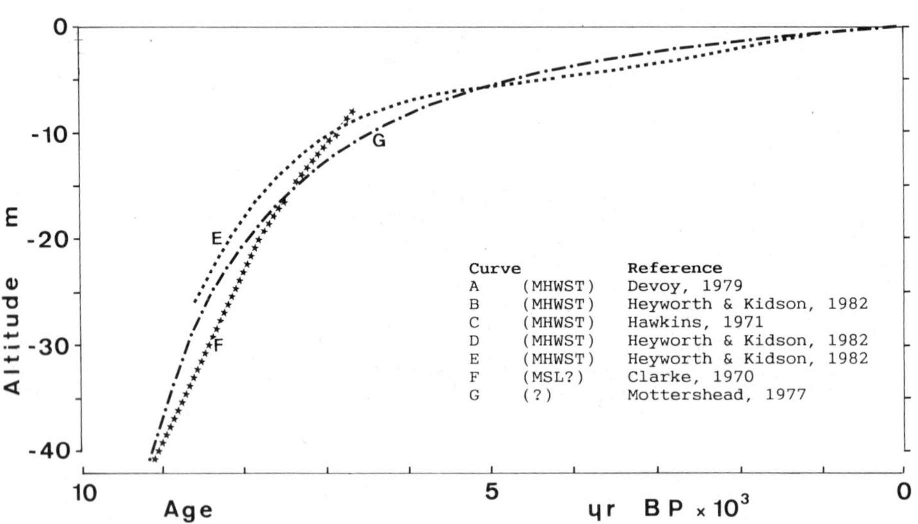

trackways in Somerset, the construction of which was considered to have been raised just above the winter flood level.

Fig. 2-6. *In situ* stump of a submerged forest appearing at low tide near Stolford, Somerset, U.K.

Fig. 2-7. On the E side of the Doge's Palace in Venice (Italy), a step (visible on the left, just above the water level) was used in 1732 as a reference sea-level mark. The datum used was the upper level of green algae, which in sheltered areas approximates the local MHW level. In 1732 the step was 17.3 cm below the upper algal marks. In 1973, due to land subsidence, increased tidal range and motor boat traffic, the upper algal marks were measured about 0.7 m above the step.

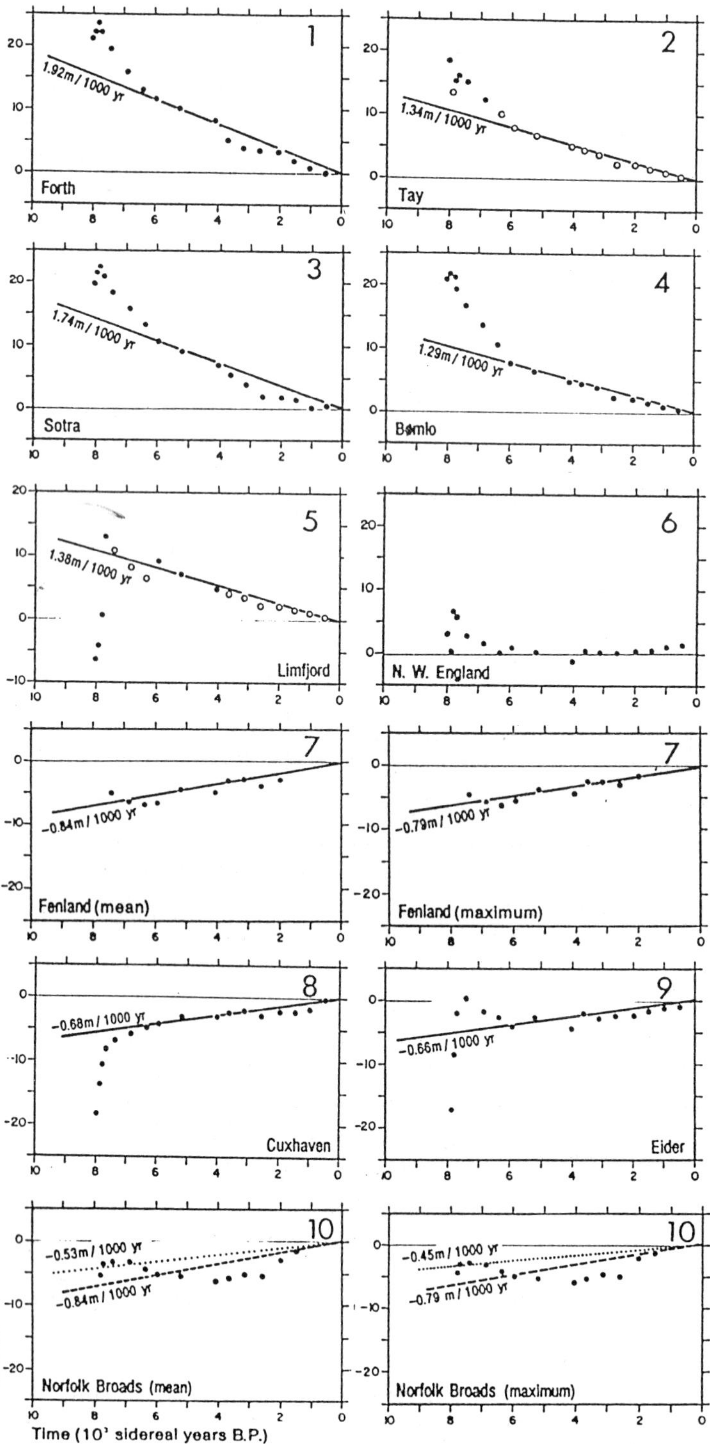
1
1.92m / 1000 yr
Forth
2
1.34m / 1000 yr
Tay
3
1.74m / 1000 yr
Sotra
4
1.29m / 1000 yr
Bømlo
5
1.38m / 1000 yr
Limfjord
6
N. W. England
7
-0.84m / 1000 yr
Fenland (mean)
7
-0.79m / 1000 yr
Fenland (maximum)
8
-0.68m / 1000 yr
Cuxhaven
9
-0.66m / 1000 yr
Eider
10
-0.53m / 1000 yr
-0.84m / 1000 yr
Norfolk Broads (mean)
10
-0.45m / 1000 yr
-0.79 m / 1000 yr
Norfolk Broads (maximum)
Metres
Time (10³ sidereal years B.P.)

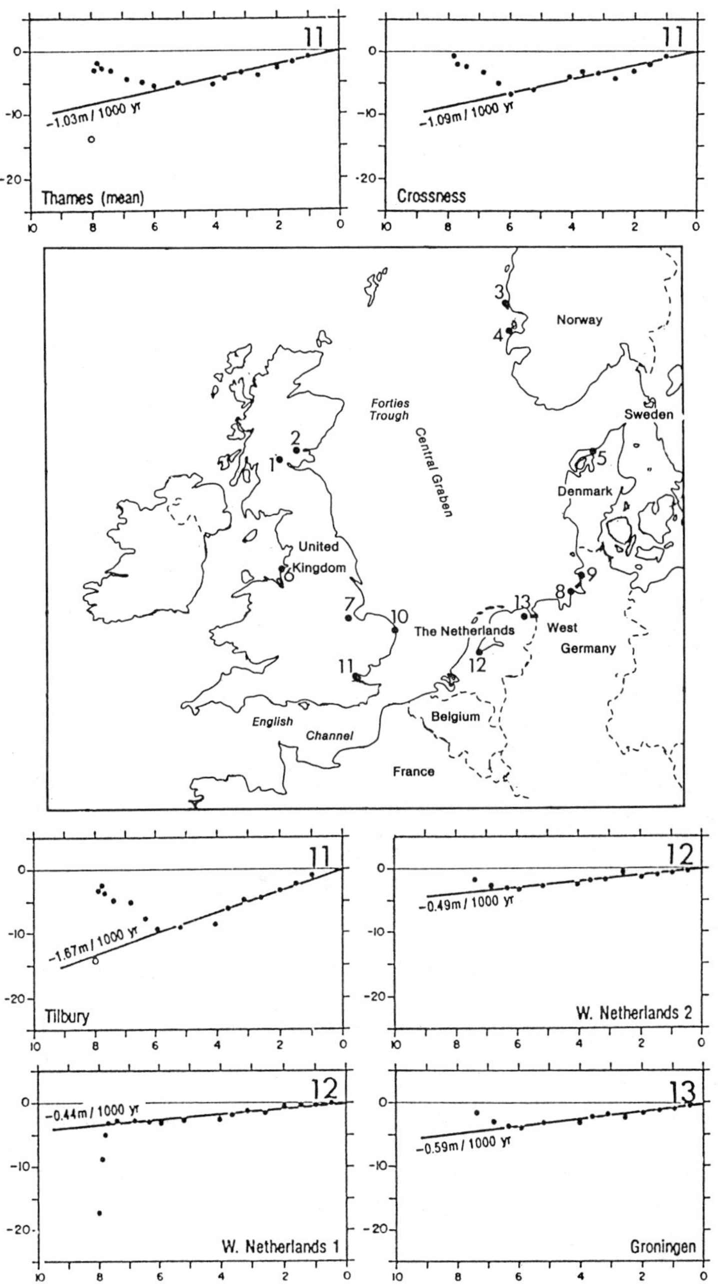

Fig. 2-5. Computed uplift and subsidence from sites around the North Sea obtained by a comparison between Holocene sea-level curves (from Shennan, 1987).

ENGLISH CHANNEL (Plate 21, below)

Heyworth and Kidson (1982) compiled the sea-level indicators available along the western part of the Channel coasts of England, obtaining, for the MHWST level, a smooth curve (**E**). More localized data were provided by Clarke (1970), using gravity cores collected from the sea floor off southeast Devon and palynological dating of intertidal muds (curve **F**). A similar curve (**F**) was reproduced by Mottershead (1977) for "sea-level" changes in Start Bay, but no details on the data and information on the reference datum used were given by this author.

BRITISH ISLES: CONCLUSION

The most reliable sea-level data from Great Britain were used by Shennan to obtain a map of estimated current rates (mm/yr) of crustal movement (Fig. 2-8). This map shows an uplift centre located in western Scotland, which is consistent with the dome-shaped isostatic recovery in this area (see Fig. 2-4) and a maximum subsidence in the Thames Estuary region.

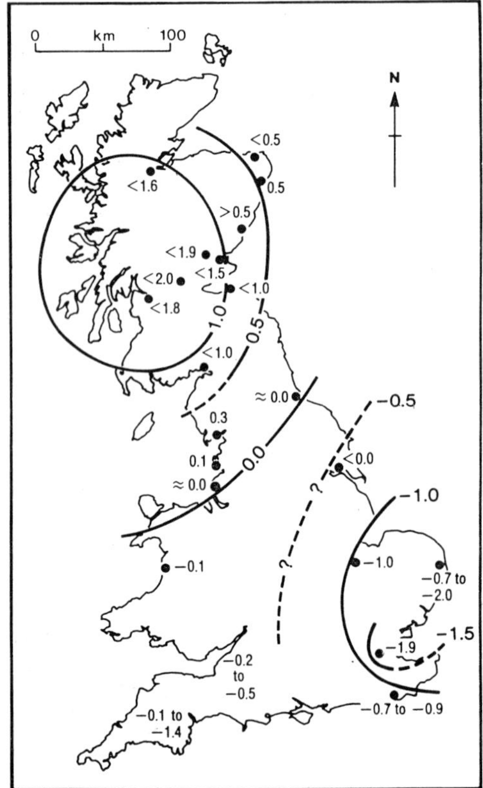

Fig. 2-8. Map of estimated current rates (mm/yr) of crustal movement in Great Britain, from Shennan (1989).

Plate 22: FRANCE I (Atlantic)

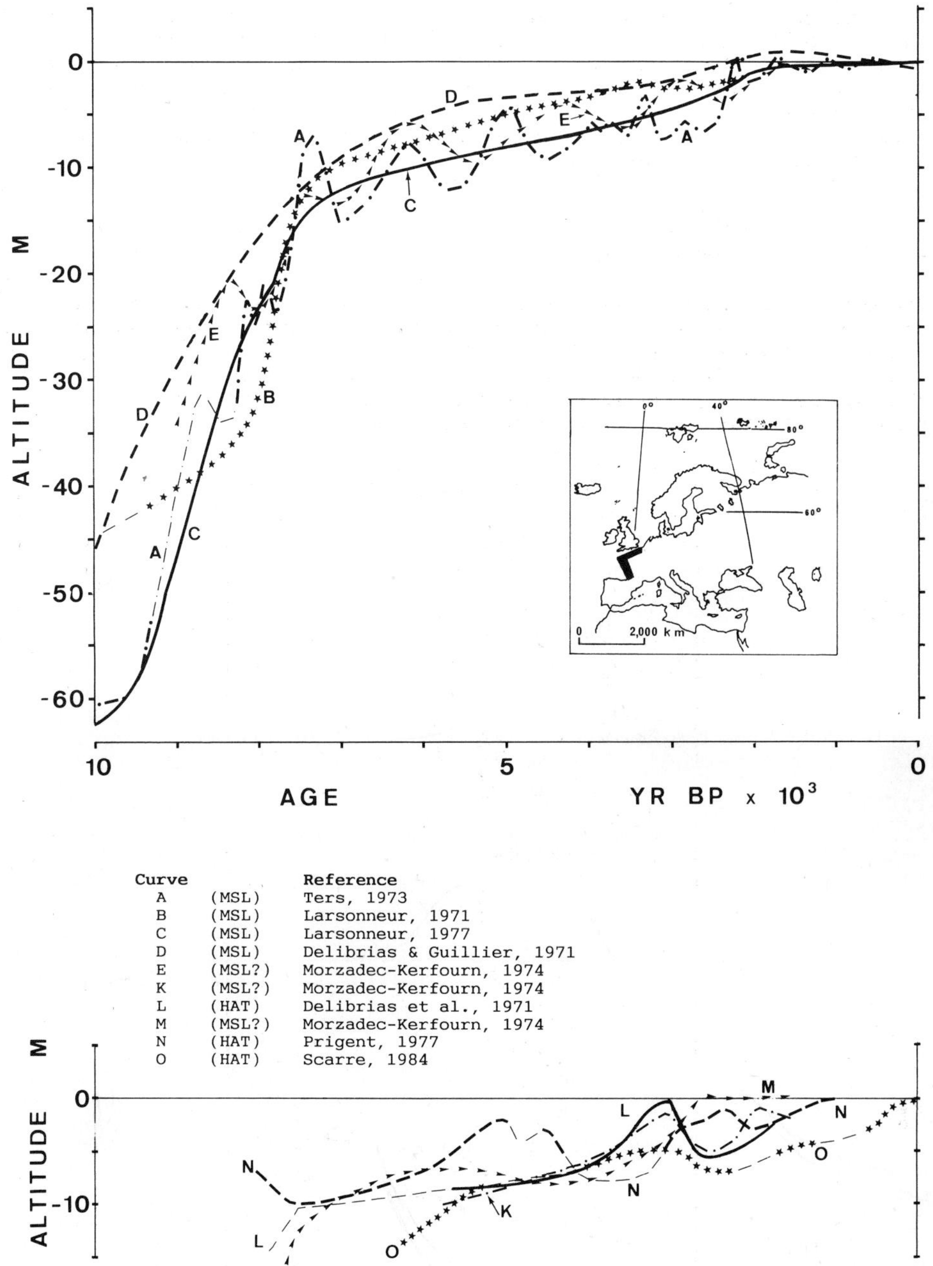

ATLANTIC FRANCE (Plates 22 and 23)

The Atlantic coasts of France belong to two main sedimentary basins (the Paris Basin to the N and the Aquitanian Basin to the SW), separated by the ancient Armorica massif stretching westwards. On the Channel coasts, limestone cliffs are frequent towards the E, whereas sand bars interrupted by capes and small marshy estuaries predominate in the W. Extensive marshes (Marais Breton, Marais Poitevin, Marais Charentais, etc.) have developed on the west coast in former bays sheltered behind sand bars or small islands. The tidal range is usually of several metres, reaching over 14 m in the Bay of Mont-Saint-Michel and changing from place to place in relation to the coastal outline. The tidal range was probably affected by important local changes during the Holocene, in relation to changes in the morphology of coastal basins, sedimentary deposition and MSL variations. The amount of tidal changes which have occurred is largely unknown, however, and this creates wide uncertainty ranges in the determination of previous sea-level positions (Fig. 1-11).

In **Plate 22** various composite curves have been assembled in the upper graph. Curve **A** by Ters (1973) is well known. It is based on data from various localities on the north and west coasts of France, under the main assumption that the tidal range did not change in the area investigated. This results in an oscillating pattern which may be explained at least in part by the assumptions made rather than by real fluctuations in MSL. The absence of uncertainty-range estimations in the Ters (1973) data prevents accurate evaluations of curve **A**.

Curve **B** by Larsonneur (1971) results from a compilation of data published by various authors and shows an estimated "average" trend along the north and west coasts of France; the occurrence of a slight sea-level oscillation is inferred between 3500 and 2500 yr BP. Subsequently however Larsonneur (1977) preferred curve **C**, which is an "average" of curve **A** by Ters (1973), to his own curve **B**.

MSL curve **D** by Delibrias and Guillier (1971) results from a compilation of over fifty radiocarbon dates of submerged peat-bogs and shell layers from sites along the north and west coasts of France. **D** is the only curve in Plates 22 and 23 suggesting a sea-level of about one metre more than the present one around 1500 yr BP; nevertheless, uncertainties on altitudes are estimated to be of the order of 1 m and are therefore of the same order as the inferred emergence.

Curve **E** was obtained by Morzadec-Kerfourn (1974) by placing side by side the parts of her curves **K**, **M** and **J** (see below) showing the most pronounced oscillations; composite curve **E** has been considered to represent sea-level changes in the Armorica region. This mixing of data is however hardly acceptable from a methodological point of view; in the absence of clearly established error boxes for each sea-level index point used, apparent oscillations may simply result from underestimated locally larger uncertainty ranges; in addition, the possibility of tidal changes has not been considered by Morzadec-Kerfourn.

While in most graphs of this Atlas the various sea-level curves are related to a vertical scale based on the present MSL, in the lower graph of Plate 22 the zero level corresponds to the

Plate 23: **FRANCE II (Atlantic)**

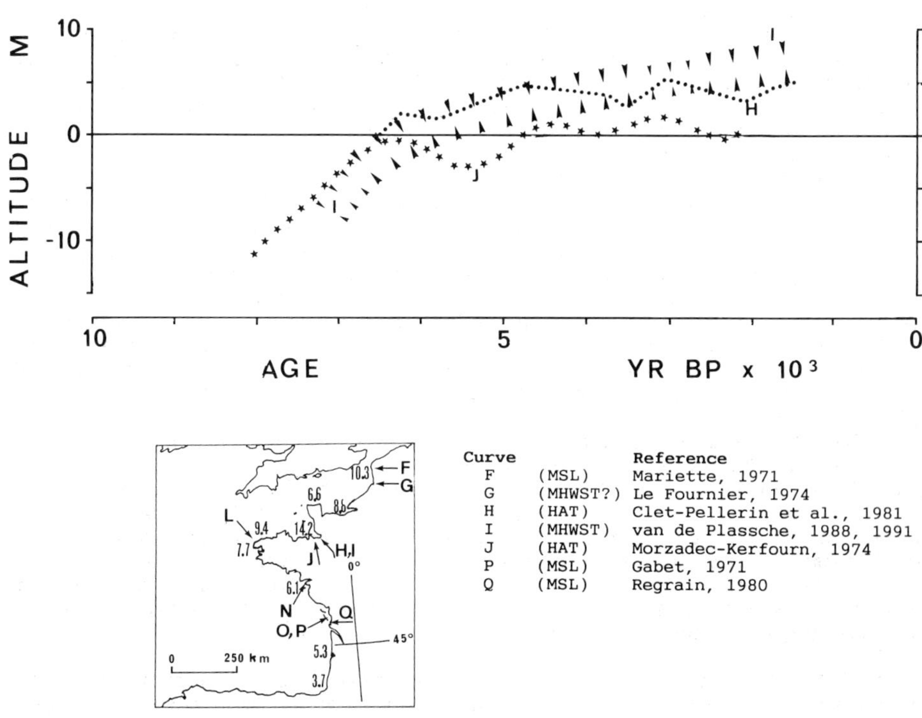

Curve		Reference
F	(MSL)	Mariette, 1971
G	(MHWST?)	Le Fournier, 1974
H	(HAT)	Clet-Pellerin et al., 1981
I	(MHWST)	van de Plassche, 1988, 1991
J	(HAT)	Morzadec-Kerfourn, 1974
P	(MSL)	Gabet, 1971
Q	(MSL)	Regrain, 1980

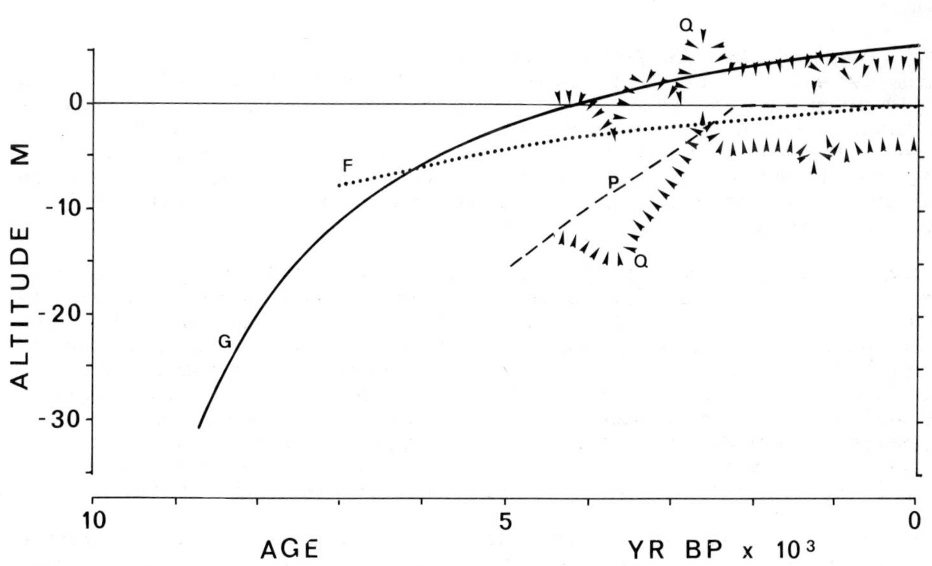

reference level used in each curve and should be read as present MSL for curves **K**, **L** and **M**, and as present HAT ("*niveau des plus hautes mers*") for curves **N** and **O**.

Curves **K** and **M** were obtained by Morzadec-Kerfourn (1974) by compiling data related respectively to "strands" and to "estuaries" in Brittany. These data were originally related to the local HAT level, then translated into MSL curves by a simple displacement in the vertical scale, in order to compare them with MSL curves by other authors; this implies an assumption that the tidal range did not change in any of the strands and estuaries considered, which seems unrealistic. On the Armorica coasts, HAT curve **L**, showing an oscillation with a sea-level peak near the present one slightly before 3000 yr BP, was deduced by Delibrias et al. (1971) from peat layers and megalithic munuments. Near the Loire River estuary, Prigent (1977) deduced from archaeological data a strongly fluctuating pattern for the HAT level (curve **N**). In west-central France, the HAT curve **O** was deduced by Scarre (1984) from evidence provided by archaeological sites. Curves **N** and **O** are however devoid of uncertainty-range estimations which would have helped to verify whether the fluctuations proposed are real phenomena, or simply the result of locally larger error boxes.

In the upper graph of **Plate 23**, all curves correspond to localities in the Bay of Mont-Saint-Michel area, where the tidal range is the largest in Europe. Curve **H** corresponds to the minimum HAT level changes proposed by Clet-Pellerin et al. (1981), based on data obtained by Lautridou in the eastern part of the Bay. **I** is a preliminary envelope for the rise of the MHWST level obtained by Van de Plassche (1988) (see also Van de Plassche, 1991) from about 300 cores collected along 11 cross-sections in the Marais-de-Dol; by comparing envelope **I** with the MSL-trend curves for the western and northern Netherlands (see Plates 16 & 17), Van de Plassche (1988) concluded that a "large" tidal range already existed in the Bay of Mont-Saint-Michel in the middle Holocene. Lastly curve **J** was proposed for the HAT level changes by Morzadec-Kerfourn (1974) in the Marais-de-Dol, using pollen analyses and radiocarbon dating of peat-layer occurrences; according to Morzadec-Kerfourn (1979-80) the vertical uncertainty range of curve **J** may be estimated between ±3 m and ±4 m and its relatively low position explained by compaction phenomena.

The gradually rising MSL curve **F** was obtained by Mariette (1971) using archaelogical and stratigraphical data from the Boulogne area, whereas curve **G** (Le Fournier, 1974) seems to correspond to variations in the MHWST level in the region near Berck.

Curve **P**, suggesting a gradual rise in the MSL from 3000 to 300 B.C. followed by relative stability along the coasts of Charentes, was proposed by Gabet (1971) on the basis of archaeological findings. Finally MSL band **Q** corresponds to the uncertainty ranges estimated by Regrain (1980) for sea-level data from the Marais Charentais. It shows clearly that curve **P** is not the only possible pattern for sea-level changes in this area.

In conclusion, many sea-level data from the Atlantic coasts of France are far from precise, the wide tidal range and the changes in the coastal environment due to heavy Holocene sedimentation being the main sources of inaccuracies. With this type of coast presenting difficulties for sea-level studies, the establishment

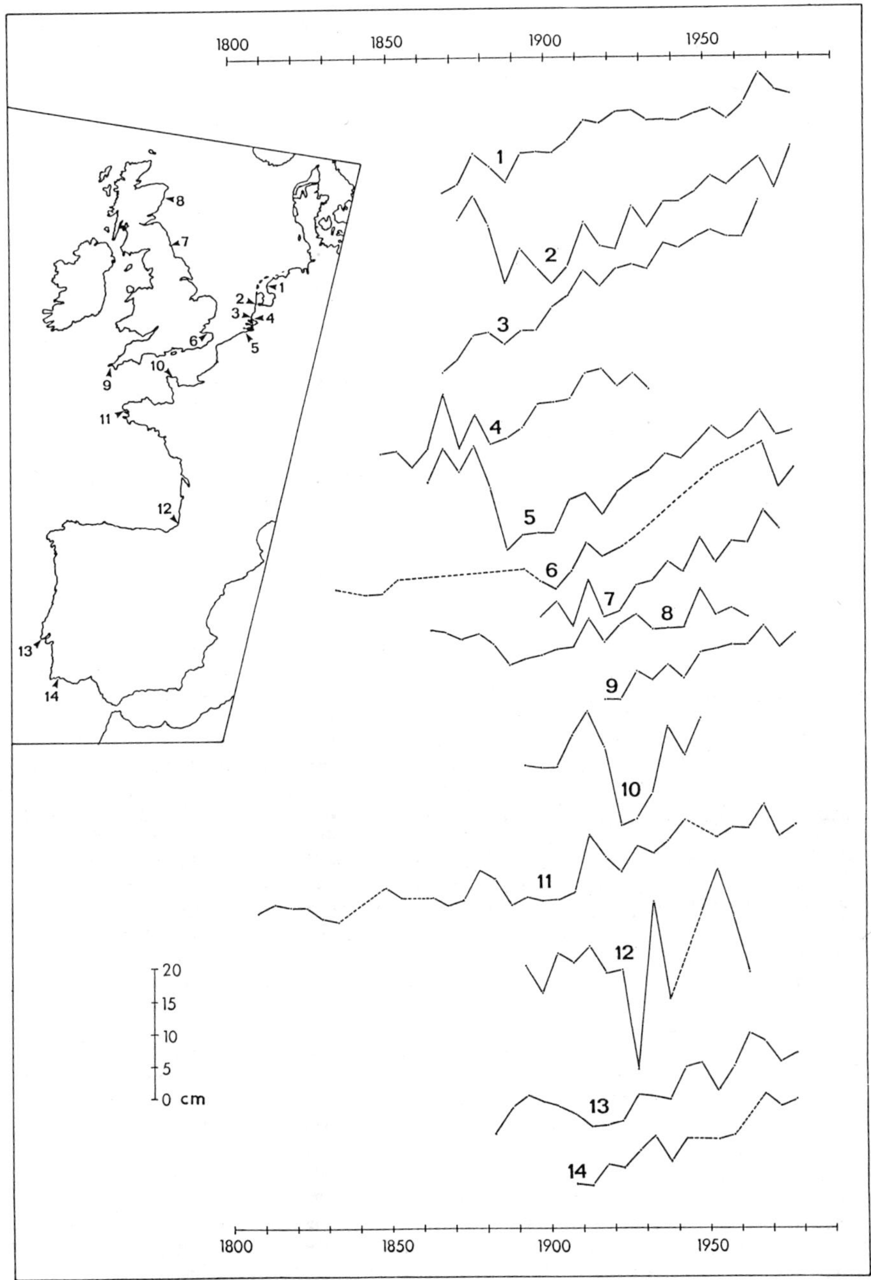

Fig. 2-9. Average 5-yr MSL variations on the Atlantic coasts of western Europe as recorded by tide-gauge stations. 1) Harlingen; 2) Ijmuiden; 3) Hoeck; 4) Maasluis; 5) Vlissingen; 6) Sheerness; 7) North Shields; 8) Aberdeen; 9) Newlyn; 10) Cherbourg; 11) Brest; 12) St. Jean-de-Luz; 13) Cascais; 14) Lagos. (From Pirazzoli, 1986b).

of accurate ranges of uncertainty for all the sea-level data available is essential, but only a very few authors have provided sufficient information. When a realistc estimation is made (e.g. band **Q**), the uncertainty range is wide enough to make unsound most of the fluctuations proposed by authors for the same area. Narrow uncertainty bands (e.g. **I**) are more difficult to obtain, since they require systematic specialized fieldwork, often repetitive, but they would be most useful.

On the whole, a gradual sea-level rise, having become probably very slow during the last 2000 years, predominated along all the coasts of north and west France during the last 7000 years, though possible uncertainty ranges prevent more accurate estimations. No reliable evidence suggests the occurrence of a sea level higher than at present.

Fig. 2-9 summarizes the present-day relative sea-level trends indicated by tide-gauge records in the Atlantic coasts of western Europe, which is situated on the glacio-isostatically subsiding belt of the Fennoscandian ice cap. All the stations show a rise of the relative sea level, with secular trends varying from 0.3 mm/yr in Aberdeen (8) to more than 2.0 mm/yr in several stations.

THE MEDITERRANEAN: INTRODUCTION

The Mediterranean Sea consists of a series of deep sedimentary basins, stretching some 4000 kilometres between Gibraltar and the Black Sea, which occupy the junction between the African-Arabian and the Eurasian continents. Together with the Black Sea and the Caspian Sea, these basins are the remnants of an ancient ocean, Tethys.

For the last 50 million years, Africa has been rotating slowly round the focal point of Gibraltar, but the relative displacements between Africa and Europe remain slight (<1 cm/yr) indicating a collision blocking the system. This collision, and the movements of the Arabian plate, are pushing Turkey in a southwestern direction, where a small oceanic space remains in the Ionian Sea. These geodynamic movements imply that great forces are at work in the lithosphere, involving considerable tectonic movements.

The Mediterranean crust is buried beneath c. 10 kilometres of sediment or more, whose weight has caused subsidence and a progressive migration of the continental margin flexure towards the continent. On the coasts the general tendency towards isostatic subsidence is overlain in places by tectonic uplift movements caused by collision or subduction phenomena.

Sea-level curves combining data from different Mediterranean regions, such as those published by Kelletat (1975), have not generally been included in the regional plates of this Atlas, since they do not refer to a single area and their fluctuations can be explained by differential tectonic movements rather than by local relative sea-level changes.

"Best fit" relative sea-level curves deduced from archaeological sites have been calculated by Flemming and Webb (1986), using 3rd and 4th degree equations, for various Mediterranean regions (e.g.: the Peloponnese, Crete, Rhodes, etc.). These are however areas of complex tectonism, with closely-spaced discontinuities in their geological structure. This kind of "best fit" curves can therefore scarcely be applied to single sites. Flemming and Webb (1986) have also proposed a "best fit" curve for all the Mediterranean archaeological data, which suggests that the "average" relative sea level has always remained above -12 m for the last 10,000 years. This is clearly an overestimation of the position of ancient sea levels, demonstrating that the results of the analysis have been biased by incorrect data.

The Black Sea was linked to the Mediterranean only during main interglacial periods. The Caspian Sea (presently at an altitude of about -28 m) may have overflowed towards the Black Sea, during certain periods when its level was higher than that of the Black Sea.

Along the coasts of the Mediterranean, detailed Holocene sea-level curves have only been provided in a few areas, mainly in southern France, the Aegean and the Levant; the major gaps include Spain, the Adriatic and most of the African shores.

CENTRAL MEDITERRANEAN (Plate 24)

On the continental shelf of Roussillon, several mollusc shell samples, collected from borings, have enabled Labeyrie et al. (1976) to construct a sea-level curve covering the last 30,000 years, the Holocene part of which (**A**) is reproduced in **Plate 24.** The most detailed study on the Gulf of Lions is probably the one by Aloisi et al. (1978), which includes data from many borings in the coastal plain and beach sites, as well as palynological observations from lagoonal sediments. Based on a set of some 80 radiocarbon dates, Aloisi et al. (1978) defined a sea-level band (**B**) showing a Holocene sea-level rise of over 50 m, culminating locally at +2 m near Cap Romarin between 5000 and 4000 yr BP, and almost no change during the last 2000 years. Curve **C** by de Lumley (1976) is a compilation of secondhand data for the entire Mediterranean coast of France and does not show any sea level higher than at present. Curve **D**, by Dubar (1987), is based on borings collected in the Nice area and on new dates of vegetal remains for the period 7500 to 4000 yr BP. The older part of curve **D** is less reliable, being deduced from data obtained elsewhere by other authors; the most recent part, suggesting a 2 to 4 m emergence around 4000 yr BP was deduced from undated beach deposits.

In the Rhone Delta area, L'Homer et al. (1981) dated marine shells collected from a series of littoral bars, obtaining a fluctuating curve (**E**) for the last 7200 years. These oscillations were ascribed to changes in sea level. The construction of littoral bars depends mostly though on wave action and differences in elevation between bars may reflect the strength of isolated storms rather than eustatic changes.

At Port Cros, in the Hyères Islands, Laborel et al. (1983) studied underwater superimposed rims bioconstructed by *Lithophyllum tortuosum* (a calcareous alga only developing very near MSL, and deduced several former MSL positions during the last 3000 years (crosses in the **F** inset in the lower graph; above the inset, curve **F** summarizes the local trend at Port Cros). A very similar sea-

Plate 24: CENTRAL MEDITERRANEAN

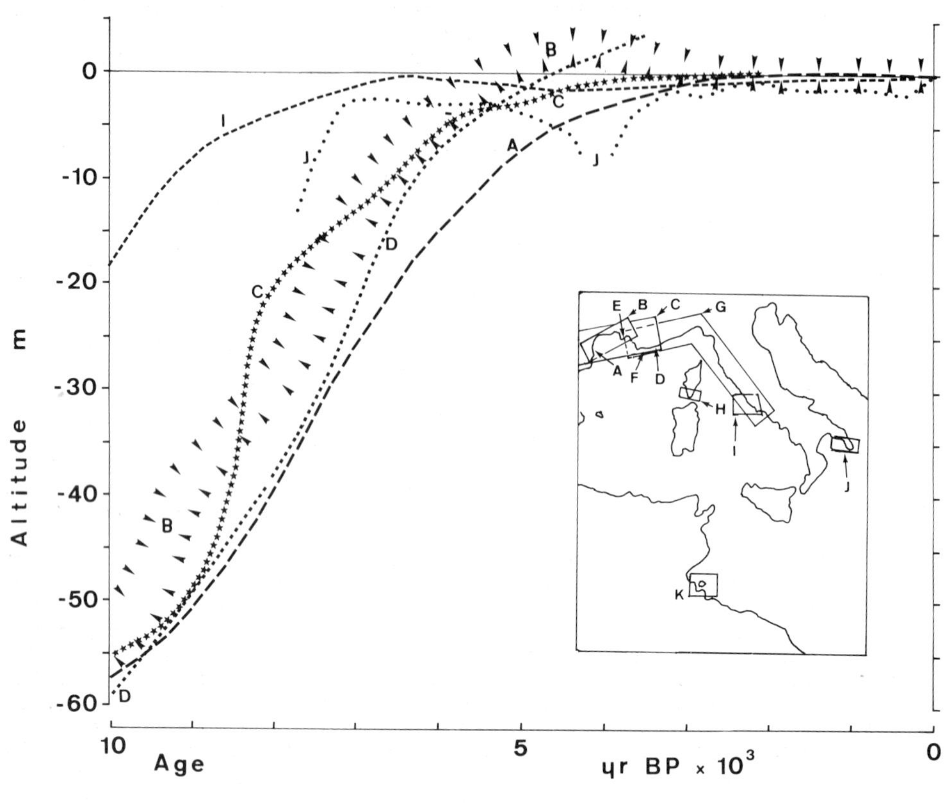

Curve	Reference
A	Labeyrie et al., 1976
B	Aloisi et al., 1978
C	de Lumley, 1976
D	Dubar, 1987
E	L'Homer et al., 1981
F	Laborel et al., 1983
G	Pirazzoli, 1976a,b
H	Nesteroff, 1984
I	Antonioli & Frezzotti, 1991
J	Dai Pra & Hearty, 1991
K	Paskoff & Sanlaville, 1983

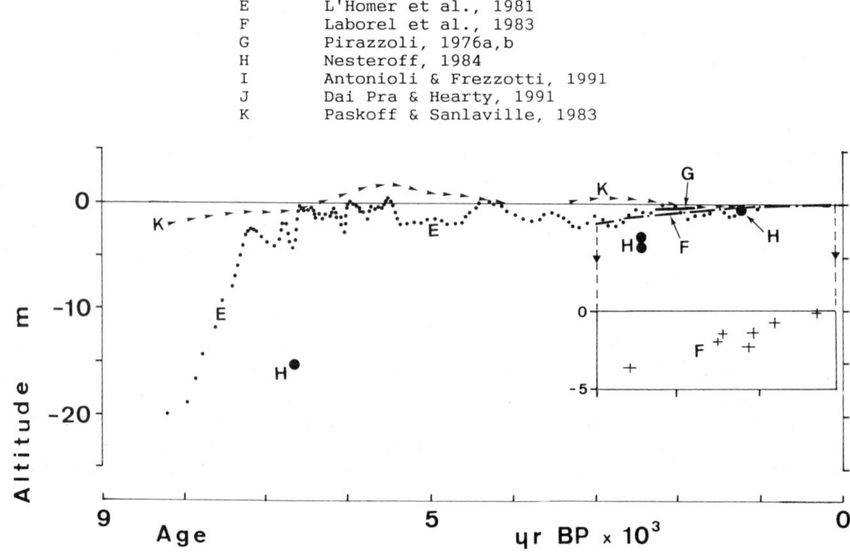

level trend (**G**), showing a relative rise of 0.75 mm/yr during the period from 300 B.C. to A.D. 100, was found by Pirazzoli (1976a,b) along the Provençal and Tyrrhenian coasts from Marseilles (France) to Formia (Italy), using archaeological remains, mainly Roman fish tanks (Fig. 1-6), though after having omitted data from some areas where major tectonic movements are evident.

In South Corsica, Nesteroff (1984) obtained four sea-level data points (**H**) by dating calcareous cements from submerged beachrock formations. It is not specified, however, whether the cements dated are all typical of a vadose environment (i.e. corresponding to the time of lithification of the beachrocks, in the midlittoral zone). The presence of phreatic micrite cements, deposited underwater after the submergence of the beachrocks, would in fact have the effect of rejuvenating the samples.

Although the Holocene transgression is often called "Versilian" in the Mediterranean, resulting from the pioneer stratigraphic work by Blanc (1936) in the Versilia Plain (Italy), few reliable detailed sea-level data based on radiometric or palynological datings have yet been obtained on the Italian coasts.

On the south Lazio coast, Central Italy, Antonioli and Frezzotti (1991) proposed relative sea-level curve **I**, based on one dozen radiocarbon dates, some of them new (on shells and on the humic fraction of paleosoils); curve **I** suggests that a level similar to the present one was reached by the sea between 7000 and 5400 yr BP, followed by a slight regression-transgression oscillation. The sea-level positions indicated by curve **I** seem however poorly constrained in the early Holocene. On the Salentine Peninsula, curve **J** was inferred by Dai Pra and Hearty (1991) from isoleucine epimerization ratios "calibrated" by using three radiocarbon dates; it suggests that at any moment sea level was higher than the present one during the Holocene, though relatively high levels between 7000 and 5000 and between 4000 and 2000 yr BP and various small oscillations were deduced from intercalations of paleosoils in aeolian deposits.

An attempt was made in the northern Adriatic region by Bortolami et al. (1977). Due to the absence of sufficiently well established relationships between the samples analysed and sea level, and to the probable existence of differential compaction movements in the wide area considered, the data obtained by Bortolami et al. (1977) appear too scattered to be used in the construction of local relative sea-level curves.

In Venice, a comparison of present-day water markings on certain buildings in the city (Fig. 2-7), with similar data recorded in historical documents, showed that some buildings indicated a relative sea-level stability from A.D. 1360 to A.D. 1872, whereas other buildings with insufficient foundations suggest the occurrence of a relative sea-level rise, though changing from place to place (Pirazzoli, 1974).

Lastly, on the southern side of the Mediterranean, Paskoff and Sanlaville (1983) proposed for the south coasts of Tunisia sea-level curve **K**, which shows slight fluctuations around the present MSL during the last 8000 years. The data provided to justify the troughs of curve **K** come however from other areas of Tunisia, which are devoid of marks of Holocene emergence. In a later study (Paskoff and Oueslati, 1991), submerged ruins of Roman times were

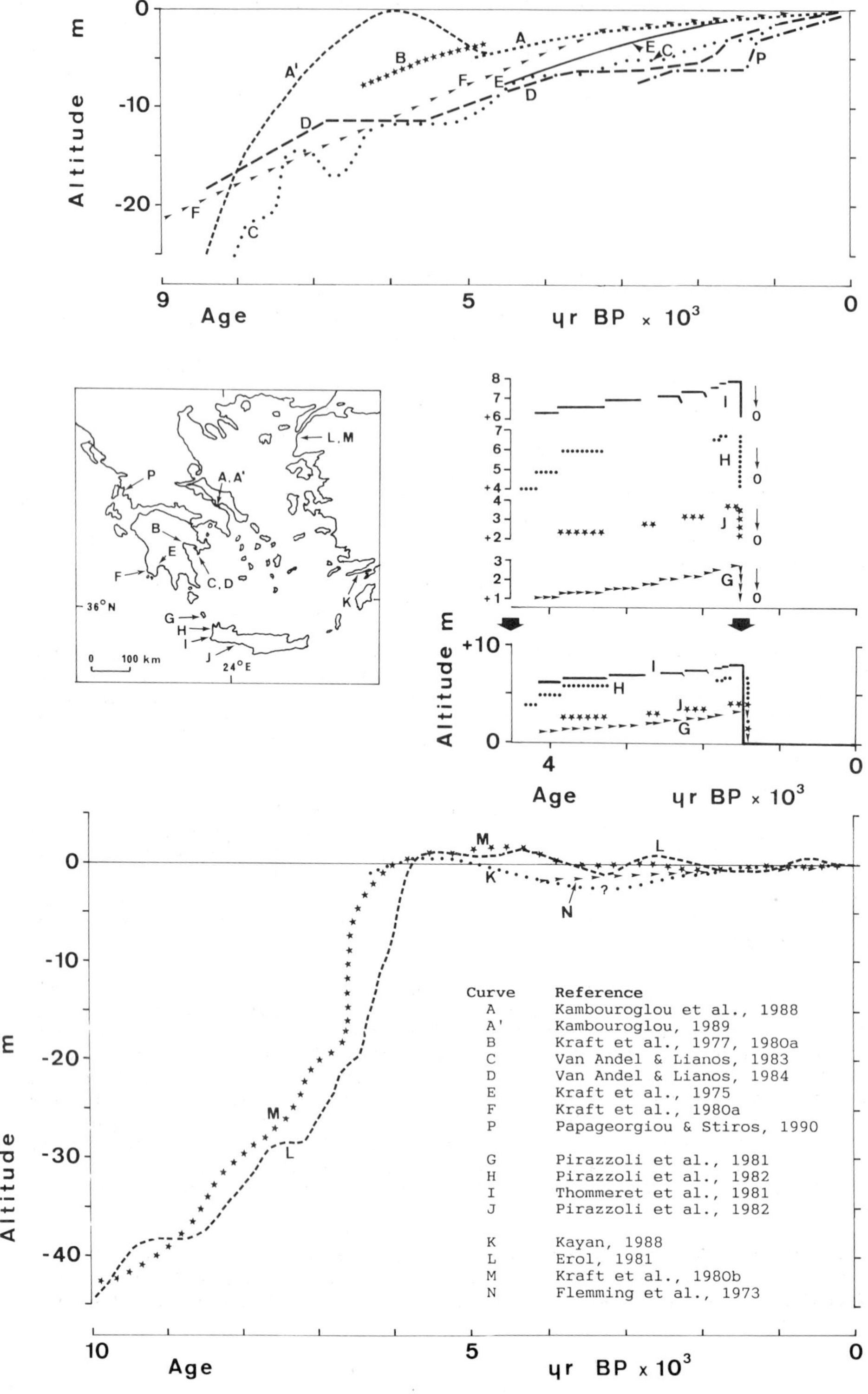
Plate 25: AEGEAN
Altitude m
Age
yr BP x 10³
A
A'
B
C
D
E
F
P
G
H
I
J
K
L
M
N
L, M
A, A'
C, D
36°N
24°E
0 100 km
Curve Reference
A Kambouroglou et al., 1988
A' Kambouroglou, 1989
B Kraft et al., 1977, 1980a
C Van Andel & Lianos, 1983
D Van Andel & Lianos, 1984
E Kraft et al., 1975
F Kraft et al., 1980a
P Papageorgiou & Stiros, 1990
G Pirazzoli et al., 1981
H Pirazzoli et al., 1982
I Thommeret et al., 1981
J Pirazzoli et al., 1982
K Kayan, 1988
L Erol, 1981
M Kraft et al., 1980b
N Flemming et al., 1973

reported also from various sites along the Gulf of Gabes, though at depth changing from a few tens of centimetres in the south part of the Gulf, to about 2 m in the Kerkennah archipelago.

Regarding the recent relative sea-level trends, an almost stability or a very slight sea-level rise (of the order of 0.2-0.3 mm/yr) seems predominant in the few localities studied in the central Mediterranean region.

THE AEGEAN (Plate 25)

Along the coast near Khalkis (**A**), Kambouroglou et al. (1988) gathered archaeological and geomorphological information from 13 sites, which they used to construct relative sea-level curve **A**, showing a gradual rise of the sea during the last 5000 years. Kambouroglou (1989) added two new points to almost the same data, deduced from an offshore peat, c. 25-m deep, and from a 6000-yr old coastal deposit near the present sea level at Eretria, to construct curve **A'**, extending **A** to before 5000 yr BP, and he suggested that curves **A** and **A'** may correspond to eustatic changes in the area. However the area considered is probably tectonically active, since evidence of a late Holocene seismotectonic uplift can be observed about 20 km northwest of Khalkis and along the north coast of Euboea Island (Pirazzoli et al., in preparation) and differential motions up to 3 m (2 m subsidence and 1 m uplift), which occurred along a distance of less than 20 km in the last 2000-3000 years in the same area (Stiros et al., in preparation), clearly cannot be considered as eustatic effects.

In the plains of the Argolid (**B**) Kraft et al. (1977) used two cores and three radiocarbon dates to describe the subsurface changes in the environment. The data were subsequently interpreted by Kraft et al. (1980a) and relative sea-level curve **B** was proposed in this area for the period 6200 to 4900 yr BP.

In the framework of a study aimed at mapping submerged shores on the shelf of the southern Argolid, using high-resolution seismic profiling, Van Andel and Lianos proposed successively two relative sea-level curves: **C** (1983) and **D** (1984). The difference between these curves, which are based on the same archaeological data, seems to result from the fact that undated palaeoshore features, revealed by seismic profiling, were interpreted as being due to sea-level fluctuations in curve **C**.

In the coastal plain of the Gulf of Messenia (**E**), Kraft et al. (1975) deduced a maximum relative sea-level curve between 5600 and 2000 yr BP from four datings of plant debris collected by boring. Indications exist here suggesting that the relative sea level moved sharply upwards about 5000 yr BP, possibly reaching an extreme of 9 m within 350 years (not shown in **Plate 25**). In the Bay of Navarino (**F**) the relative sea-level changes defined by Kraft et al. (1980a) from an environmental interpretation of the subsurface strata, covers the last 9000 years; a smooth curve was suggested, though the possibility of fluctuations was not excluded.

In northwest Greece, Papageorgiou and Stiros (1990) deduced from archaeological data of the eastern coast of Levkas Island and the nearby continental shore, a relative sea-level band (**P**) for the

Fig. 2-10. Ripple notches indicate at least two slightly raised Holocene shorelines above the present sea level at Strongilo Islet, S-E Crete, Greece. (From Pirazzoli, 1979-80a).

Fig. 2-11. A series of ripple notches appear on the limestone cliffs near Cape Ladiko (Rhodes Island, Greece), making it possible to distinguish at least five elevated Holocene shorelines between 1.6 and 3.75 m above the present sea level. The local spring tidal range is less than 0.3 m (see also Fig. 2-16, crustal block B).

Fig. 2-12. A thin white rim bioconstructed by live vermetids *Dendropoma* and calcareous algae *Neogoniolithon notarisii* appears at sea level fringing the limestone coasts of Antikythira Island (Greece). This rim is a very precise sea-level indicator. The local spring tidal range is less than 0.3 m.

Fig. 2-13. Stepping raised rims of *Dendropoma* and *Neogoniolithon* near Moni Khrisoskalitisas (East coast of Crete, Greece) enabled a series of small late Holocene uplift movements, probably of co-seismic origin, to be dated by radiocarbon (see Plate 25, curve I).

Fig. 2-14. *Dendropoma* vermetids have built a series of pools at sea level on the south coast of Crete. The almost horizontal upper limit of the pools is a very precise sea-level indicator. Arrow: fossil *Dendropoma* pools raised at about 4.5 m above the active ones (see Fig. 2-15).

Fig. 2-15. Detail of the raised *Dendropoma* pools visible in Fig. 2-14. The very good state of preservation of the pools in such an abrasive environment indicates a very rapid (probably sudden, co-seismic) uplift.

last 3000 years, showing a long-term gradual rising trend, interrupted however by a period of relative stability between 300 B.C. and 600 A.D., and a rapid, possibly sudden 1-2 m subsidence movement completed by 750 A.D.

In Antikythira (**G**) and western Crete (**H, I, J**) a number of emerged stepped shorelines, clearly visible on the limestone cliffs (Fig. 1-18, 2-10, 2-12, 2-13, 2-14, 2-15) have made possible a detailed reconstruction of late Holocene crustal movements in this area. It appears that, between 4000 and 1700 yr BP, a series of ten small movements of rapid subsidence affected without noticeable tilting a huge block of lithosphere about 150 km long. This sequence of subsidences was followed about 1530 yr BP by a sudden tremendous uplift, reaching about 9 m in the southwest corner of Crete, and a northeastward tilting (Thommeret et al., 1981; Pirazzoli et al., 1981, 1982). Curves **G to I** correspond to the relative sea-level changes recorded in various parts of the tilted lithosphere block: **G** in Antikythira Island; **H** in Phalasarna, where an ancient Greek harbour was uplifted more than 6 m above present sea level; **I** in Moni Khrisoskalitisas, near the area of maximum uplift; lastly **J** in Timios Stavros. Each emerged shoreline can be identified with very good altimetric accuracy (up to ±0.1 m) from erosional ripple notches carved in the limestone rock and from superimposed rims, bioconstructed mainly by the calcareous alga *Neogoniolithon notarisii* and the vermetid *Dendropoma petraeum*, which characteristically develop in rims at sea level, thus providing excellent material for dating former shorelines. Similar sequences of subsidence movements, followed by a sudden huge uplift, are not unknown in areas of crustal subduction, where they are related to isostatic and crustal rebound processes. The case of Crete and Antikythira, in which a complex sequence of vertical movements of a decimetric order could be identified and dated in detail, is however quite unique in the literature. This case study has been made possible by several very favourable attendant circumstances: almost negligible tide amplitude, the presence of encrusting organisms developing in calcareous rims at sea level, a hard rock basement difficult to erode and moderate exposure. The strong crustal displacements observed in Crete and Antikythira around 1530 yr BP are the most spectacular result of a major tectonic event, called by Pirazzoli (1986) the Early Byzantine Tectonic Paroxysm, which affected many areas of the eastern Mediterranean, leaving emerged shorelines in an area at least 1200 km wide, including parts of Anatolia, Syria, the Lebanon, Cyprus and possibly other areas now under study in Greece.

The assumption that earth movements occur at a constant rate (always upwards or always downwards) was used by N.C. Flemming to interpret local evidence of relative sea-level changes in several Mediterranean areas (e.g. Flemming, 1972, 1978; Flemming and Webb, 1986). Such an assumption, which may be acceptable in some cases, could lead however to misleading interpretations if applied to areas and/or to time periods characterized by reversed tectonic movements (e.g. in western Crete since 4000 yr BP). In the island of Rhodes, for instance, according to Flemming and Woodworth (1988), vertical land movements deduced from a short period of tide gauge records and from a former shoreline assumed to be 5000 years old, would indicate the occurrence of uplift movements at a constant rate. As a matter of fact, as shown by Pirazzoli et al. (1989), the long-term uplift trend of Rhodes Island takes place in a much more complex manner than Flemming and Woodworth believed (Fig. 1-15, 2-11, 2-16).

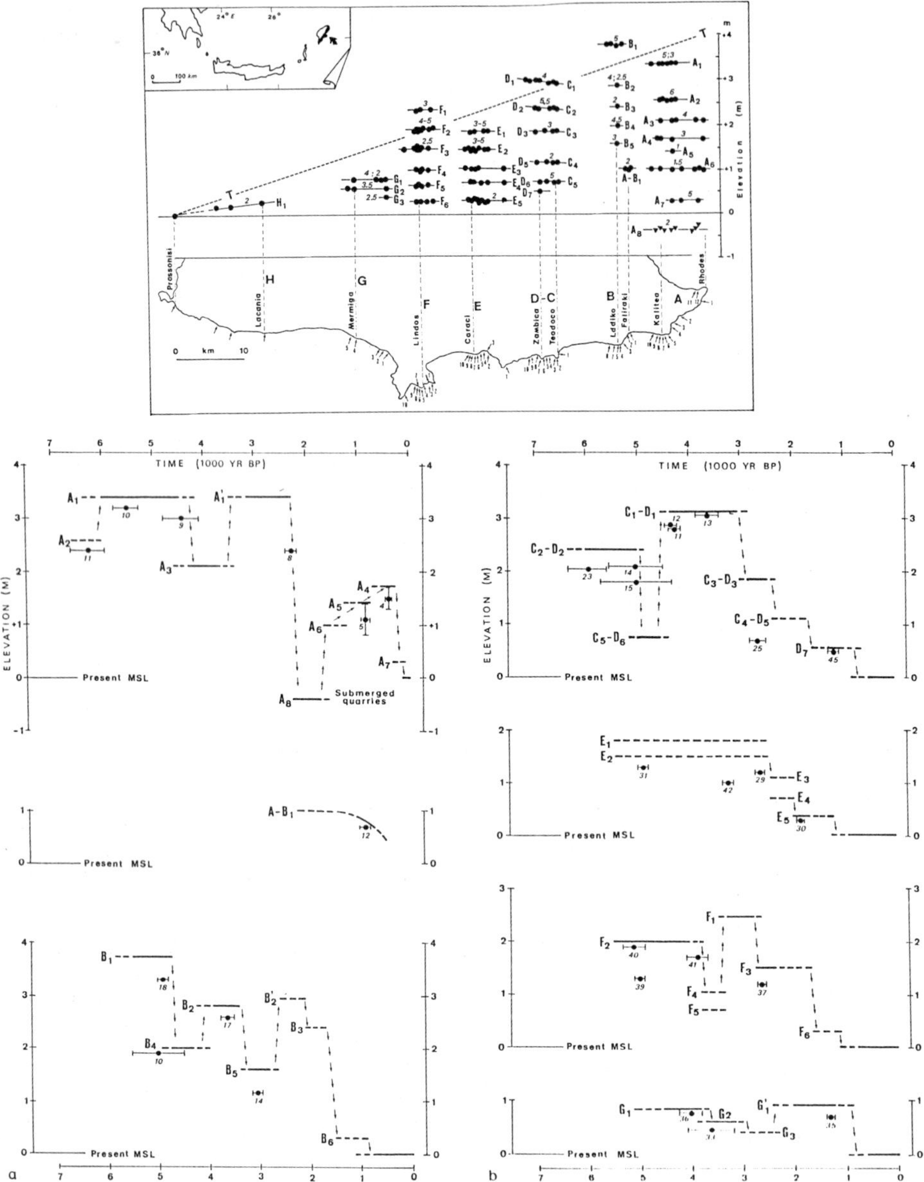

Fig. 2-16. General pattern of the Holocene shorelines and relative sea-level changes along the east coast of Rhodes Island (Greece). Above: projection of shorelines on a vertical plane; dots: emerged marks of shorelines; triangles: position of submerged quarries; T-T: average uplift trend; numbers above shorelines: approximate ages x 1000 yr BP. Middle: location of the projected shorelines. Below: relative sea-level changes: independent up and down movements appear in crustal blocks A to G. (From Pirazzoli et al., 1989).

In the Dalacak area, on the site of Old Knidos (**K**), the relative sea-level curve proposed by Kayan (1988) shows a pattern in which the tectonic jerks, so frequent in the nearby island of Rhodes, seem to be apparently completely absent. According to archaeological and geomorphic data, sea level in Dalacak reached the present position (or even 0.5 m higher) at a period assumed to be about 6000 yr BP, then perhaps descending as much as 2 m, until again reaching the present position some 1500 years ago.

Flemming et al. (1973), after having surveyed archaeological remains of three dozen ancient coastal sites between Elaea (Kazikbaglari) and Andriace-Myra (Kale), provided an average displacement curve for southwest Turkey (**N**) showing a gradual relative sea-level rise of almost 2 m during the last 4000 years. However, several sites show a submergence greater than average and other sites no submergence at all, so it is not clear in which locations the average curve **N** could be applied.

Erol (1981) and Kraft et al. (1980b) published two relative sea-level curves (**L** and **M**) for the Biga Peninsula in the Dardanelles area, which seem to be based on the same data (a few radiocarbon dates of unspecified material, Greek and Roman sherds, and undated traces of pumice on a marine terrace at +2 m). According to these curves, sea level would have risen up to 2 m above the present one in approximately 6000 years BP, with later signs of several fluctuations up to 1.5 m above and below present sea level. The evidence provided to support these views (and explain the differences between the two curves **L** and **M**) is however slight.

THE LEVANT, THE BLACK SEA, THE CASPIAN SEA (Plates 26 and 27)

For the Eastern Mediterranean several sea-level curves are available. In southeastern Cyprus, Gifford (1980) provided a sea-level trend in Larnaca, based on six radiocarbon dates, showing a relative rise of about 2.5 mm/yr between 6000 and 0 B.C. (**Plate 26**, curve **A**). In the Hatay area (Turkey), Pirazzoli et al. (1991) dated two elevated superimposed organic rims, consisting of a veneer of *Dendropoma* vermetids capping oyster shells and *Corallina* algae (Fig. 1-19), on both sides of the Orontes Delta plain, showing that the relative sea level dropped here in two sudden steps, probably at the time of two major earthquakes, around 2500 and 1500 yr BP. The latter crustal movement, which corresponds to a 80-cm uniform uplift affecting most of coastal Hatay, including the ancient harbour af Antioch, Seleucia Pieria, is thought to be related to the Early Byzantine Tectonic Paroxysm, which has already been mentioned regarding the data from Crete.

In the Lebanon, a double emerged rim, similar to and of the same age as those mentioned in Hatay, were interpreted by Dalongeville (1983), according to a previous interpretation by Sanlaville (1977), as corresponding to smooth changes of eustatic origin. However, a double uplift movement of tectonic origin seems a more likely explanation than eustatic changes for the elevated Holocene shorelines in the Lebanon (Pirazzoli, 1986a; Pirazzoli et al., 1991).

Along the coast of Israel, Galili et al. (1988) investigated submerged archaeological sites on the continental shelf between Haifa and Atlit. Their relative sea-level curve (**D**) shows a smooth

Plate 26: THE LEVANT - BLACK SEA I - CASPIAN SEA

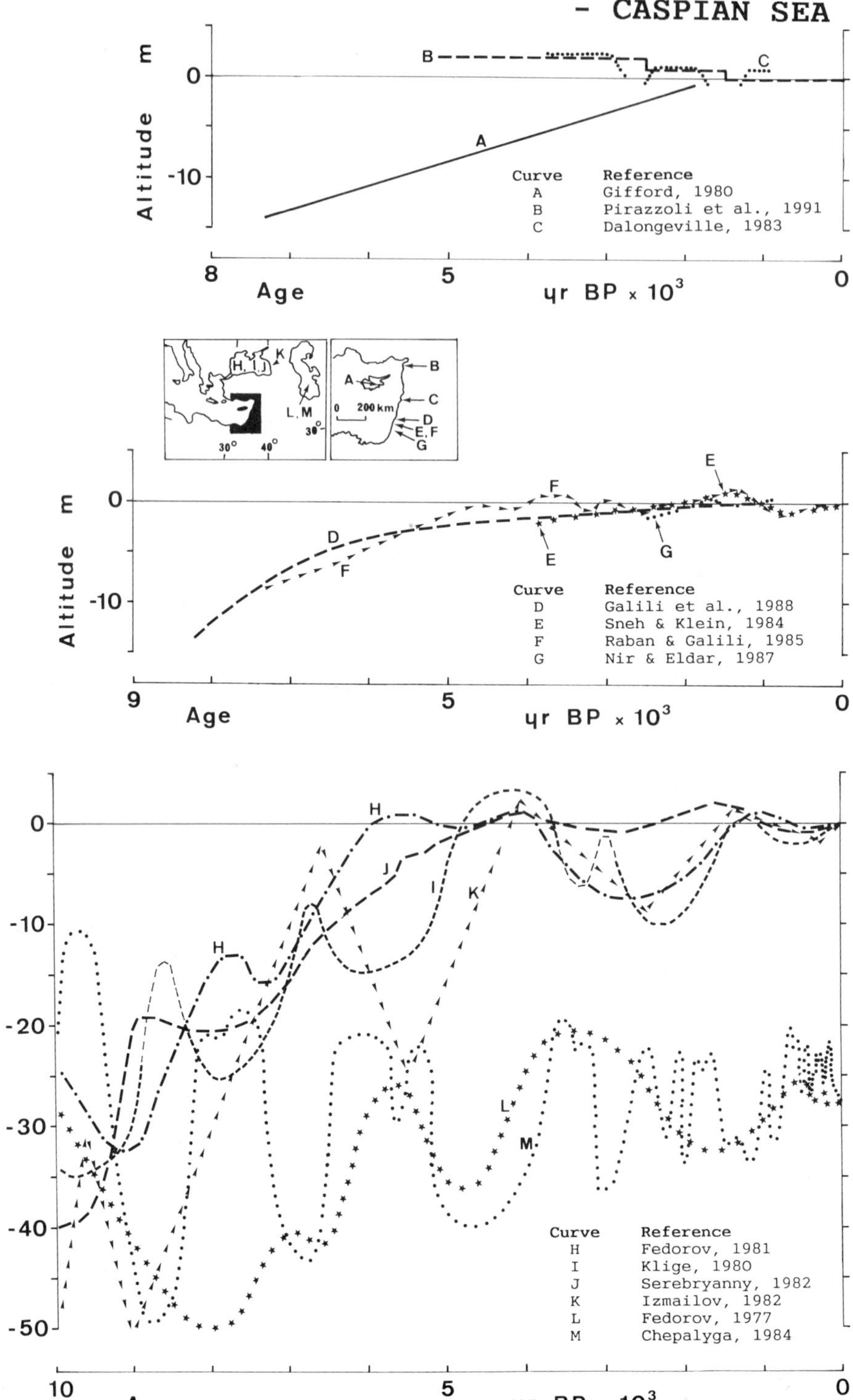

Plate 27: BLACK SEA II

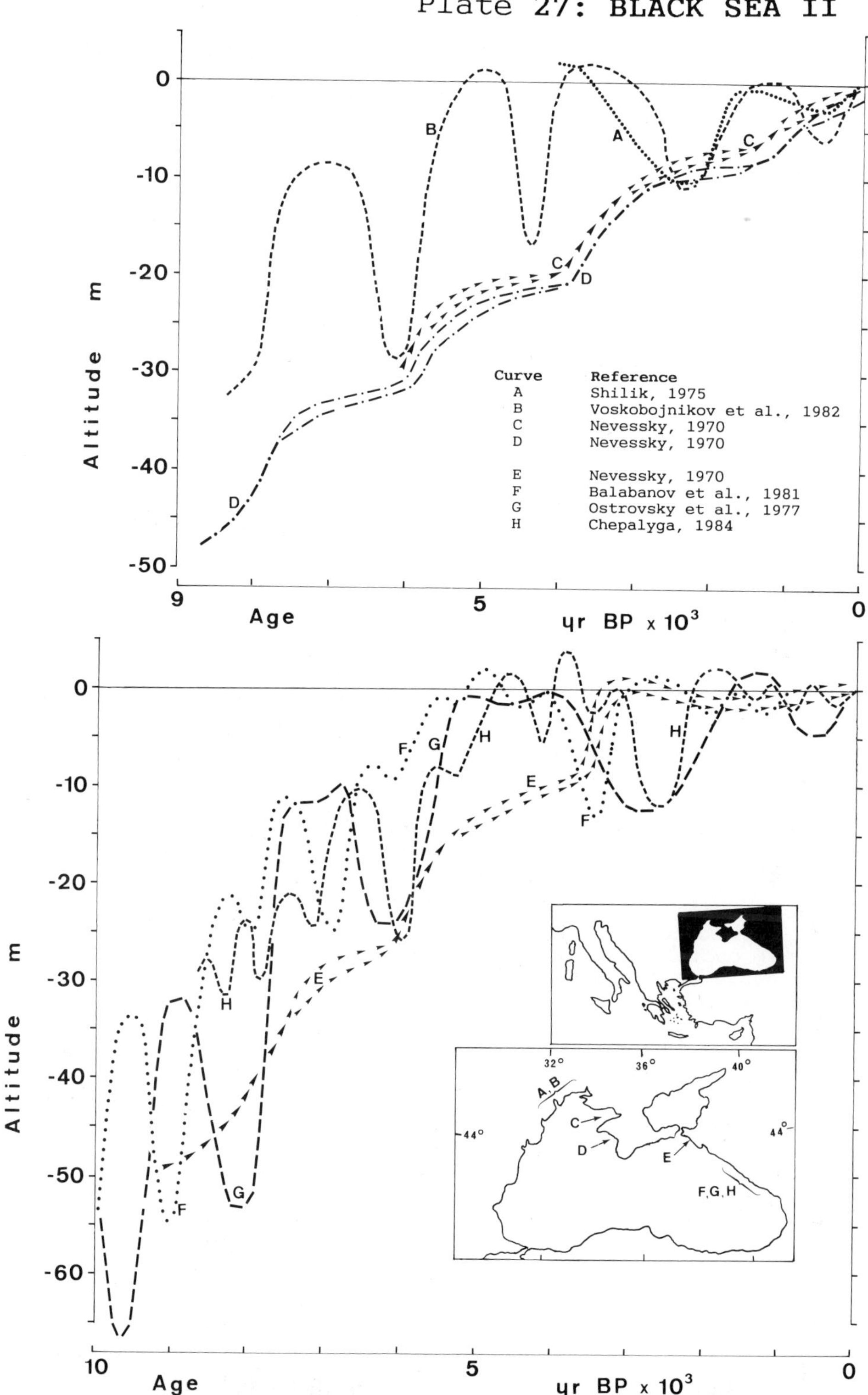

rise, reaching the present sea level slightly earlier than 1000 yr BP and is consistent with most of the data points reported previously by Flemming et al. (1978). At Dor, curves **E** by Sneh and Klein (1984) and **F** by Raban and Galili (1985) are almost the same for the last 2500 years, showing a sea-level fluctuation of over 2 m in amplitude, but differ for earlier periods, when additional fluctuations are suggested by curve **F**, at a higher level than the gradual rise proposed by curve **E**. Lastly, Nir and Eldar (1987) deduced from water table marks in different wells re-excavated along the coastal plain of Israel a curve (**G**) for the last 2500 years showing small oscillations which differ slightly from those proposed by the curves **E** and **F**.

The trends of relative sea-level change shown by curves **D**, **E**, **F**, **G** can be considered very moderate if compared to those proposed by Neev et al. (1987), who suggest the occurrence of rapid up-and-down movements, reaching 10 to 40 m in amplitude, almost everywhere in Israel. Nevertheless, a comparison with other plates of this Atlas concerning tectonically active areas makes the interpretation by Neev et al. (1987) doubtful.

Water exchanges between the Mediterranean, the Black Sea and the Caspian Sea were mainly controlled by the variations in their surface levels, due to eustatic, climatic and tectonic changes and to the elevation of the Bosphorous sill (at about -35 m, whereas the present-day level of the Caspian Sea is about -28 m). In late Glacial times the Black Sea and the Caspian Sea were still isolated from the Mediterranean, the Bosphorous discharging brackish Black Sea water into a lower Mediterranean. The hydrologic balance of the Caspian Sea differed from the one of today, as suggested by presently inactive river beds in central Asia, such as the Ouzboï, which were once flowing into the Caspian.

At certain times, when the level of the Caspian was higher than that of the Black Sea, a discharge system was activated in the Manych-Kuma area, decreasing the differences in level between the two basins. This means that sea-level fluctuations in the Black Sea may have been different from those in the Mediterranean or in the Caspian Sea, but were in some way linked to each other beyond certain altimetric limits. After the Mediterranean waters had overflowed into the Bosphorous sill in the early Holocene and a stable connection with the Black Sea was established, several sea-level oscillations have been claimed to have occurred in the latter basin, although authors disagree on their amplitude, age, and even on their existence.

According to Fedorov (1971, 1981) (**Plate 26**, curve **H**), after a period with a level near the present sea level between 6000 and 4000 yr BP, a sea-level drop of 5-7 m (Fanagorian regression) could have taken place, enabling several ancient Greek sites (Dioskouria, Gorgippia, Phanagoria, Panticapaeum, Olbia) to be established 3-5 m below present sea level. Later, a new sea-level rise no doubt occurred, surpassing the present level by 0.5-1.0 m in the middle of the first millennium A.D. Wide sea-level fluctuations are suggested also by Klige (1980) (curve **I**), Serebryanny (1982) (curve **J**) and, for the Caucasus coast, by Izmailov (1982) (curve **K**). Other sea-level curves for the Black Sea are given in Plate 27.

In the Caspian Sea, which was probably at a higher level than the Black Sea in the early Holocene, four major oscillations of decametric order have been proposed by Fedorov (1977) for the last 10,000 years (curve **L**) and as many as ten oscillations by Chepalyga (1984) (curve **M**). Detailed information on the evidence of these oscillations is not available to the present author. During historical times, according to Julian et al. (1987), the level of the Caspian Sea was high around 1700 yr BP, low (-30.5 m) at 700 yr BP and high again (-25.3 m) in the 18th century. The last peak of high sea level (A.D. 1929) was followed by a rapid 3 m drop, stabilization in 1975-77 and a new 1 m rise during the last decade. The shores of the Caspian Sea are in fact a key area where some potential effects of a global sea-level rise, which has been predicted by climatic models for next century, could be tested at the present time.

Curves **H**, **I** and **J** shown in **Plate 26** seem to be composite, i.e. based on compilations of data reported from various localities around the Black Sea, which may have experienced different tectonic movements. Their fluctuations may therefore be only apparent.

In the northwestern part of the Black Sea, Shilik (1975) used data from various archaeological remains, submerged at different depths between 0 and -3 m, to construct a sea-level curve (**A**, **Plate 27**) in which he includes though a 10 m oscillation between 4000 and 2000 yr BP which is not documented locally. In about the same area, Voskobojnikov et al. (1982) deduced from sedimentary sequences several 20 to 30 m wide fluctuations, which seem quite unlikely. More realistically, Nevessky (1970) published three relative sea-level curves, deduced from sequences of coastal shelf deposits, for the Karkinit Gulf (**C**), the Kalamit Gulf (**D**) and the Anapa town region (**E**) respectively. Nevessky (1970) explained the vertical differences between these curves, which may reach a dozen metres, by differential tectonic movements.

On the tectonically active Caucasian coast, most of the sea-level curves available show a number of strong oscillations, which vary however with each author considered. This can be seen, for instance, by comparing curve **F**, obtained by Balabanov et al. (1981) near Pitzunda, using several borehole data and a dozen radiocarbon dates, with curve **G** by Ostrovsky et al. (1977), reproduced from Badyukov and Kaplin (1979, Fig. 4), based on 34 radiocarbon dates of mollusc shells and peat bogs and on archaeological and historical data. Ostrovsky et al. (1977) believe that the decrease in sea level accompanying the Fanagorian regression reached -10 to -15 m and even that a -2 to -5 m "Medieval regression" might have occurred in the 14th and 15th centuries A.D.

Lastly, curve **H** by Chepalyga (1984), compiling data from the two preceding curves **F** and **G** and other data, show even more different trends. In fact much field work is necessary in order to clarify the sea-level fluctuations which may have occurred in the Black Sea during the Holocene. One first useful step would certainly be a more critical approach to the sea-level indicators chosen and the systematic use of realistically estimated error margins for each sea-level index point. This would probably help prove that several of the sea-level oscillations claimed are in fact unsound.

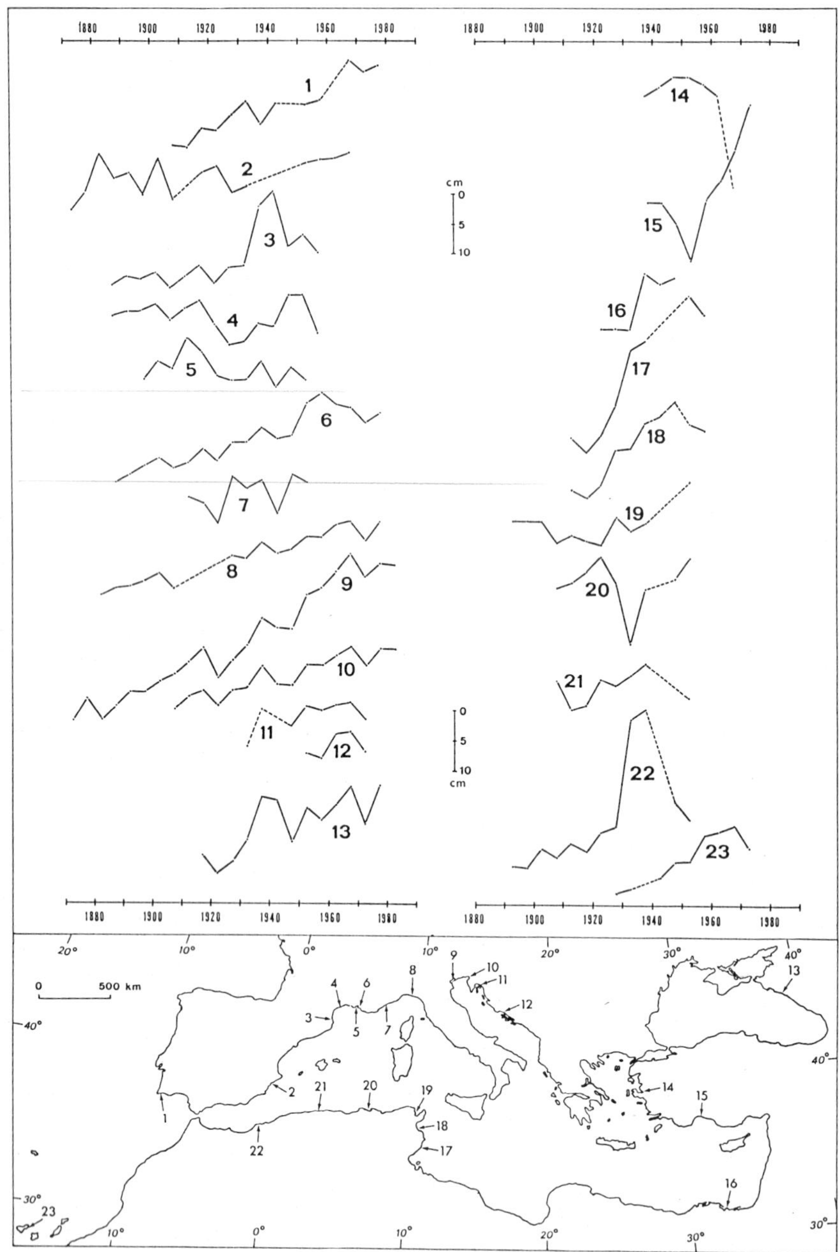

Fig. 2-17. Average 5-yr MSL variations in the Mediterranean as recorded by tide-gauge stations. 1) Lagos; 2) Alicante; 3) Port Vendres; 4) Sète; 5) Martigues; 6) Marseille; 7) Villefranche; 8) Genova; 9) Venezia; 10) Trieste; 11) Bakar; 12) Split; 13) Tuapse; 14) Izmir; 15) Antalya; 16) Port Said; 17) Sfax; 18) Sousse; 19) La Goulette; 20) Bône; 21) Alger; 22) Oran; 23) Santa Cruz. (From Pirazzoli, 1986b).

Recent relative sea-level trends shown by tide-gauge records in the Mediterranean and the Black Sea are summarized in Fig. 2-17, where most of the series examined suggest a tendency to a sea-level rise, though at rates varying from one place to another.

WEST AFRICA (Plates 28 and 29)

Being remote from Quaternary ice sheets, the African coasts have experienced only minor isostatic vertical movements. Geodetic relevellings are rare in most countries and approximate values of vertical movements can be deduced from geomorphological evidence and denudation rates calculated only over relatively long intervals of time. For the Holocene period relative sea-level changes provide therefore very useful information on active vertical movements, whatever their origin. Unfortunately, detailed field investigations have been carried out only in a limited number of coastal sectors and many areas remain to be studied.

In **Plate 28**, curve **A** by Delibrias (1974) corresponds to the Holocene part of a 26,000 yr-long curve aimed at synthetizing sea-level variations on the west coasts of Africa. It consists of two parts: (1) earlier than 7500 yr BP, the solid part of curve **A** was obtained by mixing data collected at depths between -43 m and -100 m by boring or dredging from the continental shelf of the Ivory Coast and Congo (these data were also used to construct curves A and K in Plate 29); (2) later than 7500 yr BP, two dashed curves define a band which envelops a number of local sea-level curves from Morocco to Ghana (see also curves **B**, **C**, **F**, **J** and **Q**). Such a mixing of data from areas remote from each other is however less meaningful than local relative sea-level curves, and unacceptable from a methodological point of view.

Curve **B**, from Morocco, was deduced by Delibrias (1974) from three data suggesting a 2 m sea-level drop during the last 6000 years. For the coasts of Mauritania, several sea-level curves have been published. Curve **C** by Delibrias (1974) suggests a sea level 1 to 2 m higher than at present from about 6000 to 4000 yr BP. Curve **D** by Einsele et al. (1974) shows a 6.5 m sea-level fluctuation, from +3 m about 5500 yr BP to -3.5 m about 4100 yr BP, which is based on *in situ* emerged *Arca senilis* shells for the higher sea level, and on stromatolites considered to have developed in the former intertidal zone for the lower sea level. According to Barusseau et al. (1989), however, the accuracy of stromatolitic carpets as bathymetric markers would be much less precise than assumed by Einsele et al. (1974) and the possibility of a sea-level drop greater than 5 m around 4100 yr BP is excluded on the Mauritania coast. Sea-level band **R** by Barusseau et al. (1989), which is based on 16 new radiocarbon dates of the main sedimentary evolution stages in the Banc d'Arguin area, shows a clear reduction in the amplitude and number of fluctuations in relation to most previous studies in this area. For the period prior to 7000 yr BP, Einsele et al. (1977) deduced curve **E** from beach deposits collected by coring from the continental shelf off Nouakchott.

Various relative sea-level fluctuations have been proposed in the areas between Cape Blanc and Cape Timiris (curve **F**) and between Nouakchott and Sebkha Ndrhamcha (curves **G** and **H**) by Faure and Hebrard (1977) and Faure (1980). Altimetric differences of about 4 m between curves **G** and **H**, which seem based on almost the same

Plate 28: WEST AFRICA I

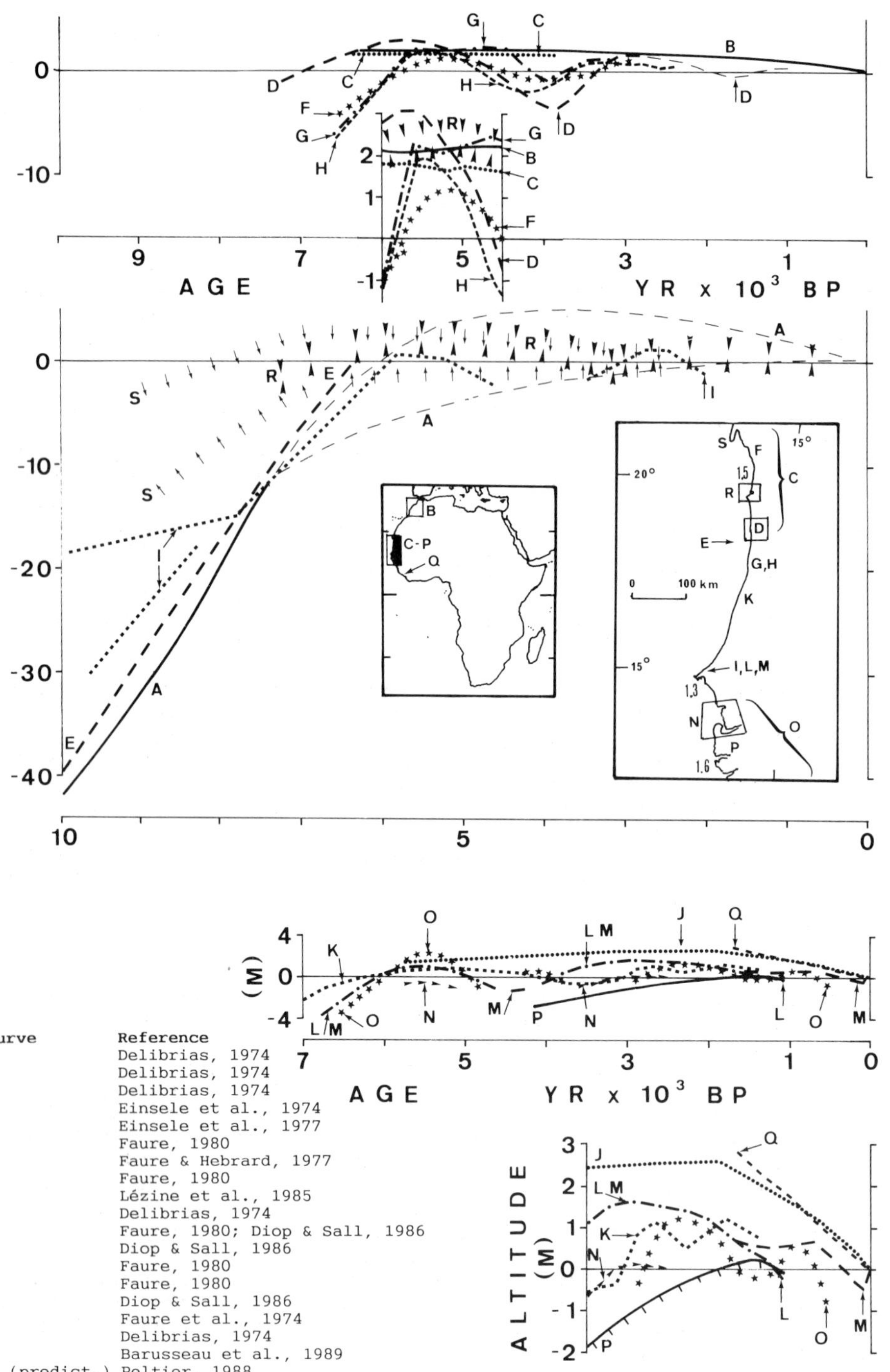

Curve	Reference
A	Delibrias, 1974
B	Delibrias, 1974
C	Delibrias, 1974
D	Einsele et al., 1974
E	Einsele et al., 1977
F	Faure, 1980
G	Faure & Hebrard, 1977
H	Faure, 1980
I	Lézine et al., 1985
J	Delibrias, 1974
K	Faure, 1980; Diop & Sall, 1986
L	Diop & Sall, 1986
M	Faure, 1980
N	Faure, 1980
O	Diop & Sall, 1986
P	Faure et al., 1974
Q	Delibrias, 1974
R	Barusseau et al., 1989
S (predict.)	Peltier, 1988

data, e.g. near 4500 yr BP (see upper inset in Plate 28), remain however unexplained. At Cap Blanc, sea-level predictions by the Peltier (1988) model are delimited by band **S**.

For the area including Aftout es Saheli and the Senegal River delta, a recent sea-level curve (**K**) by Diop and Sall (1986) is very similar to a previous curve published by Faure (1980). Curve **J** was obtained by Delibrias (1974), by mixing data from the Senegal River delta, the Dakar region and the volcanic Cape Verde Islands, 500 km off the coast of West Africa.

On the coasts of Senegal, a comparison of present-day altitudes of ancient marine shorelines in the Lake Tanma area has enabled Lézine et al. (1985) to suggest possible trends of local relative sea-level changes (curve **I**), which imply the occurrence of two periods during which Lake Tanma was connected with the sea. Near Dakar (Fig. 2-18), curve **L** by Diop and Sall (1986) differs from curve **M** by Faure (1980) only between 5000 and 4000 yr BP, when the sea level position is considered undetermined by Diop and Sall.

Curve **N** by Faure (1980) consists of only two segments, suggesting that sea level was twice near its present position in the Saloum-Gambia area, around 5500 yr BP, and between 3500 and 2500 yr BP. Curve **O** suggests four sea-level stands higher than at present (details on the data used were specified by Diop, 1990), but these have been obtained by Diop and Sall (1986) by mixing data from Saloum-Casamance and Guinea Bissau; yet in Casamance no evidence of clear emergence was reported by Faure et al. (1974) (curve **P**). Curve **Q** by Delibrias (1974), lastly, which is based on three dates of emerged vermetids, suggests a 2.5 m sea-level drop since 1400 yr BP on some islets facing Freetown (Sierra Leone).

The differences between some of the above curves have been interpreted by Faure (1980) as due to hydro-isostatic phenomena, and possibly to geoidal changes, e.g. a long wave length vertical deformation of the coast. A continental flexure at right angle to the coast has also been active, as shown by Faure et al. (1980), though its effects were smaller than those predicted by the global isostatic model by Clark et al. (1978). Barusseau et al. (1989), on the other hand, ascribe certain sea-level fluctuations, their lateral shift and change in amplitude along the coast, to diachronic phenomena induced by the morphosedimentary evolution of the coast.

In **Plate 29**, curve **A** was constructed by Martin and Delibrias (1972) and Martin (1973) using six peat samples and twelve calcareous algae nodules collected from the continental shelf of the Ivory Coast. Curve **B**, reproduced by Giresse et al. (1986), is probably the same as **A**, though it appears slightly shifted to the right. For the period earlier than 8000 yr BP, a variant in the same region (curve **C**) was proposed by Fredoux and Tastet (1976), on the basis of a palynological analysis of two additional peat samples. Always in the same area, curve **D** by Tastet (1981) seems however slightly different, though some divergence may result from inaccurate drawing. A sea-level band (**E**) was proposed for the last 5000 years in the Ivory Coast by Pomel (1979), using vermetids, peats and shell deposits collected not far from the present shore. Curve **F**, lastly, was sketched by Delibrias (1974), assembling one date from the Ivory Coast and three dates from the coasts of Ghana, and curve **G** by Paradis (1976), using eight sea-level data from the coast of Bénin.

Plate 29: WEST AFRICA II

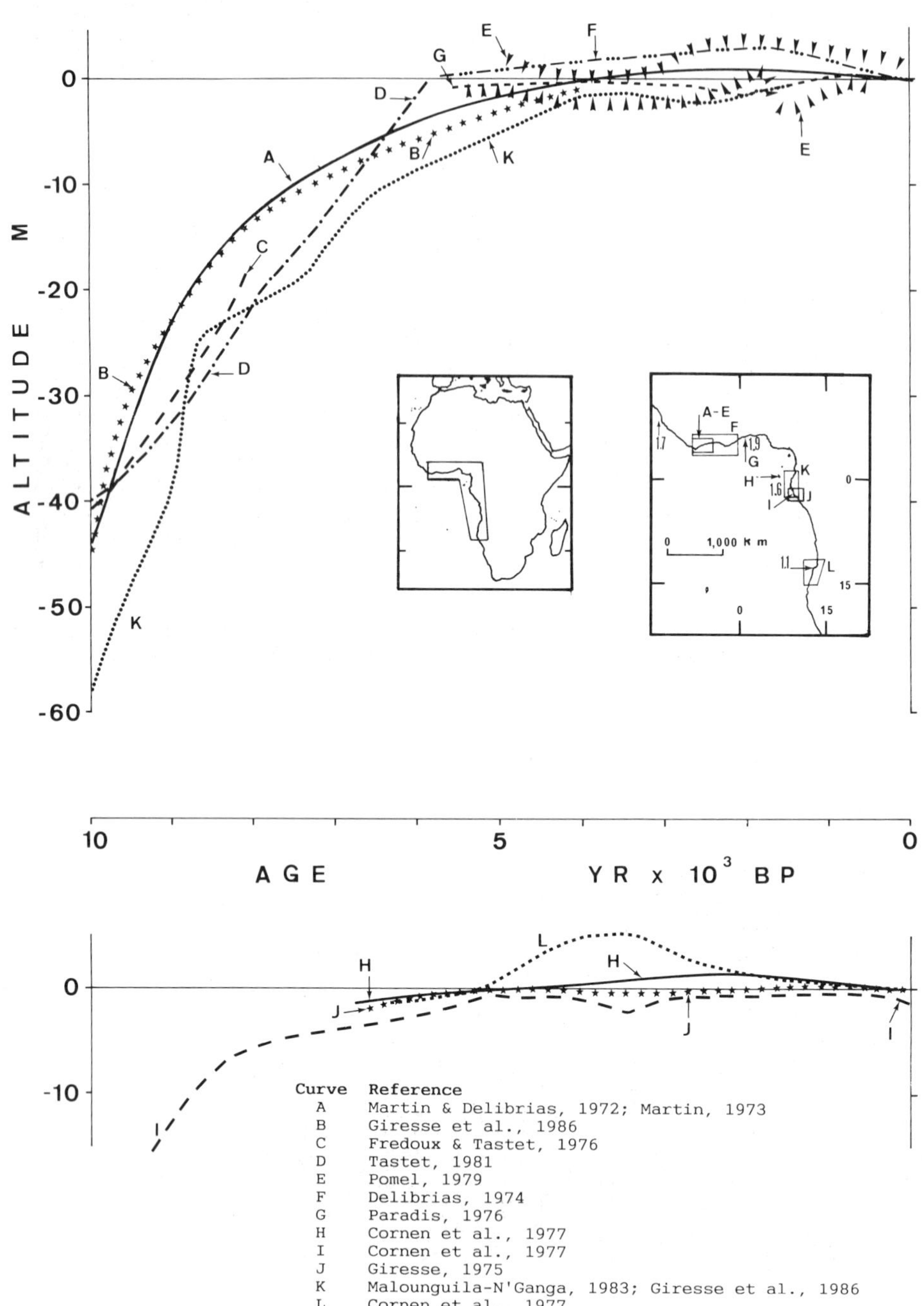

Curve	Reference
A	Martin & Delibrias, 1972; Martin, 1973
B	Giresse et al., 1986
C	Fredoux & Tastet, 1976
D	Tastet, 1981
E	Pomel, 1979
F	Delibrias, 1974
G	Paradis, 1976
H	Cornen et al., 1977
I	Cornen et al., 1977
J	Giresse, 1975
K	Malounguila-N'Ganga, 1983; Giresse et al., 1986
L	Cornen et al., 1977

Fig. 2-18. Emerged holes hollowed by sea urchins during a slightly higher late Holocene sea-level stand. Cape Almadies, Senegal.

Fig. 2-19. Erosion bench cut by the Holocene sea into a raised Pleistocene coral reef formation at Mombasa (Kenya). Mean spring tidal range is 3.2 m, water level is 1.3 m below MSL.

In São Tomé and Annobon Islands, three dates of slightly emerged beach deposits were interpreted by Cornen et al. (1977) as demonstrating a relative sea level stand higher than at present since 5000-4000 yr BP, with a peak at about +1.5 m about 2300 yr BP (curve **H**). On the Congo coast on the other hand, according to the same authors, no marks of Holocene emergence appear (curve **I**); this confirms earlier results proposed in the same area by Giresse (1975) (curve **J**). On the continental shelf of Gabon and Congo, the relative sea-level curve **K** has been obtained by Malounguila-N'Ganga (1983) and Giresse et al. (1986) using mangrove peat beds and plant remains. The shift between curves **K** and **B** was interpreted by Giresse et al. (1986) as resulting, at least in part, from an epirogenic trend of the continental basement, whose rigidity would be greater in Congo and Gabon than in the Ivory Coast or Angola. Cornen et al. (1977) reported from the Lobito-Mossamedes area in Angola a relative sea-level curve (**L**) which suggests a 5-m emergence during the last 3500 years. For more detailed data on relative sea-level changes in Angola, see Giresse et al. (1976, 1984).

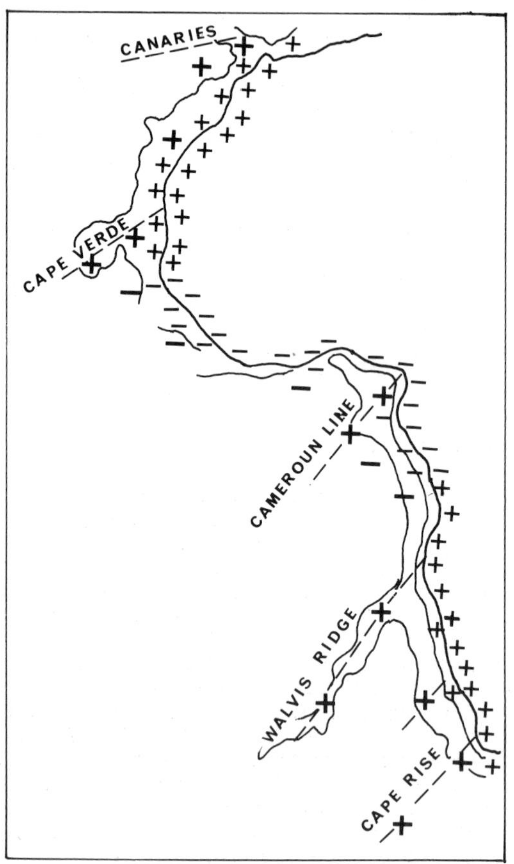

Fig. 2-20. Simplified anticlines (+) and synclines (-) on the west side of Africa, according to Cornen et al. (1977) (adapted).

Trends of epirogenetic emergence (+) and submergence (-) along the west coasts of Africa are summarized in Fig. 2-20, adapted from Cornen et al. (1977). Quaternary emergence is reported from northwest Africa (Morocco, Mauritania, north Senegal), southwest Africa (south Angola and Namibia) and from the line of volcanic islands situated off Cameroun, whereas submergence characterises most African coasts between Guinea and North Angola.

WEST INDIAN OCEAN (Plate 30)

On the east side of Africa the only Holocene sea-level curve available to the writer is that produced in coastal Mozambique between 15° and 19° S by Jaritz et al. (1977); based on some twenty radiocarbon dates, curve **A** shows that the relative sea level was at about -60 m at the beginning of the Holocene, reached the present sea level about 7000 yr BP and a peak at +2.5/+3.0 m slightly later than 6000 yr BP, then dropped slowly to the present level, which was reached between 2000 and 1000 yr BP. Nevertheless the tidal range is not negligible along the Mozambique coasts. As Jaritz et al. (1977) do not specify the "sea level" they are speaking about, the accuracy of curve **A** is unknown.

In Réunion Island, a series of dates obtained from coral samples from a core drilled into the fringing reef was used by Montaggioni (1976, 1978) to construct a sea-level curve (**B**) under the assumption that the upper members of the living coral colonies were closely controlled by the level of spring low tides throughout the Holocene ("keep up" reef). Similar methods and assumptions were used by Montaggioni (1981) in Mauritius Island (curve **C**). Curves **B** and **C**, which show a continuous sea-level rise, though at progressively decreasing rates during the last 7000 years, are however closely dependent on reef development rates. In a recent paper, Montaggioni (1988) recognized however that maximum depths of water above the growing reef may have been as great as 10 m in Réunion and 14 m in Mauritius. Accordingly, revised sea-level curves **B'** and **C'** appear steeper than **B** and **C**. Predictions in Réunion Island by the Peltier (1991) model are shown by curve **B"**, which is systematically above field data.

Sea-level data from the Persian Gulf have been assembled in the lower graph of **Plate 30.** At Failaka Island (Kuwait) a fluctuating MHW curve (**D**) was proposed by Dalongeville and Sanlaville (1987), based on three sequences of marine deposits, archaeological data and several radiocarbon dates. In the Lower Mesopotamia area, however, Sanlaville (1989) deduced from archaeological and drilled core data different sea-level oscillations (curve **D'**), which would have caused considerable advances and retreats of the Euphrates-Tigris delta area. Though the above sea-level oscillations had been "general enough to affect the whole Gulf area or at least its northern part", as stated by Sanlaville (1989, p. 24), it is not clear to what causes we should ascribe the fact that around 5500 yr BP the sea level could have been at +2 m in the Lower Mesopotamia area (**D'**) and at least -1 m at Failaka Island (**D**). Near Al Jubail, Saudi Arabia, Ridley and Seeley (1979) reported archaeological evidence interpreted as showing about 5000-6000 years ago a relative sea level 4 m higher than present "sea level" (curve **E**); again near Al Jubail, elevated beach deposits at +2.8 m, dated about 3800 yr BP by Ridley and Seeley (1979), correspond to a low sea level of curve **D'**.

Plate 30: INDIAN OCEAN I (West)

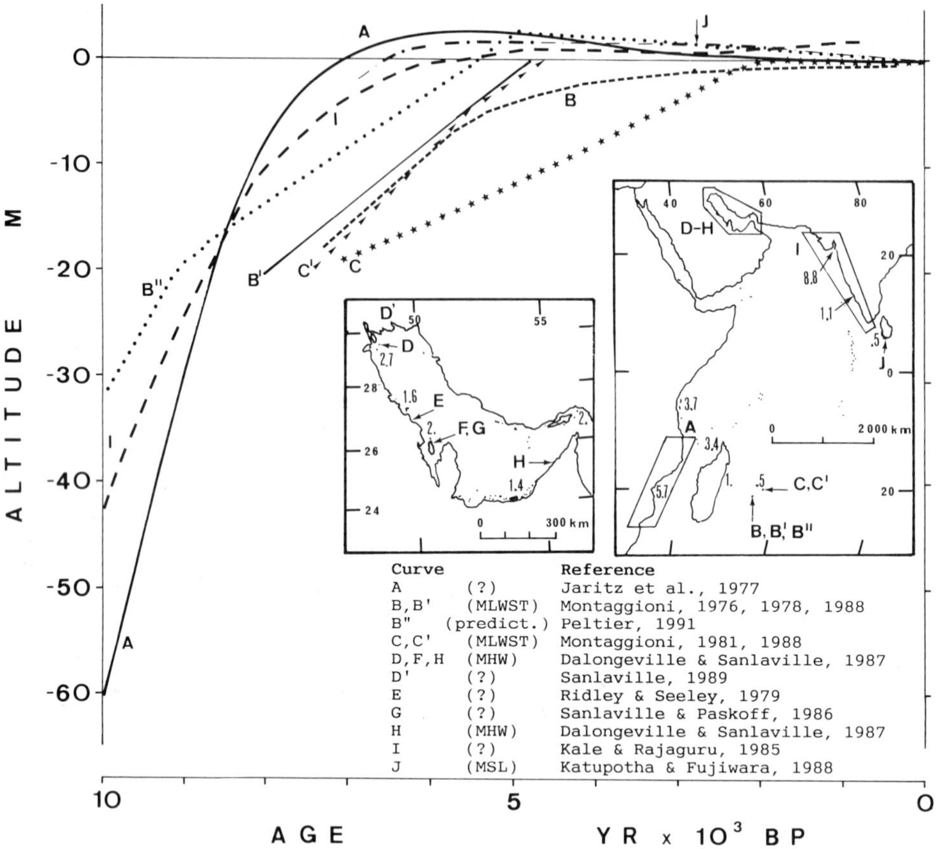

ALTITUDE M
0
-10
-20
-30
-40
-50
-60
A
B
B'
B''
C
C'
I
J
10
5
0
AGE YR x 10³ BP
D-H
40
60
80
20
8.8
1,1
.5
J0
3.7
A
3.4
5.7
1.
.5
C,C'
B,B',B''
0 2000 km
D'
50
55
D
2.7
28
1.6
E
2.
F,G
0
26
H
1.4
24
0 300 km
2.
Curve Reference
A (?) Jaritz et al., 1977
B,B' (MLWST) Montaggioni, 1976, 1978, 1988
B" (predict.) Peltier, 1991
C,C' (MLWST) Montaggioni, 1981, 1988
D,F,H (MHW) Dalongeville & Sanlaville, 1987
D' (?) Sanlaville, 1989
E (?) Ridley & Seeley, 1979
G (?) Sanlaville & Paskoff, 1986
H (MHW) Dalongeville & Sanlaville, 1987
I (?) Kale & Rajaguru, 1985
J (MSL) Katupotha & Fujiwara, 1988

ALTITUDE M
0
-10
-20
F
G
E
E
G
F
H
D
H
D
D'
2
1
0
-1
-2
G
F
D
G
D'
H
D
D
F
8
6
4
2
0
AGE YR x 10³ BP

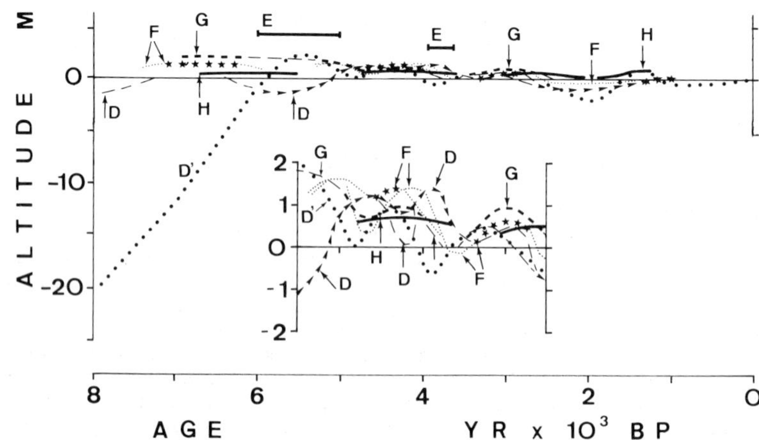

However most of the sea-level oscillations proposed by the various authors are usually of the same order as the local tidal range, eventually increasing owing to differences in level induced by changes in river runoff or by certain meteorological conditions, whereas vertical uncertainty ranges in the determination of former sea levels are not specified. This means that these sea-level oscillations need to be confirmed by more reliable data.

At Bahrain, curves **F** (Dalongeville and Sanlaville (1987) and **G** (Sanlaville and Paskoff, 1986) seem to correspond to two fluctuating variants based on almost the same data, suggesting that the relative "sea level" was 1.5 to 2 m above present from as early as 7000 to about 5000 yr BP, at about +1 m (or +1.5 m) about 4500-4000 yr BP, and still between +0.5 and +1.0 m some 3000 yr BP; the troughs of the fluctuations suggested by curves **F** and **G** appear however poorly established. Near Sharjah, according to Dalongeville and Sanlaville (1987), the relative MHW level was less than 1 m higher than at present, though with minor fluctuations, from about 6000 to 1000 yr BP (curve **H**).

On the coasts of India few data are available. Kale and Rajaguru (1985) plotted radiometric dates of beachrocks and beach dunes from Konkan, other shore data from Goa, Karwar and Kerala and offshore data from Bombay to Karwar, to obtain curve **I**, which is intended to show the trend of "sea-level" changes on the west coast of India. Nevertheless, as the area considered is very wide, curve **I** can hardly be considered as more than a promising working assumption, which still has to be tested at more local scales with detailed sea-level investigations.

On the south coast of Sri Lanka, an accurate relative MSL curve **J**, deduced from slightly emerged corals collected in growth position, was produced by Katupotha and Fujiwara (1988); it shows a sea-level stand between 1 and 2 m above present from 6000 to 2000 yr BP.

In conclusion, most coastal areas in the west Indian Ocean have not yet been systematically studied. The few sea-level curves available indicate that slight emergence predominated during the late Holocene along the coasts of Mozambique, on the southern shores of the Persian Gulf, west India and south Sri Lanka, whereas evidence of emergence is missing in oceanic islands (Réunion, Maurice). In the east part of the Indian Ocean, on the other hand, slight Holocene emergence (at least 0.5 m since about 3000 yr BP) was reported from the Cocos (Keeling) Islands by Woodroffe et al. (1990).

NORTHEAST INDIAN OCEAN (Plate 31)

In the Bengal basin, recent excavations at four sites situated 80 to 120 km inland from the present shoreline recovered biological remains of distinct ecofacies, enabling Banerjee and Sen (1987) to produce a MSL curve (**A**) based on seven new radiocarbon dates and twenty dates published by earlier workers; the relative MSL appears to have reached the present position c. 5000 yr BP, with no evidence of change since that time. The trend shown by curve **A** is extremely surprising in an area where Holocene subsidence rates of 1 to 2.5 cm/yr have been suggested (Milliman et al., 1989). The vertical accuracy of curve **A** was not specified by Banerjee and Sen

Plate 31: INDIAN OCEAN II (Northeast)

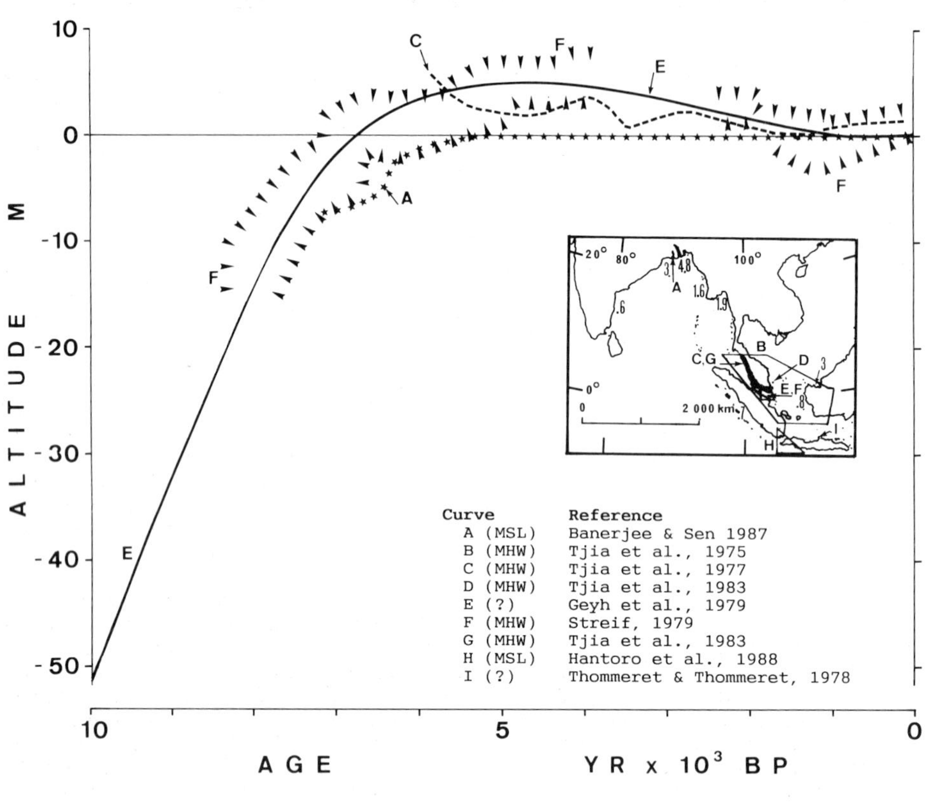

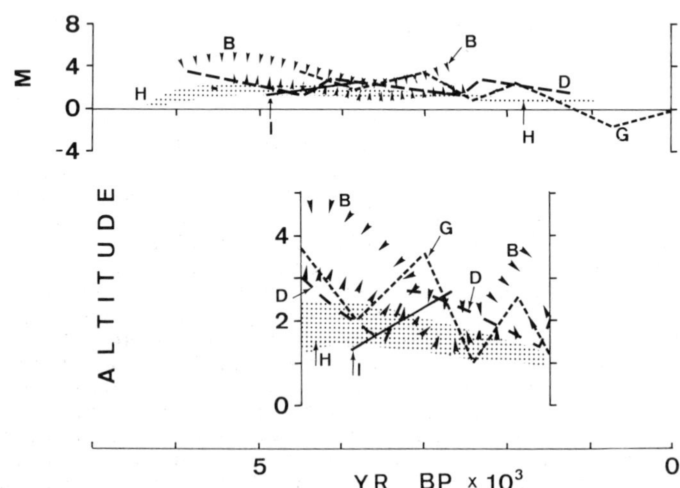

(1987); uncertainty ranges may be considerable, however, in an area where the tidal range changes from place to place, reaching c. 5 m at Diamond Harbor, and catastrophic storm surges several metres high are frequent.

Tjia et al. (1975) assembled one dozen sea-level data from "stable Sunda Land", which were used to construct a slightly fluctuating band (**B**) for the high tide level between 6000 and 2500 yr BP. Later, considering only west Malaysia, Tjia et al. (1977) produced a new composite MHW curve (**C**), again under the assumption that the area considered (which extends over 700 km along the Malacca Strait) has remained completely stable. Curve **C** shows gradual 5 m emergence (which was subsequently corrected to 3.7 m in curve **G** for "Peninsular Malaysia" by Tjia et al., 1983) since 6000 yr BP, with several oscillations which may depend on the "stability" assumption (differential hydro-isostatic vertical movements are likely to have occurred along the coast) and on underestimated uncertainty ranges affecting the sea-level indicators used.

From Tioman Island, Tjia et al. (1983) reported a tooth-shaped sea-level curve (**D**), constructed by connecting with straight lines eight dated samples for which no vertical uncertainty range had been estimated. It can be deduced from curve **D** that the relative sea level was somewhere between 1.4 and 2.7 m above the present MSL during the period 5500-1900 yr BP, though the sea-level fluctuations proposed seem insufficiently established.

The results of a combined on- and offshore survey, covering the coast of west Malaysia and the adjacent Malacca Strait along a distance of about 300 km, between Port Dickson and Singapore, were reported by Geyh et al. (1979) and Streif (1979). A MHW sea-level band (**F**) was obtained by Streif (1979) using 25 radiocarbon-dated onshore samples. Curve **E**, by Geyh et al. (1979) is based on offshore data, and extends the sea-level trends indicated by band **F** from 8000 to 10,000 yr BP. For the period later than 8000 yr BP the precision of curve **E** is however illusory, since it is based on the same data as band **F**. Even when all sea-level data from the Malay Peninsula are plotted together, as done by Bosch (1988), the character of sea-level change (smooth, marked by steps, or by erratic high and low stands) remains uncertain.

In the area including the Sunda Strait and the south coast of west Java, Hantoro et al. (1988) proposed a preliminary MSL band (**H**) for the past 7000 years, showing a slight mid-Holocene emergence (Fig. 2-21). On the north coast of Java, at Jepara, 700 km east of Jakarta, Thommeret and Thommeret (1978) dated seven in situ shell and coral samples from a stratigraphical series of emerged marine deposits, obtaining consistent indications that the local relative "sea level" had been rising from +1.3 to +2.5 m between 5000 and 3600 yr BP (curve **I**). According to Tjia et al. (1983), curve **I** would reflect a combination of vertical crustal movements and real sea-level rise. In the lowest part of **Plate 31**, the vertical scale of bands **B** and **H** and of curves **D, G** and **I** has been enlarged to facilitate comparisons between 5500 and 2500 yr BP. In conclusion, slight late Holocene emergence seems to predominate in most places examined in Plate 31, with the exception of the Bengal basin, where recent relative sea-level stability is surprising (subsidence phenomena are usually predominant in delta areas) and needs therefore to be confirmed by more precise data.

Fig. 2-21. Holocene *Porites* corals outcropping on the south coast of Sumatra (Indonesia). Scale is 2 m. Mean spring tidal range is about 1 m.

Fig. 2-22. Emerged holes, bored into an *in situ* mudstone formation by *Lithophaga*, still contain their shells in growth position. Near Himi, Toyama Prefecture, Japan.

Plate 32: VIETNAM - SOUTH CHINA I

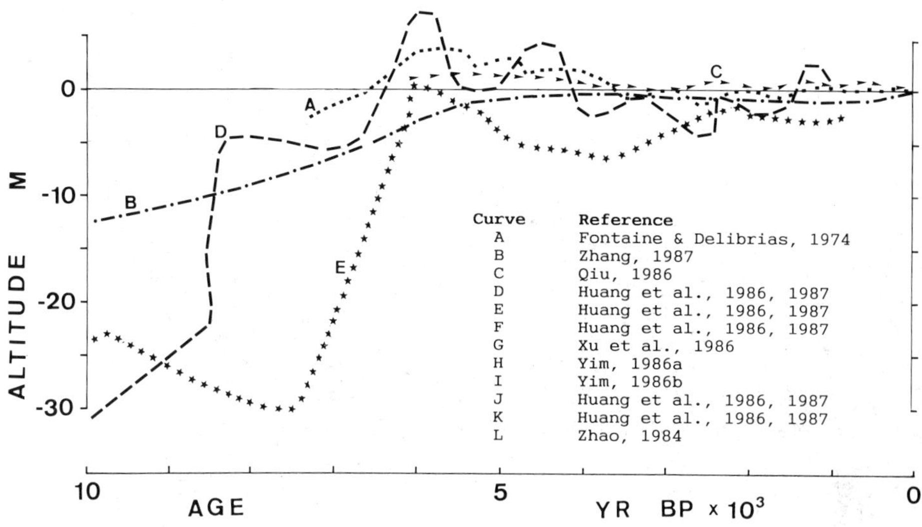

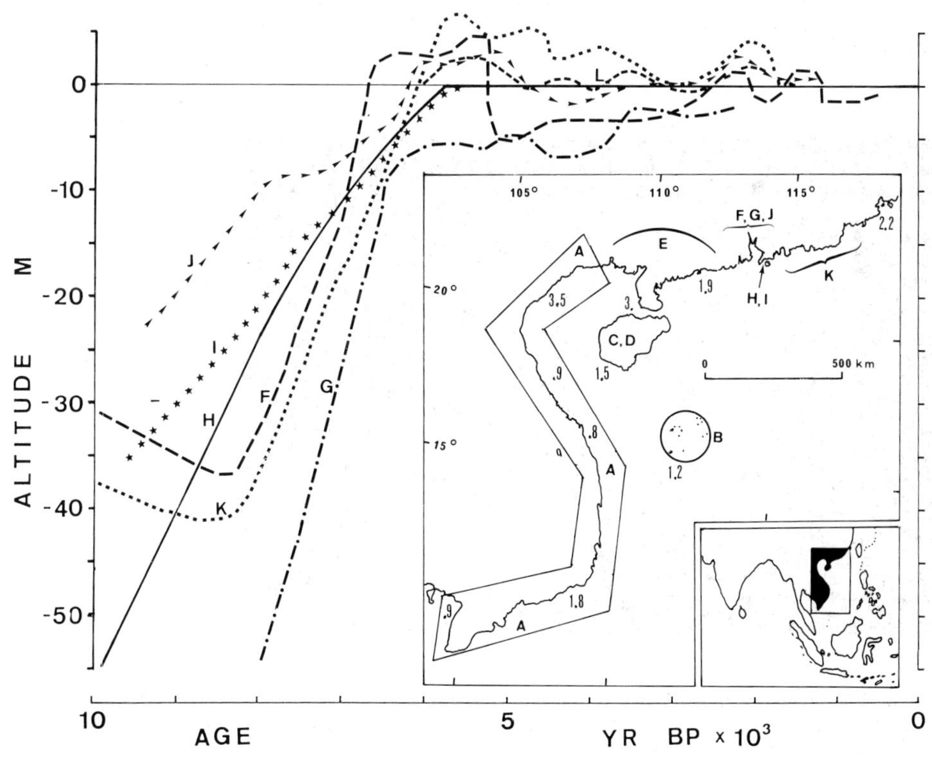

Relative sea-level trends shown by tide gauge records in Africa and South Asia are summarized in Fig. 2-23. These records are inadequate for regional statements.

VIETNAM - SOUTH CHINA (Plates 32-33)

Fontaine and Delibrias (1974) have produced a "sea-level" curve (**A, Plate 32**) using data scattered all along the Vietnam coast (which is over 2000 km long, with spring/diurnal/tropical tidal ranges varying from 0.5 m to 3.5 m). There is evidence of sea levels decreasing with age following a maximum at c. +4 m reached about 6000 years ago. The fluctuating pattern of curve **A**, inspired by Fairbridge (1961), is however doubtful, since negative oscillations were not studied by Fontaine and Delibrias (1974).

In the Xisha region (Paracel Islands), Zhang (1987) did not find any indications of recent emergence and proposed a curve (**B**) showing a relative sea-level rise of no more than a dozen metres since 10,000 yr BP, which seems very underestimated.

In Hainan Island, Qiu (1986) used 22 radiocarbon dates, mostly of *in situ* corals, to obtain a sea-level curve (**C**) which shows 1 to 2 m emergence from 6000 to 4000 yr BP, followed by a gradual sea-level drop with fluctuations less than 1 m in amplitude, ascribed at least in part to crustal movements.

Huang et al. (1986, 1987) proposed curves of sea-level change for six coastal sectors. Five of these curves are presented in **Plate 32**: **D** for Hainan Island, **E** for Guanxi and western Guangdong, **F** for the Zhujang River delta, **J** for Shenzhen and Hong Kong and **K** for eastern Guangdong. All these curves show rather unlikely fluctuating patterns, which probably depend on the assumptions and methods used. The assumptions consisted in ascribing to each sea-level indicator an arbitrarily fixed former sea-level position [e.g. all coral samples collected *in situ* were assumed to have grown at -5 m (5 m below the contemporary sea level), all *Ostrea* shell beds at -7 m, all shell ridges to have deposited at +2 m, all "marine mud beds with shells or putrid woods" (i.e. one third of the samples) at -1.5 m, etc.]. These assumptions might have been acceptable if adequate uncertainty ranges (± several metres in some cases) were adopted, but this was not the case and even the influence of the tidal range (which varies from 1.1 to 4.5 m along the coasts in the area considered) was not taken into account. The method used by Huang et al. (1987) consists in extrapolating to the ages of the samples dated and deducting from their paleoelevation the trends of "crustal movement" deduced from recent levelling surveys (usually several mm/yr). Again these trends were considered without uncertainty ranges, though they may depend in some cases on man-induced subsidence or systematic errors, rather than on real crustal movements.

In Hong Kong two smooth sea-level curves (**H** and **I**) have been published by Yim (1986a,b), suggesting that the present sea level was reached some 6000 years ago and that no change occurred after that time. As curves **H** and **I** seem to be based on almost the same data, the 15-m departure they show in the early Holocene remains unexplained.

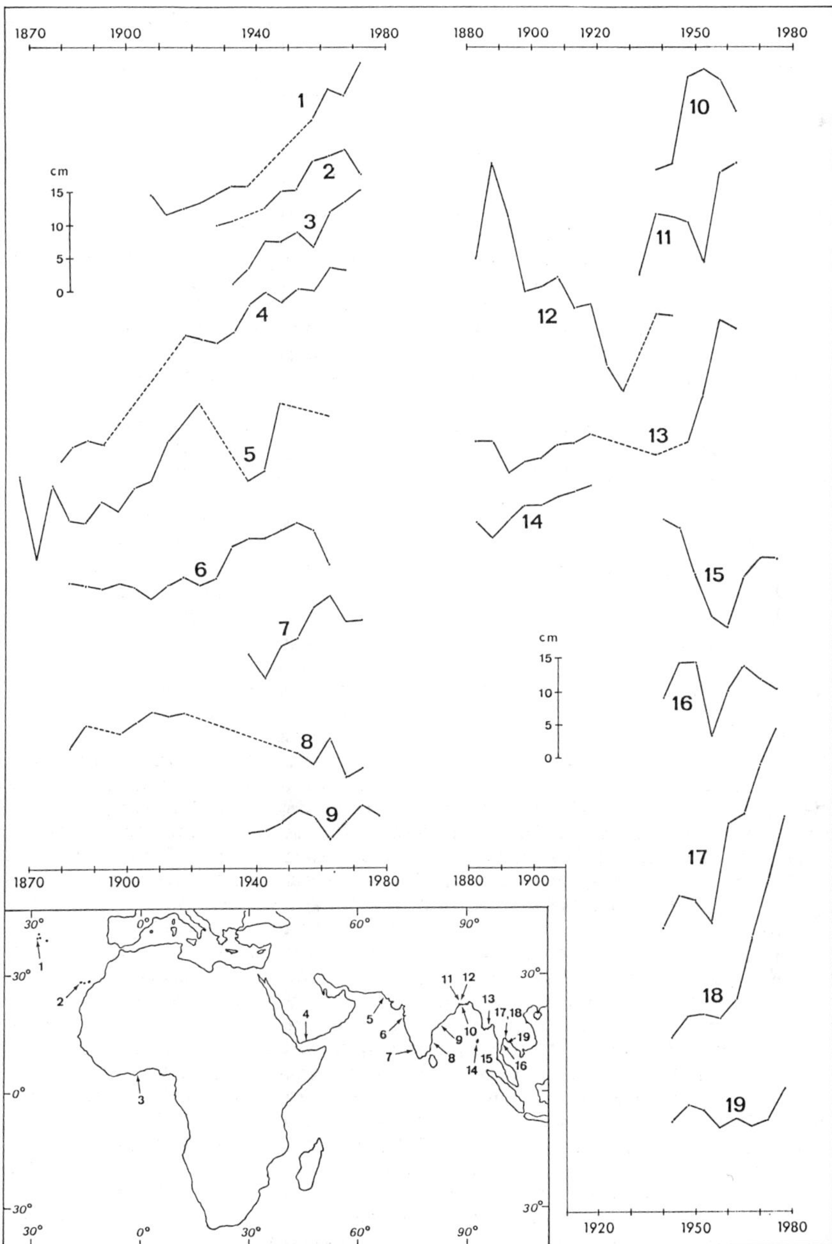

Fig. 2-23. Average 5-yr MSL variations in Africa and South Asia as recorded by tide-gauge stations. 1) Horta (Azores); 2) Santa Cruz de Tenerife; 3) Takoradi; 4) Aden; 5) Karachi; 6) Bombay; 7) Cochin; 8) Madras; 9) Vishakhapatam; 10) Saugor Island; 11) Calcutta; 12) Kidderpore; 13) Rangoon; 14) Port Blair; 15) Ko Taphao Noi; 16) Phrachwap Kirikhan; 17) Bangkok Bar; 18) Fort Phrachula; 19) Ko Sichang. (From Pirazzoli, 1986b).

Plate 33: SOUTH CHINA II

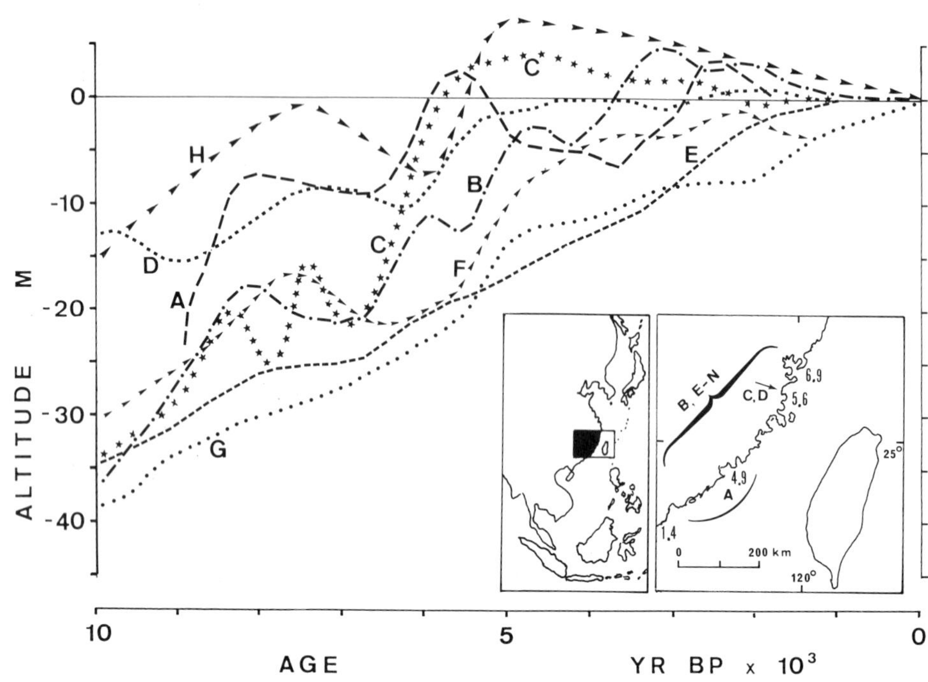

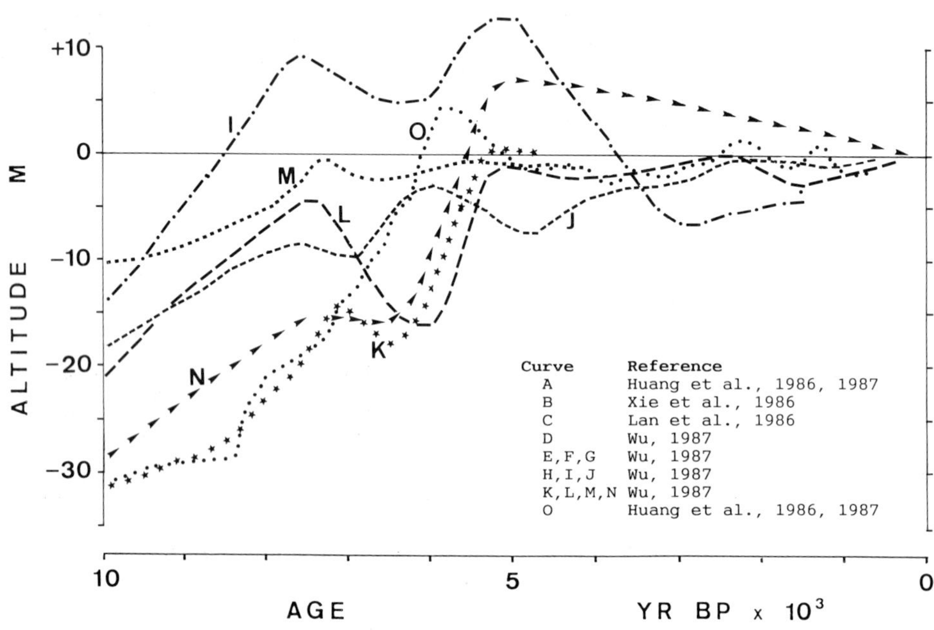

Curve	Reference
A	Huang et al., 1986, 1987
B	Xie et al., 1986
C	Lan et al., 1986
D	Wu, 1987
E,F,G	Wu, 1987
H,I,J	Wu, 1987
K,L,M,N	Wu, 1987
O	Huang et al., 1986, 1987

Plate 34: EAST CHINA I

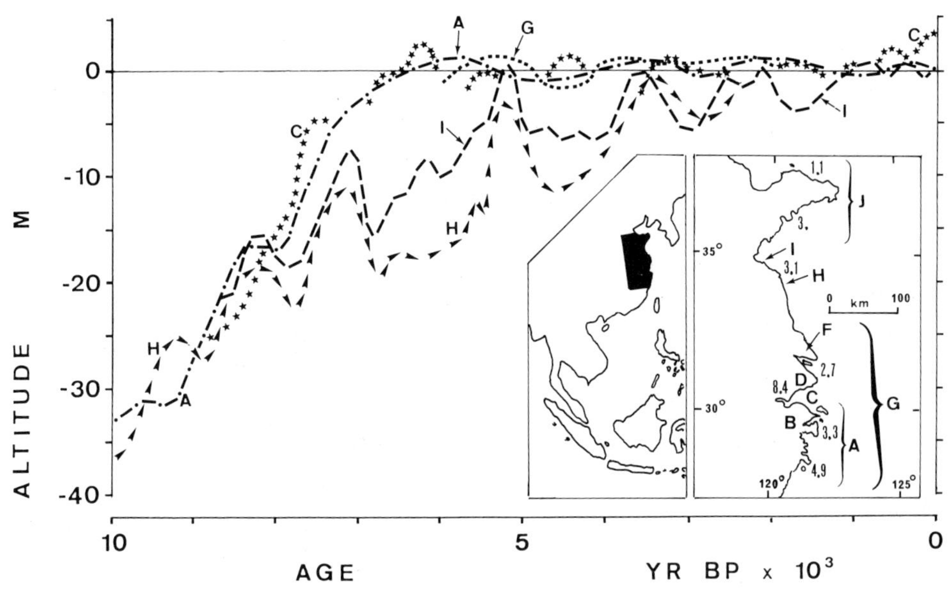

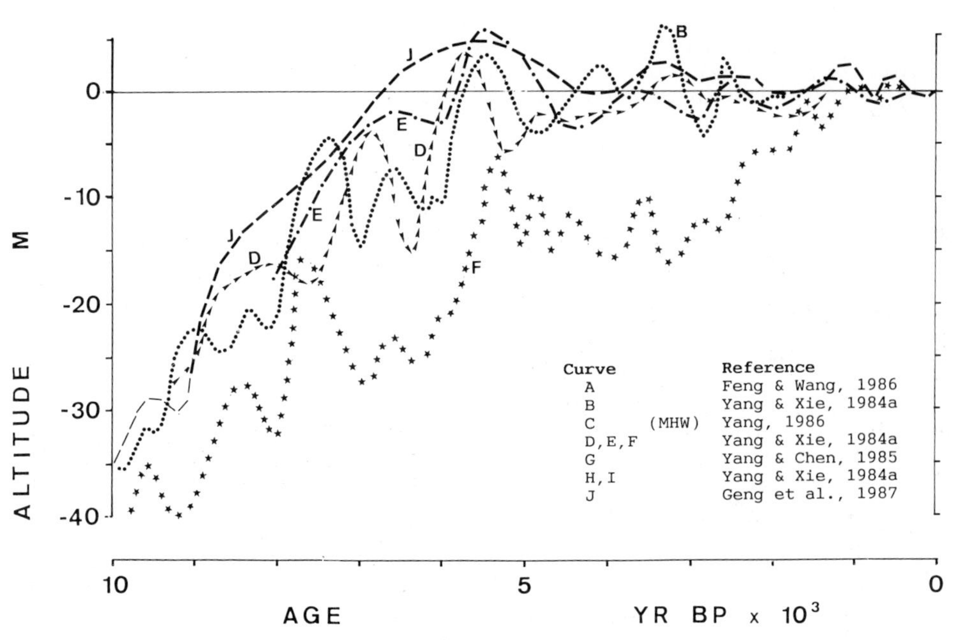

Curve		Reference
A		Feng & Wang, 1986
B		Yang & Xie, 1984a
C	(MHW)	Yang, 1986
D,E,F		Yang & Xie, 1984a
G		Yang & Chen, 1985
H,I		Yang & Xie, 1984a
J		Geng et al., 1987

In the Zhujang River delta, the **G** sea-level curve proposed by Xu et al. (1986) suggests a rapid rise until 6000 yr BP, followed by some fluctuations. For the "South China Sea coasts" Zhao (1984) proposed a sea-level curve (**L**) with four stands 1 to 2 m higher than present since 6000 yr BP.

In **Plate 33**, curve **A** was proposed for southern Fujian by Huang et al. (1986, 1987), using assumptions and methods already discussed above. On the Fujian coast, Xie et al. (1986) were able to obtain a sea-level curve (**B**) "similar to Fairbridge's curve", though slightly shifted in time. In the Fuzhou basin, Lan et al. (1986) deduced from sporopollen and diatom analysis of drilled cores a sea-level curve (**C**) suggesting two rather unlikely wide fluctuations between 9000 and 7000 yr BP, followed by a peak slightly above the present sea level between 6000 and 5000 yr BP and a gradual fall to the present situation until c. 2000 yr BP. In the same area, at the mouth of the Minjang River, curve **D** obtained by Wu (1987) from analysis of samples collected from the core CK7 suggests a sea-level rise of no more than 15 m since 10,000 yr BP, which is probably underestimated and a Holocene peak at +1 m between 3000 and 2500 yr BP.

Relative sea-level curves **E** to **N**, published by Wu (1987), seem to have been deduced from 10 cores drilled in "basins and plains along the coast of Fujian Province", the location of which is not given. Curve **E** corresponds to core ZK4352, **F** to core ZK4752, **G** to core ZK1039, **H** to core ZK1039, **I** to core ZK51, **J** to core ZK2495, **K** to core ZK0124, **L** to core ZK1027, **M** to core ZK1026 and **N** to core ZK0539. Information provided by Wu (1987) on curves **E** to **N** consists only of a very summarized stratigraphic column for each core and is therefore insufficient to assess the curve's reliability; nevertheless, differences in level between the curves is often so great that they may be suspected to result from rough underestimation of uncertainty ranges.

Curve **O**, which is considered by Huang et al. (1987) to be a "typical sea-level curve for South China", was obtained "through synthesizing and averaging the variations of heights of sea levels reflected by the six curves" **D, E, F, J, K** (Plate 32) and **A** (Plate 33), which have been commented on above.

EAST CHINA - KOREA (Plates 34-35)

In the Zhejiang Province, a detailed sea-level curve (**A, Plate 34**) was established by Feng and Wang (1986), using 21 radiocarbon dates of sea-level indicators collected from shelf and coastal areas. Various lines of evidence support the interpretation of a rapid sea-level rise, from -50 m around 12,000 yr BP to -10 m around 7500 yr BP, apparently interrupted by two short stillstands of the sea (around -30 m about 9300 yr BP, and -16 m 8300 yr BP), which reached a culminating point in the middle of the Holocene, followed by a regression trend. On the other hand the sea-level oscillations proposed by Feng and Wang (1986) after 5000 yr BP seem less well established, in spite of several radiocarbon dates, and their amplitude may be of the same order of the vertical uncertainty ranges of the data used.

Yang and Xie (1984a) proposed curves of relative sea-level change for the coastal plains south (curve **B**) and north (curve **D**) of

Plate 35: EAST CHINA II - KOREA

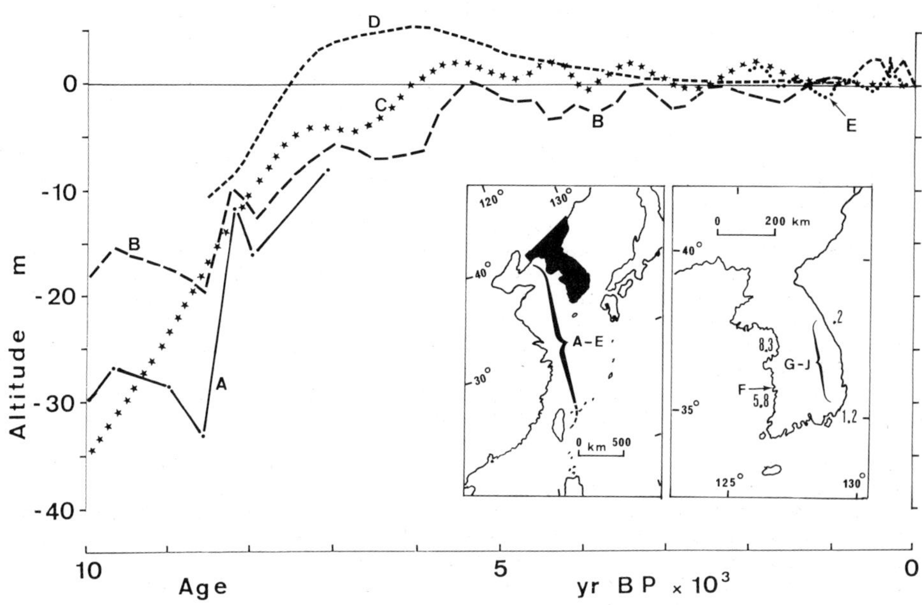

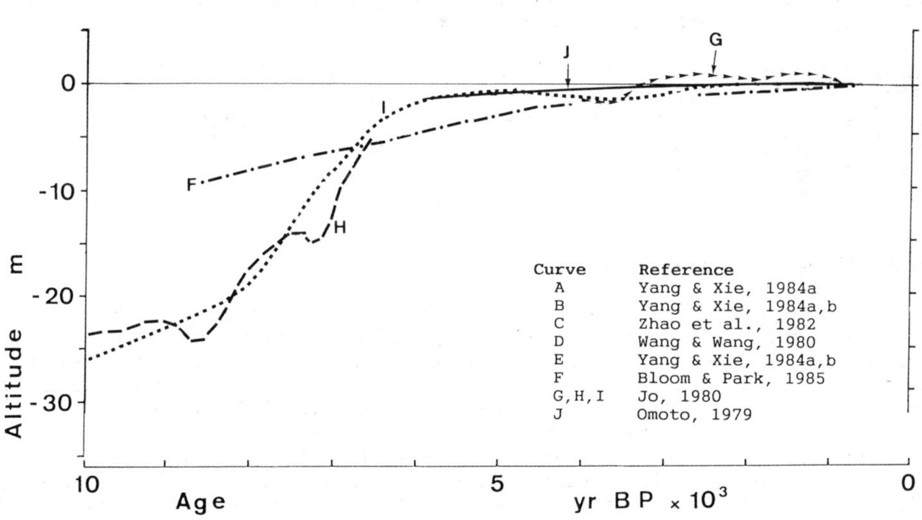

Hangzhouwan, for the western part of the Taihu lacustrine plain (curve **E**) and for the Changjiang delta plain east of Nantong (curve **F**). No information was provided by Yang and Xie (1984a) to explain the shifts between these curves, though compaction and subsidence effects seem likely to have occurred in some cases and underestimation of vertical uncertainty ranges may explain several apparent but unlikely oscillations.

In the Changjiang estuary, Yang (1986) determined, from fossil deposits, marks of ten relative peaks of the high-water level during the last 8000 years (curve **C**). As the tidal range seems to have been decreasing since 7000 yr BP in the Changjiang estuary, the present sea level is the highest one reached during the Holocene.

Curve **G** by Yang and Chen (1985) summarizes changes in elevation between 6000 and 2000 years ago of the lowest level of dwelling sites in ancient Chinese history in the wide area comprising the plains of Jiangsu, Zhejiang and the Changjiang River delta.

In the Subei coastal plain, two relative sea-level curves have been proposed by Yang and Xie (1984a) for the areas south (**H**) and north (**I**) of the mouth of the Guanhe respectively, using data from a number of cores and remains of Neolithic cultures. No information is provided however to justify important oscillations such as rapid up-and-down movements of one dozen metres around 7000 and 5000 yr BP, which seem rather unlikely.

Around the Shandong Peninsula, 67 radiocarbon-dated sea-level indicators, deduced from beach, chenier and beachrock shells, lagoons, swamp and estuarine deposits, marine terraces and archaeological data, have enabled Geng et al. (1987) to produce a detailed sea-level curve (**J**). According to **J**, a relative sea-level rise occurred from -40/-30 m at the beginning of the Holocene to the present position (reached between 7000 and 6800 yr BP, and possibly to a peak at +4/+5 m around 6000 yr BP, if the development of certain abrasion terraces dates from this period. Later on, the occurrence of five fluctuations of several metres in amplitude, decreasing in range and wave-length with time, was proposed by Geng et al. (1987); vertical uncertainty ranges of the above data are not specified.

In **Plate 35** the above graph summarizes some general sea-level curves which have been proposed for "East China". Curve **A** was obtained by Yang and Xie (1984a) from shelf data of East China by joining with straight lines "samples of littoral facies" or "corrected points" deduced from "samples of shallow sea facies". Curve **A** would indeed appear smoother if adequate uncertainty ranges had been adopted.

Curve **B** was obtained by Yang and Xie (1984a,b) "after eliminating the effects of isostasy, tectonism and compaction" from seven curves (B, D, E, F, H, I in Plate 34; A in Plate 35); the method used to obtain such "elimination" is not specified however.

Curve **C** is part of a 20,000-yr sea-level curve constructed by Zhao et al. (1982) on the basis of some 40 radiocarbon dates of various sea-level indicators from the coasts and shelf of the Bohai and Yellow Seas, 12 dates from the East China Sea and even 9 dates from the north coast of Taiwan. Curve **C**, which shows several oscillations, is considered by Zhao et al. (1982) as representative of "eastern China", i.e. of a wide coastal and shelf belt extending in a north-south direction from 39°N to 25°N and including areas with different tectonic and isostatic histories.

Curves **A**, **B** and **C**, obtained by mixing data scattered over too wide areas, which are characterized by different tectonic, sedimentological, tidal, climatic and oceanographic situations, probably do not correspond to actual local sea-level histories in any place. No first hand information is available to the present author on curve **D** by Wang and Wang (1980), reproduced from Ota (1987).

For the last 2000 years, Yang and Xie (1984a,b) provided a fluctuating sea-level curve **E** based on "rich historical records", showing high levels in the 4th, 9th and 16th centuries A.D., low sea levels in the 6th, 12th and 18th centuries, and oscillations of more than 2 m in amplitude; in particular, sea level would have been 1.5 m higher than at present from the 8th to the 10th century; the evidence justifying these estimations is not provided however.

The lower graph in **Plate 35** summarizes sea-level curves reported from South Korea. On the west coast, curve **F** by Bloom and Park (1985) is based on 8 radiocarbon dates of estuarine deposits containing transgressive sedimentary facies from the Kimje, Kunsan and Baenaru areas and suggests that sea level rose gradually to the present position from about -10 m at 8500 yr BP. As noted by Ota (1987), such a high situation 8500 years ago is inconsistent with similar data from nearby regions.

On the east coast of South Korea, Jo (1980) studied several cores drilled in alluvial plains and produced three sea-level curves. Curve **G**, deduced from pollen analysis of back marsh deposits, shows a slightly undulating pattern since c. 5000 yr BP. Curve **H**, which also fluctuates, was deduced from grain size analysis for the period 10,000 to 7000 yr BP. Lastly curve **I** appears much smoother and was produced by Jo (1980) to summarize the above data. Curve **J**, proposed by Omoto (1979) and based on almost the same data as Jo (1980), suggests a gradual sea-level rise of less than 2 m during the last 6000 years.

ARCTIC & PACIFIC USSR (Plate 36)

The upper graph of **Plate 36** groups together sea-level curves reported from the Arctic coasts of the USSR, with the exception of Franz Joseph Land, for which a curve was inserted in Plate 1, and of Kola Bay, in Plate 3. Curve **A** was proposed by Badyukov and Kaplin (1979) to show "the general tendency in the level dynamics in the White Sea basin", though isostatic movements on the northwestern coasts were different from those on the southern ones. The basin was covered by ice masses or isolated from the ocean until about 11,000 years ago. At the beginning of the Holocene, the sea-level position suggested by Badyukov and Kaplin (1979) is at c. -35 m. Following a slight lowering or temporary stabilization of sea level between 9100 and 8400 yr BP, comparable with the Litorina phase in Scandinavia, the rise of sea level appears to have been rapid until 7000 yr BP, except during a short period of deceleration around 7800 yr BP, which left a thin peat layer in some areas. According to curve **A**, two slight fluctuations occurred after the present sea level was reached about 2000 years ago.

Plate 36: U.S.S.R. (Arctic & Pacific)

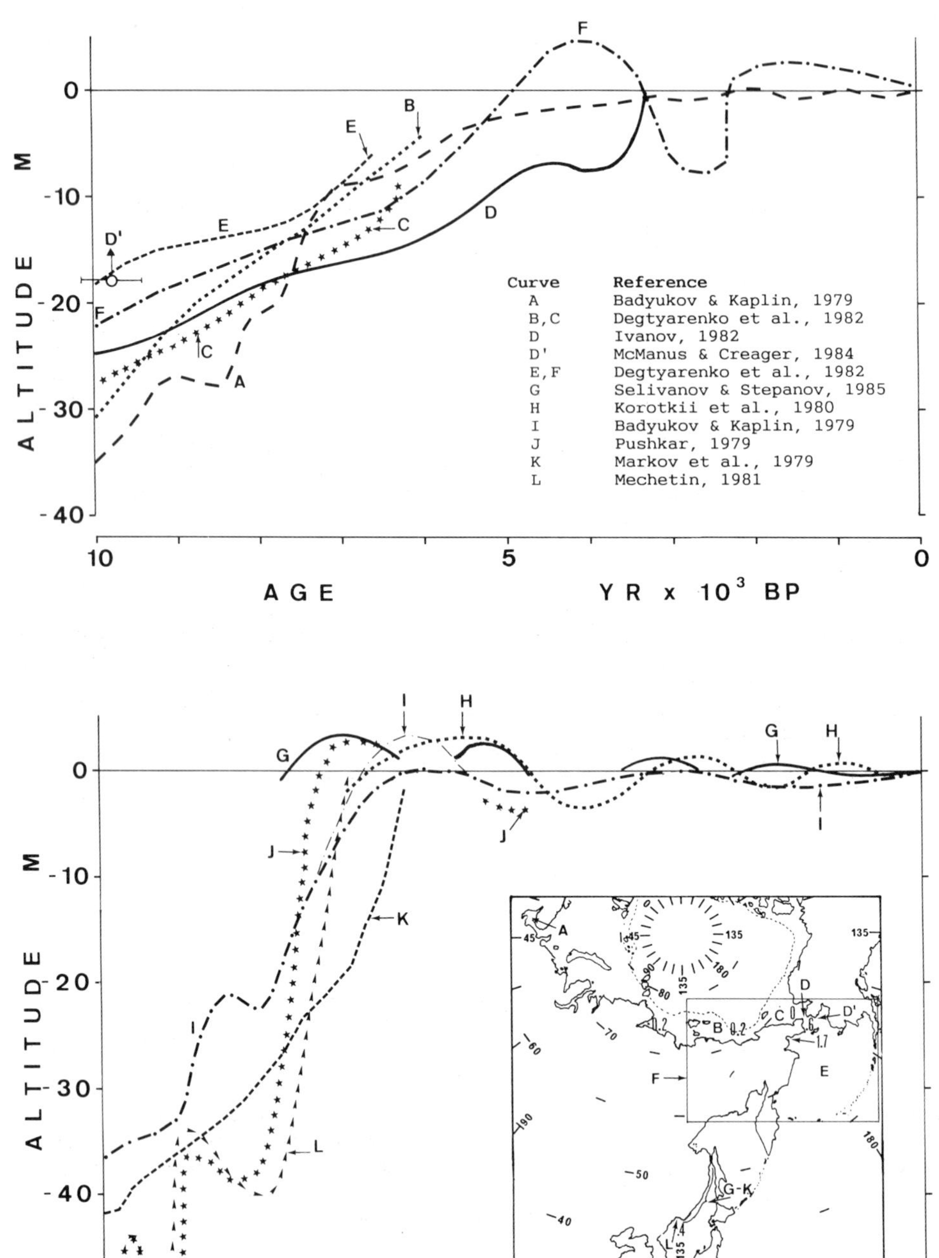

Degtyarenko et al. (1982) published three sea-level curves: **B** for the East Siberian Sea, **C** for the Chukchi Sea and **E** for the Bering Sea coasts; very little information on these curves is available to the present writer. Curve **F**, presented by Degtyarenko et al. (1982) as an average of curves **B**, **C** and **E**, suggests the occurrence of some wide oscillations since 6000 yr BP, which seem rather unlikely however. In addition, the physical meaning of such an average is unclear.

Near the Bering Strait coasts, according to Ivanov (1982), a 10-m rise in sea level occurred between 4000 and 3500 yr BP (curve **D**), following a relative stabilization between 5000 and 4000 yr BP. Near the Alaskan coast of the Bering Strait, **D'** corresponds to a peaty soil dated 9700 yr BP, reported by McManus and Creager (1984) at a depth of -18 m, which indicates a relative sea level above this level.

Several sea-level curves have been produced for the Primorye coast. Selivanov and Stepanov (1985) proposed curve **G**, showing four emergence phases since 8000 yr BP, based on geoarchaeological data from sixteen ancient sites, supported by ten radiocarbon dates; the first emergence period was dated as early as 7700 to 7300 yr BP using "algae peat" samples. Curve **H** by Korotkii et al. (1980) is based on geological and geomorphological data; it shows various oscillations, with however only three emergence phases, which in addition do not coincide with those of curve **G**. Curve **I** was obtained on the northwest coast of the Japan Sea by Badyukov and Kaplin (1979) from an interpretation of buried Holocene coastal complexes discovered by drilling under recent sandy sediments. It suggests a relative sea-level rise from -35 m to zero (with a variant to +3 m) between 10,000 and 6500 yr BP and three regression phases culminating c. 8000, 4000 and 1500 yr BP. Finally curves **J** by Pushkar (1979), **K** by Markov et al. (1979) and **L** by Mechetin (1981) have been reproduced from a graph published by Korotkii (1985), who explained discrepancies between the various curves by "different chronologic manifestations of hydro-isostasy" and considered that inferences concerning sharp oscillations of sea level in the early Holocene "seem unwarranted".

JAPAN: INTRODUCTION

The Japanese Islands comprise four island arcs, stretching along the eastern margin of the Asian continent, which have been tectonically active throughout recent geological times. As reported by Yoshikawa et al. (1981), their morphotectonic units are relatively small in area and have a complicated mosaic structure. Many sea-level curves have been reported from the Japanese islands during the last two decades. Some early curves were composite, mixing data from various areas which were often characterized by different tectonic histories. These early curves, although they were often claimed to be "eustatic", are not reproduced here and, with a few exceptions, only sea-level curves constructed with data coming from small areas have been considered. Earlier compilations of sea-level records, made by Ota et al. (1981; 1987a,b) in the framework of the activities of the IGCP Projects n° 61 and 200, have been extremely useful in the preparation of Plates 37-41.

Plate 37: JAPAN I (North)

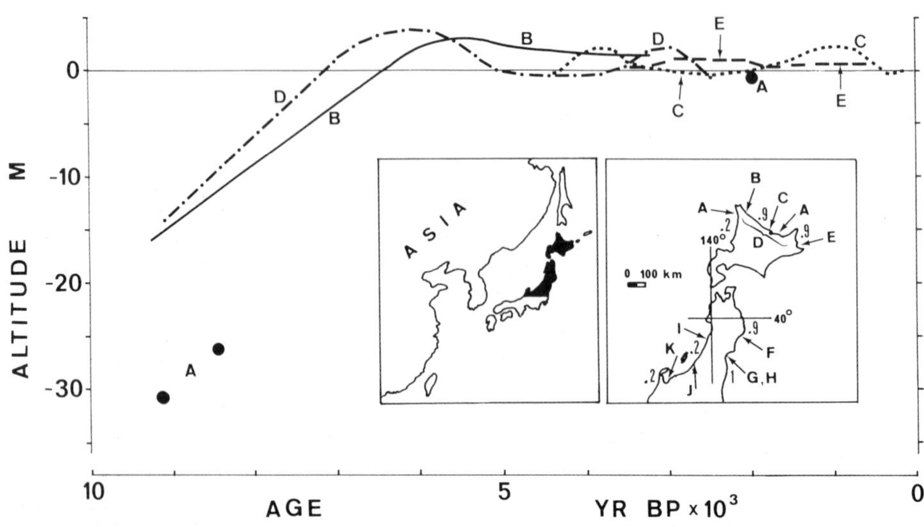

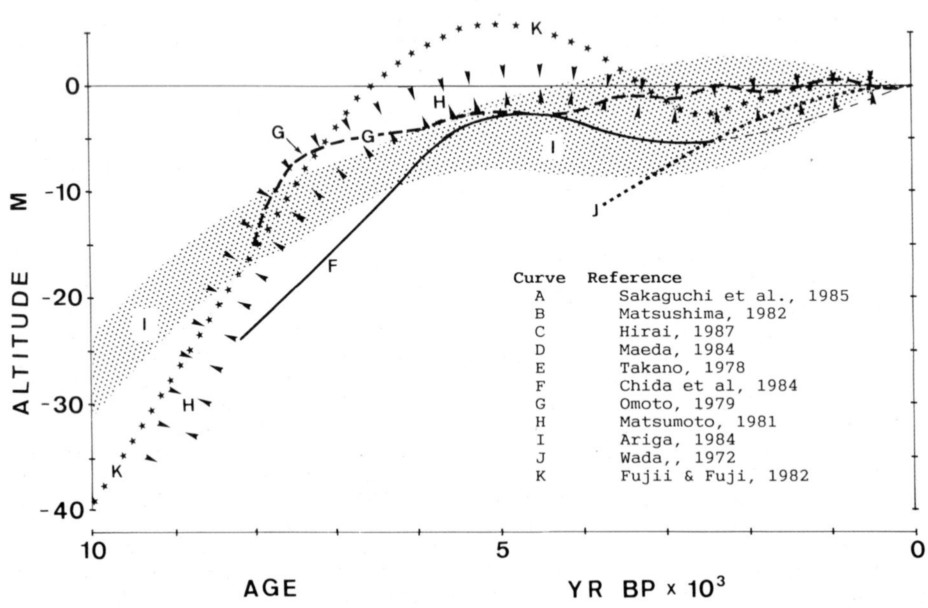

Curve	Reference
A	Sakaguchi et al., 1985
B	Matsushima, 1982
C	Hirai, 1987
D	Maeda, 1984
E	Takano, 1978
F	Chida et al, 1984
G	Omoto, 1979
H	Matsumoto, 1981
I	Ariga, 1984
J	Wada,, 1972
K	Fujii & Fuji, 1982

NORTH JAPAN (Plate 37)

In Hokkaido, most of the available sea-level data come from the north coast. According to cores bored in the Tokoro and Sarobetsu lowlands (loc. A) studied by Sakaguchi et al. (1985), the sea level was about -30 m around 9000 yr BP (circles **A**). Matsushima (1982) (curve **B**) and Maeda (1984) (curve **D**), on the other hand, locate the contemporaneous sea level some 15 m higher in the Kucharo Lake area (loc. B), on the basis of the date of a silty peat sample collected at the base of the Holocene marine deposits. They suggest that the present sea level was probably reached earlier than 7000 yr BP (curve **D**) or around 6500 yr BP (curve **B**), after which it may have fluctuated above and below zero in the former case, or remained above zero until at least 3000 yr BP in the latter case. The marine origin of the peat seems doubtful however, and in both cases the few data provided are insufficient to justify convincingly such divergent conclusions.

Since 6000 yr BP, a sea-level curve oscillating between 4 m above zero and a few decimetres below zero is proposed by Hirai (1987) in the Saroma Lake area (curve **C**), using several of the same data provided by Maeda (1984). Lastly, on the east coast of Hokkaido, beach ridges at various elevations have been interpreted by Takano (1978) as corresponding to changing sea levels (curve **E**). Although none of the above curves appears sufficiently precise to estimate the rate of recent relative sea-level changes in northern Hokkaido, slight emergence seems to have predominated during the last 6000 years. This suggests an average rate of relative sea-level drop of 0.3 ± 0.4 mm/yr for this area.

In northeast Honshu, curve **F** (Rikuzentakata coastal plain) has been constructed by Chida et al. (1984) using four radiocarbon dates of shells and organic clay samples obtained by boring. Subsidence phenomena were probably active here during the late Holocene. Remeasurement of benchmarks by levelling has found a recent subsidence rate of 0.8 mm/yr. Omoto (1979) summarized previous work in the Sendai plain by producing a sea-level curve (**G**) which shows a gradual rise with slight oscillations, most of the curve remaining however below zero. A similar (but not identical) trend is suggested by Matsumoto (1981), whose sea-level band (**H**), largely based on data from borings, confirms that the relative sea level has been rising in the Sendai area for the last few thousand years.

On the Japan Sea side, Ariga (1984) estimated a sea-level band (**I**) using borehole data from the Shonai coastal plain. As the uncertainty range widens towards 2000-3000 yr BP, it is impossible to make a precise estimation of recent trends in the relative sea-level change. In the Niigata plain (**J**), the curve outlined by Wada (1972) shows a rapid relative sea-level rise, suggesting strong subsidence movements during the late Holocene, but few details are given. Lastly, Fujii and Fuji (1982) reproduced for the Hokuriku area a sea-level curve (**K**) already proposed in earlier publications, which shows a maximum clearly above the present sea level, between 6000 and 4000 yr BP, followed by a negative oscillation, between 3000 and 2000 yr BP, and by a sea-level rise continuing until the present situation. However, as already noted by Pirazzoli (1978), these fluctuations are a consequence of the choices made by the authors, who in the same composite curve mixed data from areas showing different tectonic behaviours: emergence on both sides of Toyama Bay (Fig. 2-22),

Plate 38: JAPAN II (Central)

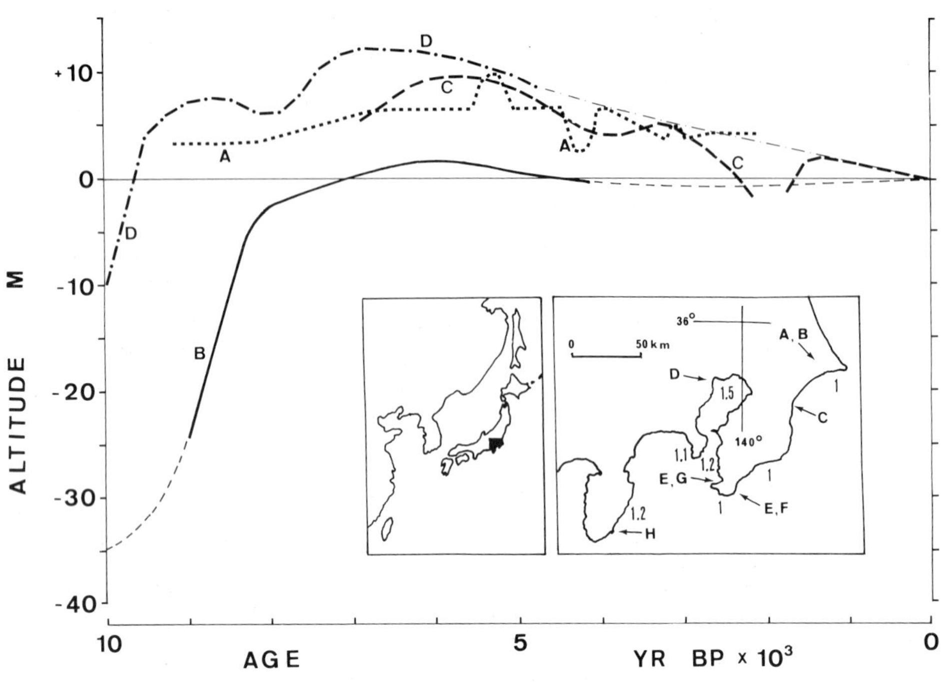

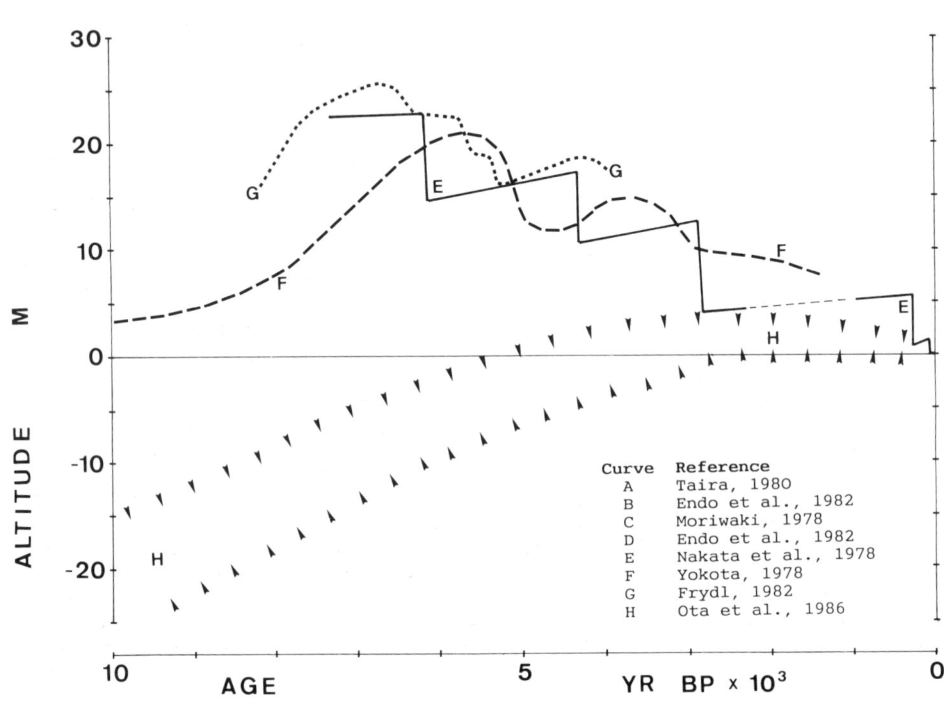

where the sea-level data above zero have been collected, and subsidence in the alluvial plain of the central part of the bay, near Toyama City, where all the sea level data are below zero. These different tectonic trends are confirmed by the 1: 500,000 Neotectonic Map published in 1985 by the Geological Survey of Japan (Sheet 10: Kanazawa), which indicates that the Toyama plain is delimited on the west and east sides by inferred active faults.

CENTRAL JAPAN (Plates 38 and 39)

The Tokyo area, which is located near a zone where three crustal plates come into contact with each other, is characterized by strong tectonic movements, which are most often part of a long-term uplift, although uplift movements usually occur by sudden jerks, at the time of major earthquakes, and moderate subsidence may prevail during interseismic periods. All the sea-level curves from the Boso Peninsula are influenced by these predominating trends, though some localized differences may appear.

Curve **A** (**Plate 38**) by Taira (1980) was obtained by linking radiocarbon dates of shell middens in the Tone River area with the altitude of the alluvial plain adjacent to the middens, assuming that gaps in the sequence of dates resulted from regressions. In fact these criteria are not sufficient proof that the fluctuations shown by curve **A** really occurred. Endo et al. (1982) used boring data to summarize the stratigraphy of several alluvial plains in the Kanto region. In the Tone River area, their curve (**B**) appears much lower and less fluctuating than in the preceding curve **A**. In the paleo-Isumi Bay area (curve **D**), the influence of uplift movement is clearly visible, the average uplift rate being about 2 mm/yr. Not far from there, in the Kujukuri coastal plain (**C**), the relative sea level has dropped gradually from over +9 m to zero since c. 6000 yr BP according to Moriwaki (1978), although the average rate is estimated to have been no more than 1 mm/yr by Moriwaki (1979). A possible drop of sea level below the present zero, around 2000 BP, is inferred by Moriwaki from data concerning other areas of Japan, though with little local support.

In the southern part of the Boso Peninsula, remnants of coral reefs in growth position are found up to +24 m, marking the northernmost limit reached by Holocene coral reefs in the North Pacific. Relative emergence has been sufficiently rapid here (c. 3 mm/yr) to expose marine sediments of the last 8000 years. Nakata et al.(1978), in agreement with previous work by Yonekura (1975) and other authors, interpreted a series of four stepped terraces as being due to sudden uplift movements occurring at the time of great earthquakes, such as those in 1703 and in 1923, when the lowest terrace was suddenly uplifted in two steps. Their curve (**E**) is based on 36 radiocarbon dates. A slight relative sea-level rise between two sudden seismic movements was inferred from interseismic subsidence, revealed by levellings and tide-gauge records. Yokota (1978), on the other hand, represented the relative sea-level changes in this area with a smooth curve (**F**) showing wide oscillations rather than sudden jerks. It seems unlikely however that postglacial emergence could have started earlier than 10,000 yr BP in this area, as suggested by Yokota, who believed that "the rate of eustatic rise of sea level was slow about 9000 yr BP". Finally Frydl (1982) deduced from the biofacies distribution of ostracod species in the same area a

Plate 39: JAPAN III (Central)

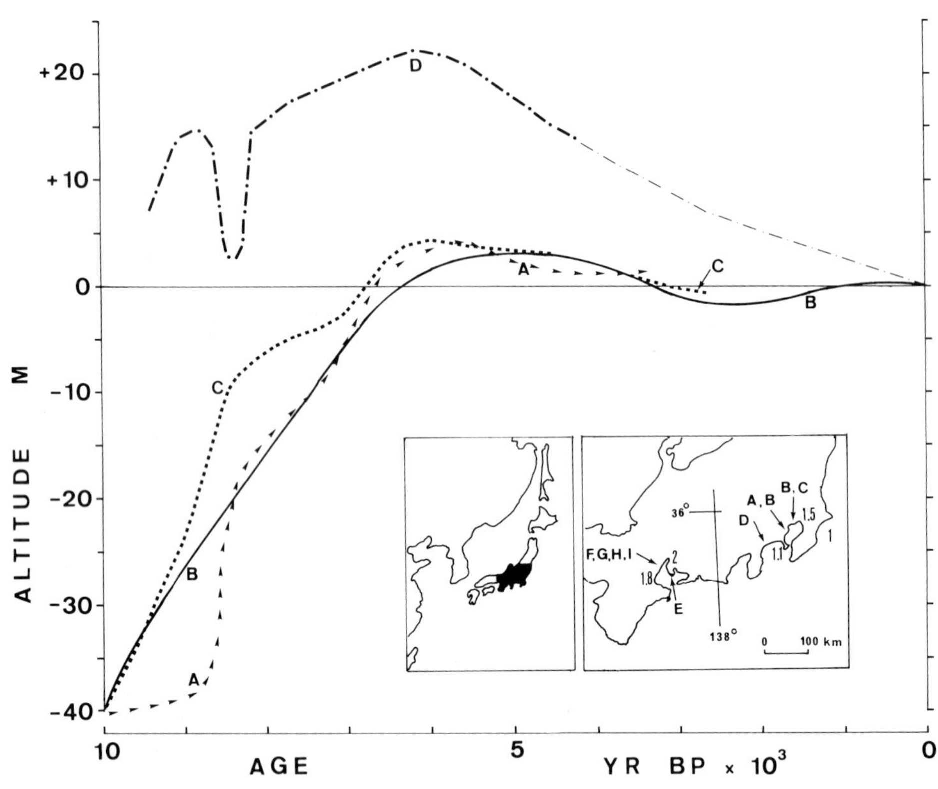

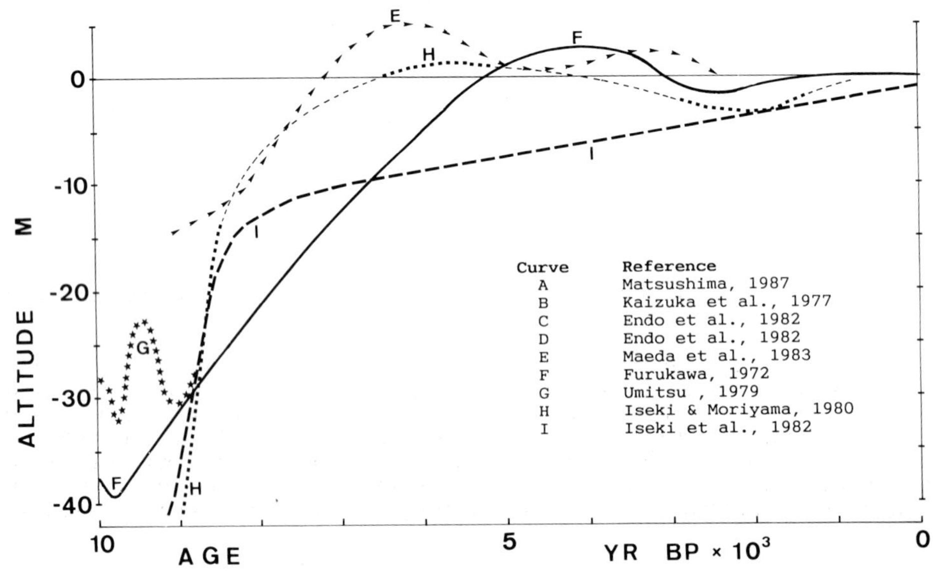

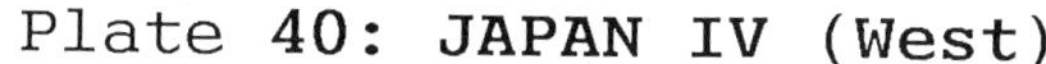

Plate 40: JAPAN IV (West)

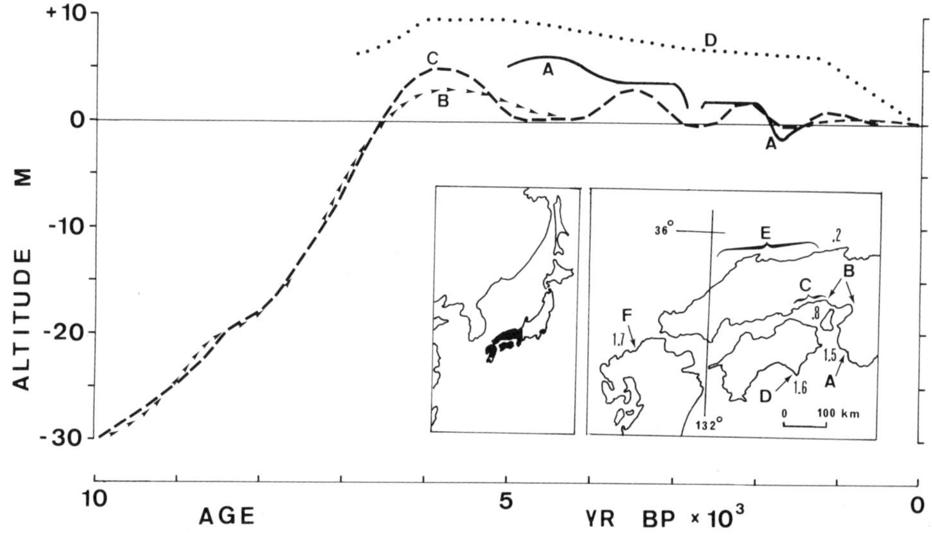
+10
0
-10
-20
-30
ALTITUDE M
A
B
C
D
36°
E
.2
F
1.7
.8
1.5
1.6
132°
0 100 km
10
5
0
AGE
YR BP × 10³

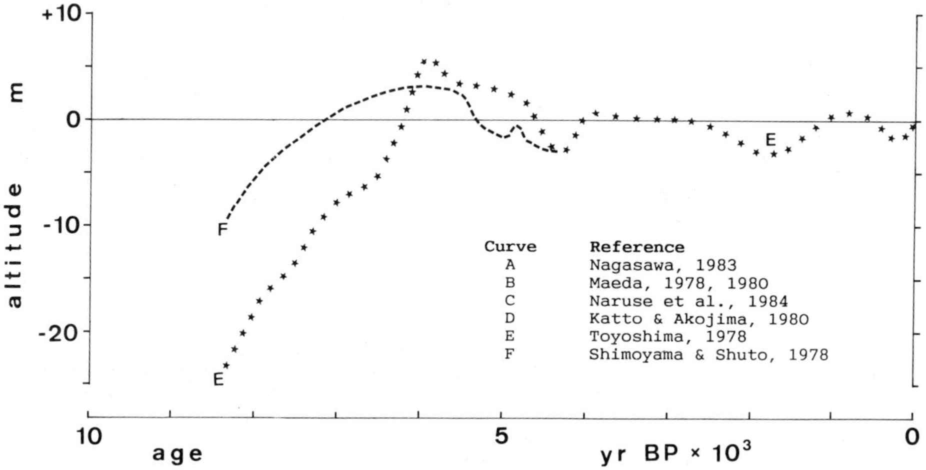
+10
0
-10
-20
altitude m
E
F
Curve Reference
A Nagasawa, 1983
B Maeda, 1978, 1980
C Naruse et al., 1984
D Katto & Akojima, 1980
E Toyoshima, 1978
F Shimoyama & Shuto, 1978
10
5
0
age
yr BP × 10³

rather irregular trend of relative sea-level movements between 8000 and 4000 yr BP (curve **G**).

In the southern part of Izu Peninsula which, as part of the Philippine Sea plate, is in collision with Honshu, the uplifting trend is much slower than in Boso Peninsula. The sea-level band (**H**) published by Ota et al. (1986) was obtained from excavations in two coastal lowlands, determining sediment facies and making an environmental assessment of fauna, and is based on about thirty radiocarbon dates. It shows an average rate of relative sea-level drop of slightly less than 1 mm/yr since c. 3000 yr BP. Ota et al.(1986) suggested however that emergence may have occurred suddenly, as a consequence of a coseismic uplift.

In the Tama River lowland, a very detailed curve (**A**, **Plate 39**) by Matsushima (1987), based on seven borehole data, has been derived from some seventy radiocarbon dates. The lowest part of this curve, showing a sea-level stand at about -40 m around 9000-10,000 yr BP, differs from the contemporaneous rapidly rising sea level shown in the same area by curves **B** (Kaizuka et al., 1977) and **C** (Endo et al., 1982), in which a sea-level standstill or regression is suggested earlier than 10,000 yr BP. Around 8500 yr BP, the highest position of curve **C**, in relation to curves **A** and **B**, seems based on only one sample. No local data are provided to justify the possibility of a regression which is suggested towards 2000 yr BP.

In Sagami Bay, at the mouth of the Oshikiri River (**D**), Holocene marine sediments can be found above 20 m in altitude. Here a major fluctuation in the relative sea level is suggested by Endo et al. (1982) between 9000 and 8000 yr BP. Since 6000 yr BP, the average rate of the local relative sea-level drop is at least 3 mm/yr. The oscillating **E** curve proposed by Maeda et al.(1983) at Utsumi City is based on the upper limit of marine sediments near shell mounds and on a few radiocarbon datings.

From the Nagoya area, four relative sea-level curves are available. Curve **F** was published by Furukawa (1972) at a very small scale with few details. Curve **G** by Umitsu (1979) is noticeable for it suggests the possibility of an unusal oscillating sea-level trend between 10,000 and 8000 yr BP. Iseki and Moriyama (1980) have proposed a sea-level drop around 2000 yr BP (curve **H**), following a maximum above zero in the middle Holocene. Curves **F** and **H** are however composite, mixing data from the subsiding Nobi plain with emergence data found in areas around the plain which may be relatively stable or even slightly uplifting. From this point of view curve **I** by Iseki et al. (1982), showing a continuous relative sea-level rise since 8000 yr BP, at an average rate slightly exceeding 1 mm/yr, seems more realistic when only the west part of Nagoya Harbour is considered. This suggests that the so-called "Yayoi Regression", which is often reported in Japanese papers for the period 2400 to 1500 yr BP, may be in some cases an artifact due to the use of composite sea-level curves mixing data from uplifting areas with data found in subsiding coastal plains.

WEST JAPAN (Plate 40)

In the Tanabe Bay area (loc. A), where several emerged features (benches, notches, sea caves, etc.) have been described by various authors between +6 m and the present sea level, Nagasawa (1983) has attempted to draw a relative sea-level curve (**A**) based on the frequence of erosional features at various levels and on a few dates of raised beaches and prehistoric sites. Although the date of a raised beach at +3 m (2025 ± 75 yr BP) corresponds to the middle of the Yayoi period, the occurrence of a "rapid regression in the Yayoi period" is suggested by Nagasawa in curve **A**, without however producing any local radiometric or archaeologic date.

A series of borings, direct observation from a caisson of the Holocene sediments under the sea floor and many radiocarbon dates have enabled Maeda (1978) to construct a detailed relative sea-level curve (**B**) for the period 10,000 to 4000 yr BP in the Bay of Osaka. An acceleration in the rate of sea-level rise appears clearly from the stratigraphy between 8000 and 6000 yr BP. Curve **B** is completed in its upper part by Naruse et al. (1984), who investigated emerged marine features along the Hamarinada, whereas the numerous oscillations of curve **C** are based on only four radiocarbon datings.

The Muroto Peninsula (loc. D) is known to have been uplifted and tilted landward, as shown by the distribution of Pleistocene marine terraces (Yoshikawa et al., 1981). At the Muroto Cape, where the terraces reach their maximum elevation, coseismic uplift movements have occurred several times during the last few centuries: +1.5 m in 1707, +1.2 m in 1854, +1.2 m in 1946. Here interseismic subsidence between two consecutive earthquakes is estimated to be about 95 cm. An average residual uplift rate of about 2 mm/yr can be deduced from the above data (for references, see Pirazzoli, 1978). Katto and Akojima (1980) suggested however that an uplift rate of about 5 mm/yr could have occurred since about 1000 ± 200 yr BP without tilting all along the west coast of the peninsula. The curve they provided (**D**) uses data from several sites along the west coast, indicating an average rate of relative sea-level drop of less than 1.5 mm/yr during the last 6500 yr. Katto and Akojima (1980) concluded that the uplift rate was variable during the Holocene, with possible periods of subsidence.

Along the San-in coast, Toyoshima (1978) has constructed a single sea-level curve (**E**) summarizing all available data. This is a composite curve however, with data coming from localities far apart. In addition, geomorphological changes occurring over short distances are typical of the block movements which characterize the Inner Zone of Japan to which the San-in coast belongs. This means that local data should be interpreted with caution and in context. Though curve **E** may not correspond to the real relative sea-level history in a single locality, it gives a useful idea of the size of crustal movements which may have occurred in this area.

In the Fukuoka City area, north Kyushu, a sea-level curve between 8000 and 4600 yr BP, showing a peak at about 6000 yr BP (**F**) was constructed by Shimoyama and Shuto (1978), with assumptions on the basis of two radiocarbon datings.

SOUTH JAPAN (Plate 41)

Koba et al. (1979) reported from Takara and Kodakara Islands, in the Ryukyus, a noteworthy example of a sudden uplift movement giving an almost rigid tilt to two islands situated on the same crustal block. According to radiometric data, the movement occurred simultaneously about 2380 yr BP in both islands, probably after a period of relative sea-level rise. At that time Kodakara Island (curve **B**) was uplifted 7.4 m (the corresponding shoreline being raised from +2.4 m to +9.6 m), whereas the uplift jerk in Takara Island (curve **A**) was 2.9 m (from +1.1 m to +4.0 m). It is not specified whether the present sea level has been reached gradually since then, or is due to another sudden jerk.

Plate 41: JAPAN V (South)

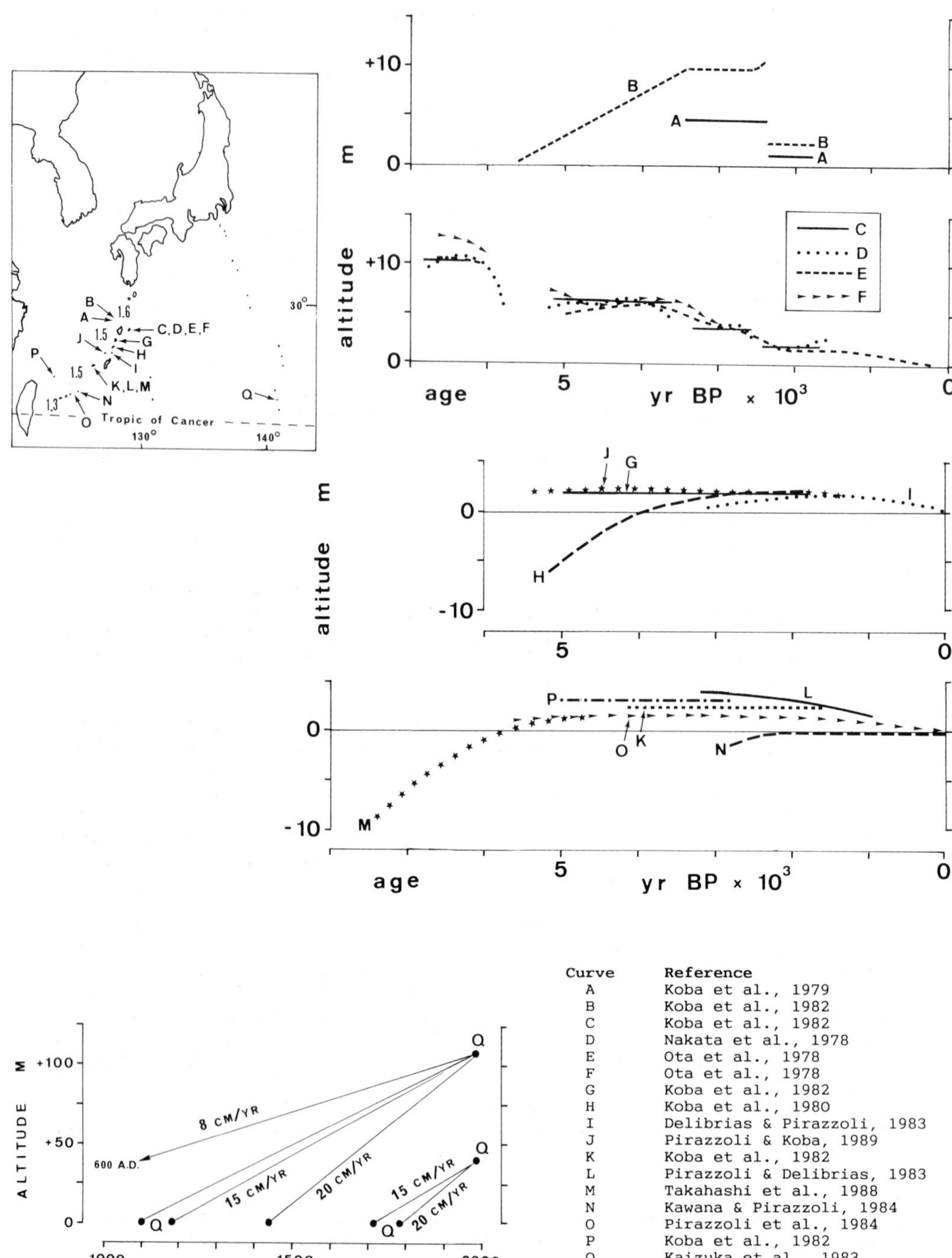

Curve	Reference
A	Koba et al., 1979
B	Koba et al., 1982
C	Koba et al., 1982
D	Nakata et al., 1978
E	Ota et al., 1978
F	Ota et al., 1978
G	Koba et al., 1982
H	Koba et al., 1980
I	Delibrias & Pirazzoli, 1983
J	Pirazzoli & Koba, 1989
K	Koba et al., 1982
L	Pirazzoli & Delibrias, 1983
M	Takahashi et al., 1988
N	Kawana & Pirazzoli, 1984
O	Pirazzoli et al., 1984
P	Koba et al., 1982
Q	Kaizuka et al., 1983

Kikai Island is completely surrounded by emerged Holocene fringing reefs, reaching 12-13 m in elevation. Six distinct shorelines have been recognized, four of which have been dated. The curves **C**, **D**, **E** and **F** correspond to different interpretations of about the same data. According to Nakata et al. (1978) (curve **D**) and to Koba et al. (1982) (curve **C**), the four main elevated coral reef terraces emerged have been uplifted by sudden jerks at the time of major earthquakes, with minor subsidence movements occurring occasionally in interseismic periods. Ota et al. (1978), on the other hand, assume that the uplift movement has been regular and that marine terraces at various levels are the consequence of eustatic sea-level fluctuations superimposed to a uniform rate of tectonic uplift. Curve **E** is obtained from data concerning the whole coastline of the island, whereas curve **D** refers to the northwestern coast only. Anyway, there is general agreement that the island has been uplifting at the average rate of 1.5 to 2.0 mm/yr since at least 125,000 yr BP.

In Toku Island (**G**), according to Koba et al. (1982), the sea level may have remained stable at +2 m from 5000 to 2000 yr BP. According to Machida et al. (1976) and Pirazzoli (1978), on the other hand, several sudden uplift movements affected the island during the late Holocene, one of them raising the +2 m shoreline at about 3000 yr BP.

A former sea-level stand at +2 m is also reported from Iheya Island (**J**), lasting from about 5500 to about 1500 yr BP (Pirazzoli and Koba, 1989). In Okinoerabu Island, according to Koba et al. (1982), curve **H** indicates a minimum position for the relative sea level from 5000 to 2000 yr BP. The Yoron Island curve (**I**), obtained by Delibrias and Pirazzoli (1983), suggests a possible continuation until the present of the various curves in this area, in the absence of sudden seismo-tectonic movements.

In Kume Island, Koba et al. (1982) proposed a stillstand of the sea at about +2.5 m (**K**). The island is tilting however, as shown by Pirazzoli and Delibrias (1983). On the western coast of the island, where uplift is most rapid, the relative sea-level history can be summarized by the minimum positions of curve **M**, obtained with boring data by Takahashi et al. (1988), then, for the more recent period, by curve **L**, obtained by Pirazzoli and Delibrias (1983) using fossil barnacles.

Miyako (**N**) is one of the rare islands in the Ryukyus where marks of Holocene emergence are lacking, as has been shown by Kawana and Pirazzoli (1984) (Fig. 2-24). This may be only a temporary situation, however, due to the fact that uplift movements accompanying great earthquakes did not occur here in the late Holocene. Although there is evidence of relative sea-level stability for the last two thousand years, a slight uplifting trend is shown to be predominant over longer periods, as indicated by the occurrence of three steps of Pleistocene marine terraces, the lowest of which, at about +15 m, has been ascribed to the last interglacial stage.

In Tarama Island (**O**), according to Pirazzoli et al. (1984), Holocene emergence is only 1.5 m and no marks of sudden crustal movements are visible. In the Uotsuri Island (**P**), Koba et al. (1982) have reinterpreted data already published by Konishi et al. (1979), suggesting that the relative sea level remained stable at +3 m, from 5000 to 2800 yr BP.

Fig. 2-24. In Miyako Island (the Ryukyus, Japan) sea-level indicators suggest relative sea-level stability during the late Holocene, in contrast with the long-term uplift trend indicated by raised Pleistocene marine terraces. Mean spring tidal range is 1.5 m. **a**) A tidal notch (left) and a beachrock (right) are well developed in the present-day intertidal zone (as shown by *Bostrychia tenella* algae, which usually reach an upper limit between MSL and HAT (Kawana, 1982). Water level is 0.4 m below MSL. **b**) Detail of a notch profile along the west coast of Miyako Island. A *Bostrychia* cover is clearly visible above the notch floor. Scale is 1 m. Water level is 0.25 m below MSL.

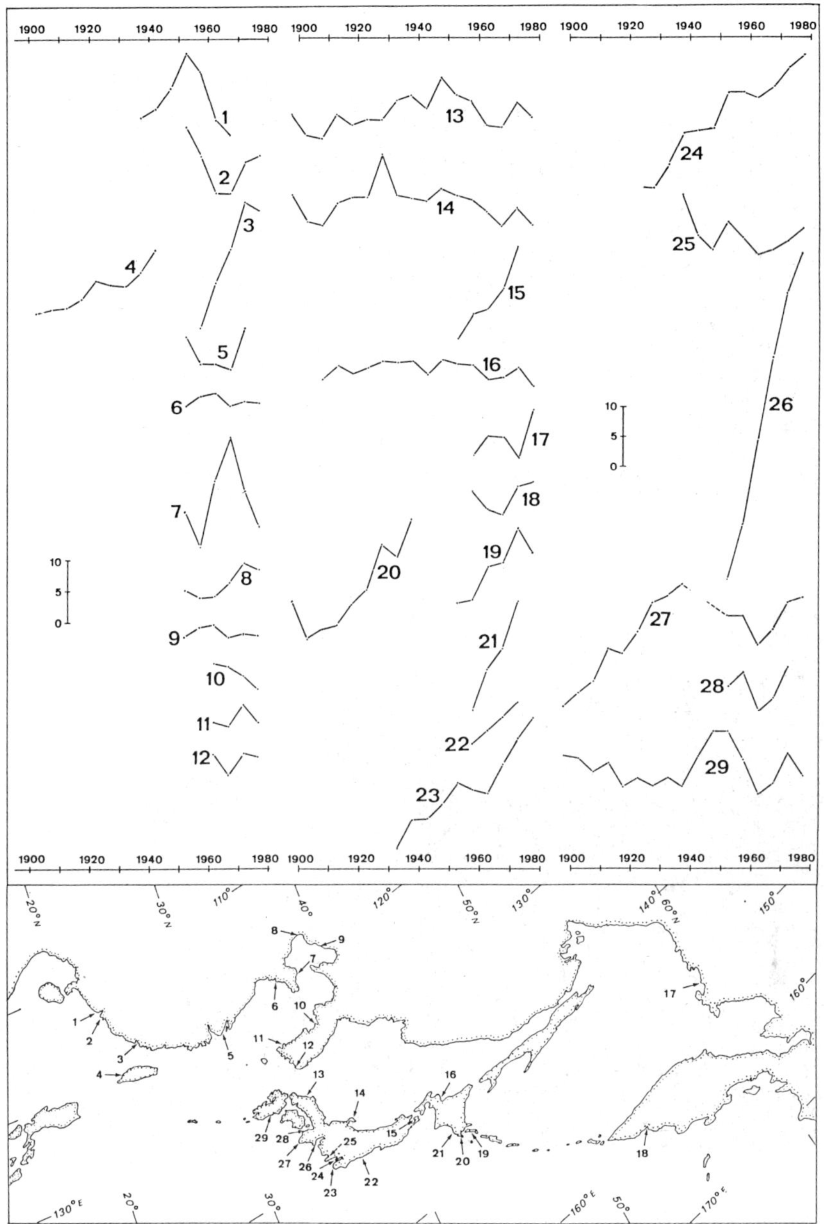

Fig. 2-25. Average 5-yr MSL variations in East Asia as recorded by tide-gauge stations. 1) Macao; 2) Hong Kong; 3) Xiamen; 4) Takao; 5) Shanghai; 6) Qingdao; 7) Yantai; 8) Tanghu; 9) Qinghuangdao; 10) Inchon; 11) Mokpo; 12) Pusan; 13) Tonoura; 14) Wajima; 15) Ominato; 16) Oshoro; 17) Nagajeva Bay; 18) Petropavlosk; 19) Yuzno Kurilsk; 20) Hanasaki; 21) Kushiro; 22) Onohama; 23) Mera; 24) Aburatsubo; 25) Uchiura; 26) Toba; 27) Kushimoto; 28) Kainan; 29) Hosojima. (From Pirazzoli, 1986b).

Lastly Iwo (Sulphur) Island, a volcano situated near the Japan Trench 1200 km south of Tokyo (**Q**), provides the most impressive example of very rapid uplifting. Over twenty steps of marine terraces, occupying nearly the whole area of the island, up to +120 m, have been recognized by Kaizuka et al. (1983), corresponding to average uplift rates of over 100 mm/yr over several centuries, reaching 200 mm/yr in some parts of the island.

Recent relative sea-level trends shown by tide-gauge records in East Asia are summarized in Fig. 2-25. Most of the series longer than 50 years come from Japan, where trends are irregular and change from place to place. Indeed tectonic and oceanographic influences are very strong in this area.

TAIWAN - MICRONESIA (Plate 42)

In Taiwan a strong tectonic activity is caused by a continent-arc collision in the transition zone between opposite movements of the nearby Philippine Sea plate: an underthrust in the Ryukyu arc and an overthrust along the volcanic islands down to central Luzon. This tectonic activity results in Holocene uplift rates which are among the highest in the world outside areas of former ice sheets.

Curve **A** (**Plate 42**), which shows six sea-level fluctuations of several dozen metres each (the minima of which are however obscure), was presented by Lin (1969) and Fujii et al. (1971) as "the apparent eustatic oscillation curve of the coastal region of Taiwan". Although this definition is inappropriate, curve **A** can be considered as a kind of discontinuous summary of the elevation of marine deposits studied by Lin (1969) around the island. Dotted areas correspond in fact to differences in elevation between marine deposits of similar age in different areas. The peak around 4000 yr BP, for example, with a dotted area extending vertically from +17 to +40 m, summarizes the range of elevations at which the upper limit of dated marine deposits was reported by Lin (1969), from the east coast of Taiwan, at c. +20 m near Tulan (22°52'N), +25 m at Ch'engkung (23°07'N), +30 m at Shihti (23°29'N) and +40 m near Hualien City (24°00'N). Though curve **A** cannot be used as a relative sea-level curve at any given locality, dotted areas may be useful to gain an idea of the amount of differential crustal movements which have occurred since a given time between coastal localities. The most recent peak in curve **A**, reaching +65 m about 1000 yr BP, is doubtful however.
Curve **B** represents, according to Peng et al. (1977), the minimum uplift movements which can be deduced from dated samples of raised corals along the east coast of Taiwan. Curve **C**, also proposed by Peng et al. (1977), corresponds to the minimum uplift along the Hengchung Peninsula and Tainan area. Curve **D**, showing three steps, was obtained by Taira (1975) from dated coral samples and is expected to represent crustal uplift in Hengchun Peninsula. As concerns uplift movements, they probably occurred in jerks at least in some cases; on the east coast of Taiwan, for example, marks of at least ten superimposed Holocene shorelines have been observed by the present writer between the present sea level and +40 m in the Ch'engkung-Tanman area.

In fact, rates of vertical movements are changing so rapidly from place to place in active tectonic areas such as Taiwan (Fig. 2-26, 2-27), that even "regional" curves like **B**, **C** or **D** are hardly applicable to a given locality. Only along the southern coast of

Plate 42: TAIWAN - MICRONESIA

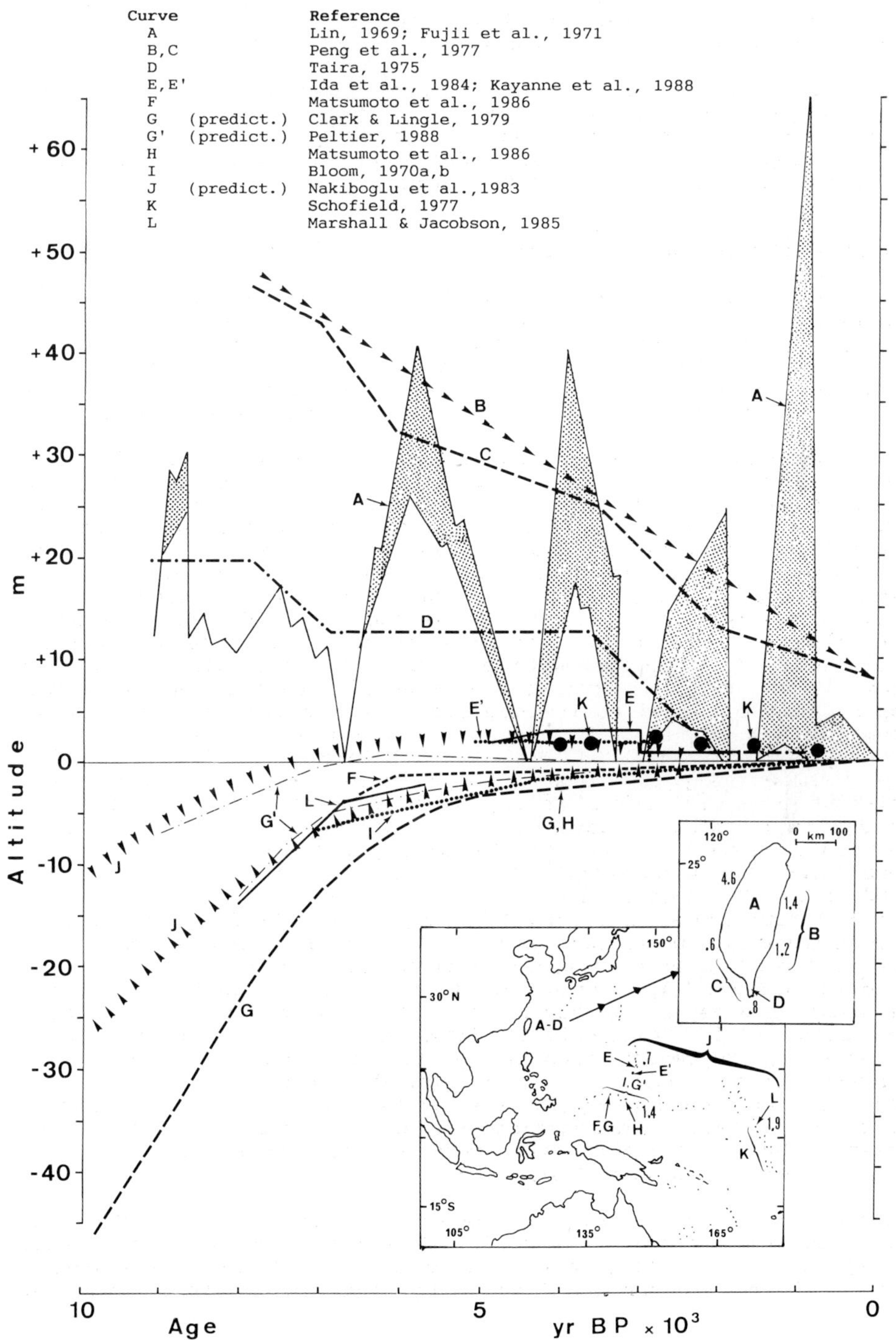

Fig. 2-26. Sea-level marks in an uplifting coral reef area (southernmost tip of Taiwan). A: upper limit of the present intertidal zone (coloured in green or light brown by living algae); A-B: bare Holocene raised reef, washed by occasional storm waves (coloured in grey by endolithic algae); B: upper limit of present-day wave action (beach deposits) and lower limit of vegetation; C: late Pleistocene raised coral reef terrace.

Fig. 2-27. *In situ* coral reefs about 3000 years old uplifted at about 25 m above MSL at Hualien, east coast of Taiwan. This rapid tectonic uplift rate (over 8 mm/yr) is of the same order as the glacio-isostatic rebound in former ice sheet areas.

Hengchung Peninsula, for example, is the inner edge of the mid-Holocene marine terrace reported by Liew and Lin (1987) to vary in altitude from +5 to +36 m, with several undulation movements, along a distance of less than 10 km.

In the southern Marianas, Ida et al. (1984) and Kayanne et al. (1988) deduced a three-step relative sea-level curve (**E**) from raised coral reefs and superimposed notches at Rota Island and a two-step curve (**E'**) at Guam Island. At Rota, where emerged coral reefs and erosional terraces are developed above the andesitic basement up to the top of the island at +496 m, the highest Holocene standstill at +4.5 m on the eastern coast, though only at +3 m on the western coast, was dated between 4000 and 3000 yr BP. A second standstill, at c. +1 m on the western coast, was dated about 2500 yr BP. In Guam, a standstill at +1.8 m lasted from 5100 to 2900 yr BP. Emergence movements in both islands have been rapid and probably occurred at the time of great earthquakes.

In Truk Island, in the Carolines, where marks of Holocene emergence seem absent, sea-level curve **F** was constructed by Matsumoto et al. (1986) using five dated samples of coral or peat; it suggests a rate of relative sea-level rise of 0.2 mm/yr during the last 6000 years. In the same area, the trends of relative sea-level change predicted by global isostatic models are shown by curve **G** (Clark and Lingle, 1979) and band **G'** (Peltier, 1988). For the last 5500 years, **G** is almost identical to curve **H**, determined in Ponape Island by Matsumoto et al. (1986), using four radiocarbon dates of peat and coral, i.e. a relative sea-level rise of about 0.8 mm/yr. Curve **I**, constructed by Bloom (1970a,b) by mixing nine dated samples from Truk, Ponape and Kusaie, is intermediate between **F** and **H**. Lastly band **J** is a sea-level envelope predicted by Nakiboglu et al.(1983) in Micronesia, which shows a trend similar to band **G'**.

Exposed platforms, consisting of coral reef deposits lithified at or above high tide, have been described in many reef islands (Fig. 2-28). In Micronesia, these were interpreted as lithified storm deposits (Shepard et al., 1967; Newell and Bloom, 1970) providing no evidence for emergence. More recent studies, however, have supported the occurrence of late Holocene emergence of at least one metre in Micronesian atolls (Tracey and Ladd, 1974; Buddemeier et al., 1975). This suggests that relative sea-level curves deduced from paludal stratigraphy (e.g. **F**, **H**, **I**) may be more or less influenced by local compaction or subsidence phenomena.

Marshall and Jacobson (1985) used cores from ten holes drilled to a maximum depth of 30 m into the reef of Tarawa Atoll (Kiribati) and two dozen radiocarbon dates to obtain a minimum sea-level curve (**L**) for the period 8000 to 5500 yr BP, suggesting a deceleration in the sea-level rise after 7000 yr BP. In Kiribati and Tuvalu (formerly Gilbert and Ellice Islands), Schofield (1977) interpreted geomorphological features and dates of emerged *in situ* corals and *Tridacna* shells as representing peaks of six "transgressions" (full circles **K**) corresponding to former sea levels between +0.6 and +2.4 m; these can be re-interpreted, however, as belonging to a smooth relative sea-level curve indicating slight, persistent, late Holocene emergence in this area.

MELANESIA (Plate 43)

The marine terraces of Huon Peninsula, Papua New Guinea, are a well known reference basis for the estimation of late Pleistocene sea-level changes (Chappell, 1974, 1983; Bloom et al., 1974; Chappell and Shackleton, 1986). In the Holocene reef, which emerges as far as 12 m, Chappell and Polach (1976) studied an exposed seacliff section, 8 m high, showing a shallow-water reef-crest facies that grew steadily upwards, between 8200 and 6600 yr BP, while the relative sea level rose approximately 7 m (curve **A**). If the uplift rate in the area considered (c. 1.9 mm/yr) is assumed uniform in time and deducted from the present altitude of the deposits, the sea-level position "around northeast Papua New Guinea" at 8000 and 6000 yr BP can be estimated at -14 m and -4 m respectively (**A'**). However, superimposed benches and notches indicate that the Holocene emergence occurred in jerks (Fig. 1-8). The assumption of a uniform uplift rate, though locally reliable on the long term, may thus introduce vertical errors of a few metres when time scales of the same order of possible recurrence periods of major vertical crustal movements are considered. In the same area, sea-level band **B** was predicted by the Peltier (1988) global isostatic model.

On the south coast of Santo Island (Vanuatu), a reef with a probable age of 6000 years is now almost 33 m above sea level on the Tasmaloum Peninsula (**C**), whereas on the nearby Tangoa Island (**D**), a reef crest 6700 years old is now at +19 m (Bloom and Yonekura, 1985). From the radiocarbon dating of numerous corals, collected between the present sea level and the emerged Holocene reef crest, Bloom and Yonekura (1985) considered that the most likely interpretation was that summarized by relative sea-level curves **C** and **D**, which suggest the occurrence of a sequence of sudden or of more or less gradual uplift movements, interspaced with quiescent intervals.

In New Caledonia Baltzer (1970) deduced from four radiocarbon-dated samples of *Rizophora* peat, collected from the west coast of the island, that the high water level was 5.15 m below its present counterpart 7300 yr BP and 0.15 m below 5600 yr BP (curve **E**). Cabioch et al. (1989) mixed the data of **E** with results obtained from cores taken from 39 holes, drilled into ten fringing reefs developed around the mainland of New Caledonia, to construct curve **F**. The shift between curves **E** and **F**, which until 5600 yr BP are based on about the same data, seems to result from the fact that curve **F**, considered to represent MSL, was constructed in fact by combining together indicators of the higher part of the intertidal zone (*Rizophora* peats) with indicators of the upper subtidal zone (corals). As a result, the elevation of the first part of curve **F** (based on peats) is slightly overestimated and the second part (based on corals) underestimated. As shown by Coudray and Delibrias (1972), sea level in New Caledonia was above +1 m between 4400 and 3000 yr BP and still at about +0.7 or +0.8 m 770 yr BP. Band **G** corresponds to sea-level changes predicted in New Caledonia by Peltier's (1988) model.

In the Fiji Islands several new results were provided by the HIPAC team. MSL curve **H** by Miyata et al.(1988), based on several coral dates and geomorphological observations on the southern coast of Vanua Levu, shows a gradual rise from c. +1 m to +2 m between 6000 and 4000 yr BP, followed by a sudden sea-level drop which occurred some time later than 3400 yr BP. On the southern coast of Viti

Plate 43: MELANESIA

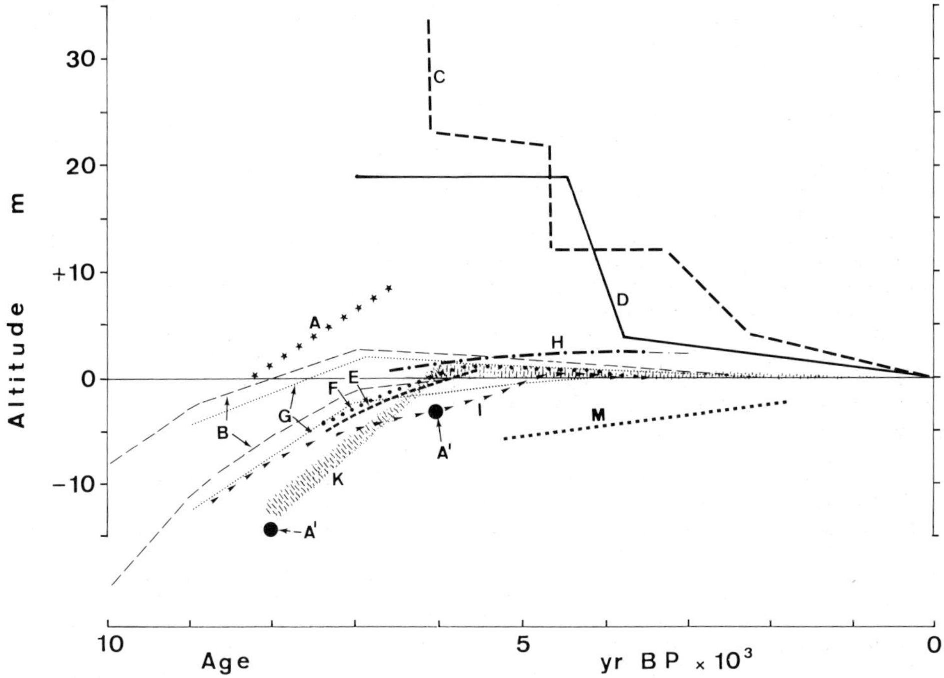

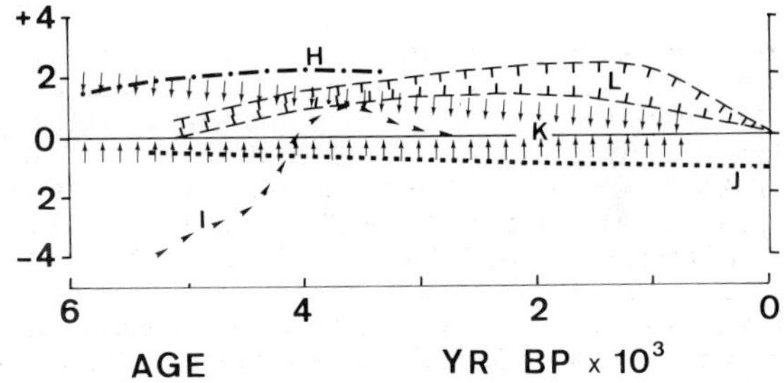

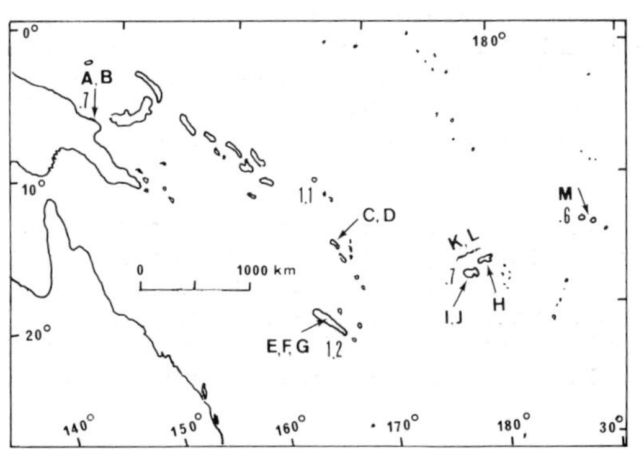

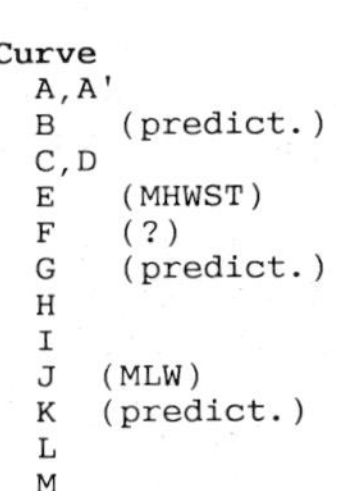

Curve		Reference
A,A'		Chappell & Polach, 1976
B	(predict.)	Peltier, 1988
C,D		Bloom & Yonekura, 1985
E	(MHWST)	Baltzer, 1970
F	(?)	Cabioch et al., 1989
G	(predict.)	Peltier, 1988
H		Miyata et al, 1988
I		Sugimura et al., 1988a
J	(MLW)	Ash, 1987
K	(predict.)	Nakada & Lambeck, 1989
L		Nunn, 1990
M		Bloom, 1980

Fig. 2-28. Coral-rich conglomerate flagstones, corresponding to elevated reef flats. The boundary between the lower part of the conglomerates, usually consisting of an *in situ* reef framework, and the upper part, often made of reworked reef material lithified in the intertidal or supratidal zone, can be specified through petrological analysis (Montaggioni and Pirazzoli, 1984). **a**) Temoe Atoll, Tuamotu Islands; **b**) Bora Bora barrier reef, Society Islands.

Levu, relative sea-level curve **I** was obtained by Sugimura et al. (1988a) from borings carried out in the Lobau Lowland. It shows a gradual rise from 8000 to 4000 yr BP, when sea level reached a peak at about +1 m, followed by a fall to the present situation probably before 3000 yr BP and an almost stable state since that time. Sea-level band **K** corresponds to the predictions of Nakada and Lambeck (1989) model in the Fijis. Curves **H**, **I** and **K** are reproduced at an enlarged scale in the lower graph of Plate 43, together with low-water trend **J** proposed for the north coast of Viti Levu by Ash (1987) and sea-level band **L** propounded for the Fiji archipelago by Nunn (1990).

In the Western Samoa Islands, where curve **M** by Bloom (1980) suggested the occurrence of rapid submergence since 5500 yr BP, and sea level would still have been 2 m below the present sea level 1500 yr BP, Sugimura et al. (1988b) and Rodda (1988) summarized various indications of possible 1-2 m late Holocene emergence.

POLYNESIA (Plate 44)

In the Cook Islands, Yonekura et al. (1984) proposed MSL curve **A** for Rarotonga and Aitutaki Islands, deduced from dated coral samples, suggesting a slow 1-m rise from 6000 to 3000 yr BP, when the present sea level was reached, followed by relative stability. In Mangaia Island, the sea-level curve reported by Yonekura et al. (1988) (**B**) is parallel to **A**, but situated 2 m above **A** until 3150 yr BP, when a sudden 2-m uplift movement is claimed to have occurred. Sea-level curve **C**, which is identical to **A** after 6000 yr BP, was predicted by Nakada and Lambeck (1989) in the Rarotonga area.

Badyukov (1982) proposed two sea-level curves for Pacific Ocean atolls: **D** between 10°N and 10°S, and **E** south of 10°S. If the great variety of tectonic situations existing in the areas considered is taken into account, such generalizations are unrealistic; nevertheless, it would be interesting to know on which kind of local data the 3-m negative oscillation shown by curve **D** around 4000 yr BP, or the sea-level above -5 m suggested by **E** between 8000 and 7000 yr BP have been based; unfortunately, no details are provided by Badyukov (1982).

Minimum sea-level rise curves deduced from drillholes were provided by Montaggioni (1988) for Tahiti (**F**) and Moorea (**G**) Islands. Curve **F**, in particular, was deduced from a 24 m-long drillhole in which shells of the gastropod *Serpulorbis annulatus*, inhabiting water depths less than 3 m, were found incrusting *in situ Acropora* corals at intervals systematically less than 1 m. Curve **F**, which is therefore expected to provide reliable indications on the sea-level position to within ±3 m during the last 7000 years (Pirazzoli and Montaggioni, 1988a), shows a rise at the rate of about 15 mm/yr between 7000 and 5500 yr BP. After the present sea-level position was reached, about 5500 yr BP, curve **H** by Pirazzoli et al. (1985) indicates that slight emergence, reaching about 0.5 m between 2000 and 1000 yr BP, predominated in Tahiti and Moorea during the late Holocene; this is in spite of local subsidence phenomena, due to the load of volcanic masses, which can be assumed to be 0.15 mm/yr greater in Tahiti and Moorea than in nearby islands. Curve **I** shows the sea-

Plate 44: POLYNESIA

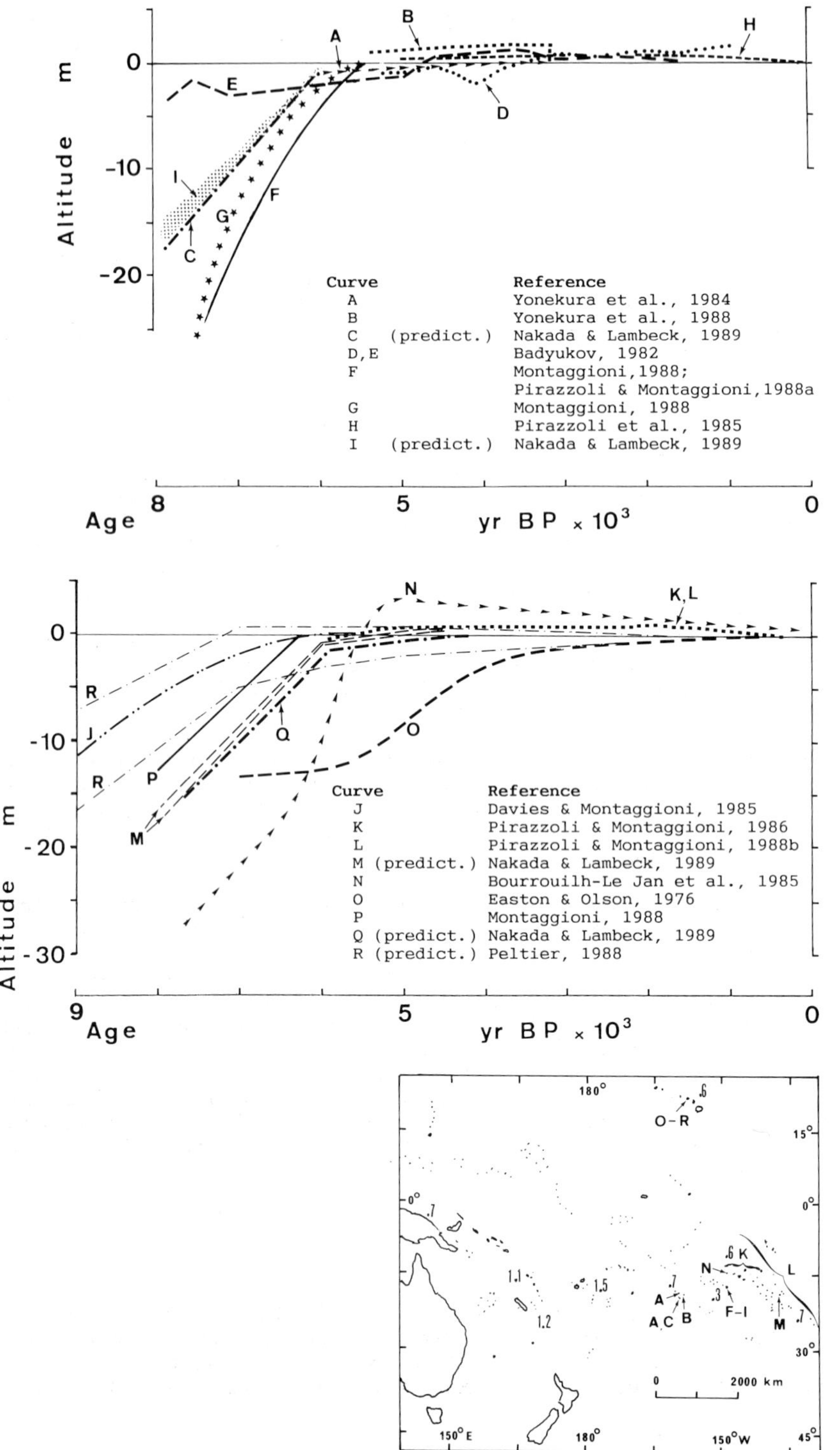

level history predicted near Tahiti by the Nakada and Lambeck (1989) model.

According to Davies and Montaggioni (1985), **J** is a "Central Pacific" sea-level curve inspired by the "composite South Pacific" curve of Adey (1978). Even if the purpose of **J** is to provide merely an evaluation of reef growth, this curve differs from most local ones proposed for the "Central" or "South Pacific" and can therefore hardly be used for local applications.

Pirazzoli and Montaggioni (1986) deduced from surface observations and boreholes carried out in six northwest Tuamotu atolls, supported by some forty radiocarbon dates, a local curve of MSL variations (**K**), showing that sea level remained above the present situation from between 6000 and 5500 yr BP until at least 1200 yr BP, reaching a maximum elevation at approximately +0.9 m. Field work was subsequently extended to 24 other islands and atolls in the Tuamotu, Society, Gambier and Austral archipelagos, and over 70 new radiocarbon dates were obtained (Pirazzoli and Montaggioni, 1988b), providing almost consistent results over an intraplate area as wide as western Europe (curve **L**). With the exception of a few islands located in two anomalous areas (around the hot spot of the Society Islands, and near Rurutu in the Austral Islands) the pattern shown by curves **K** or **L** was confirmed in most islands. It can be considered therefore that these curves are representative of the regional eustatic pattern in the French Polynesia area (curve **L** is indiscernible from **K** at the scale used in Plate 44), though a slight increase of emergence southwards remains a possible phenomenon, as predicted by the model by Clark et al. (1978). At Vahitahi Atoll in the eastern Tuamotus, where the sea-level data available are consistent with curve **K** (Pirazzoli et al., 1987), Nakada and Lambeck (1989) have predicted sea-level curve **M**, which fits field observations perfectly.

Bourrouilh-Le Jan et al.(1985) claimed that sea level in Mataiva Atoll (northwest Tuamotus) would have reached a peak at +2 to +3 m about 5000 yr BP (curve **N**), but no evidence was provided to support this view, which is inconsistent with all the data available in the same area.

In Oahu, Hawaii, curve **O**, deduced from ten core holes drilled through an active fringing reef and supported by 64 radiocarbon dates, was considered by Easton and Olson (1976) to represent sea-level changes; **O** is in fact a reef growth curve, certainly sea-level dependent, but postdating the sea-level rise. According to Montaggioni (1988), reef growth in Oahu would have occurred under water depths of 2-3 m on the central part of the reef flat, 8 m on the backreef zone and 15 m on the reef edge; the corresponding sea-level curve suggested by Montaggioni (1988) is **P**. In the same area, sea-level predictions by rheological models are summarized by curve **Q** (Nakada and Lambeck, 1989) and band **R** (Peltier, 1988).

NEW ZEALAND (Plates 45-47)

On the Weiti River estuary, in North Island, 18 radiocarbon-dated shell samples, collected from former cheniers, beach ridges and tidal flats, were referred by Gibb (1986) to the height of their modern counterpart and used to construct sea-level band **A** (**Plate**

45), which does not suggest any unequivocal departure from the present sea level during the past 6300 years. This site was considered tectonically stable by Gibb (1986) and employed, together with Blueskin Bay, about 1090 km apart in South Island (see locality F, Plate 47) as a zero datum in relation to which uplift of subsidence trends could be deduced in other areas.

In the Firth of Thames, Schofield (1960) interpreted constructional and erosional morphologies of beach ridges and tidal flats as due to sea-level fluctuations 0.5 to 1.5 m in amplitude (curve **B**), that he attempted to correlate with changes in the European climate. According to Schofield, sea level fell over 2 m between 4000 and 2000 yr BP and remained relatively stable since that time. Woodroffe et al. (1983), on the other hand, concluded that in the same place 3600 years ago sea level was only 0.7-0.9 m above present and that it fell gradually, without any oscillations until 1200 years ago, when it was close to present.

Again in the Firth of Thames, Gibb (1986) provided relative sea-level curves at Kellys Beach (**C**) (where a rapid apparent sea-level rise of several decimetres was reported around 2200 yr BP), Kaiaua (**D**) and Miranda, where he reinterpreted the data used by Schofield (1960) (**B'**) and by Woodroffe et al. (1983) (**E**). From a comparison with curve **A**, he deduced a local uplift trend in the Firth of Thames of 0.1 mm/yr, though a much more rapid uplift trend (0.45 mm/yr) had been estimated previously by Gibb (1983) at Miranda. In the same area, predictions by the rheological model of Nakada and Lambeck (1989) are summarized by band **F**.

The eastern coast of North Island, which is located 50-150 km west of the Hikurangi Trench, is subjected to the stress of oblique convergence as the Pacific plate is subducted beneath the Australian plate. This causes frequent uplift movements accompanied by landward tilting, with local patterns which may change sharply over short distances. Several approximate relative sea-level curves, based on one hundred radiocarbon dates, were deduced from transgressive deposits exposed at sites of former estuaries. In the Pakarae River area, where a flight of Holocene terraces reaching 24 m in elevation were interpreted by Berryman (1987b) as indicating five co-seismic uplift events of 2 to 10 m each since 7000 yr BP (curve **G**), the **H** curve by Ota et al. (1988) shows that marine deposits as old as 10,000 yr BP are exposed in the same locality. According to Ota (1987a), the next uplift movement is expected to occur in the Pakarae River estuary area within 150 to 500 years. In Sponge Bay, emerged marine deposits suggest that the relative sea level was rising between 8000 and 6000 yr BP (curve **I**, by Ota et al., 1988).

In Mahia Peninsula, where Ota et al. (1988) deduced a rising sea-level trend between 8000 and 7000 yr BP from marine shells in growth position at elevations attaining +13 m (curve **J**), Berryman (1987a,b) reported five Holocene terraces, attaining +15 m and recording episodic shifts in relative sea level; these were interpreted as indicating that co-seismic uplift took place with a recurrence interval of 400 to 1500 years (curve **K**).

In the north part of the Waimarama coast, the occurrence of two uplift movements, the first attaining 2.5 m slightly before 4000 yr BP, the second 2.0 m before 2000 yr BP (curve **A, Plate 46**) were suggested by Ota (1987a). Near Aramoana, on the other hand, where

Plate 45: NEW ZEALAND I

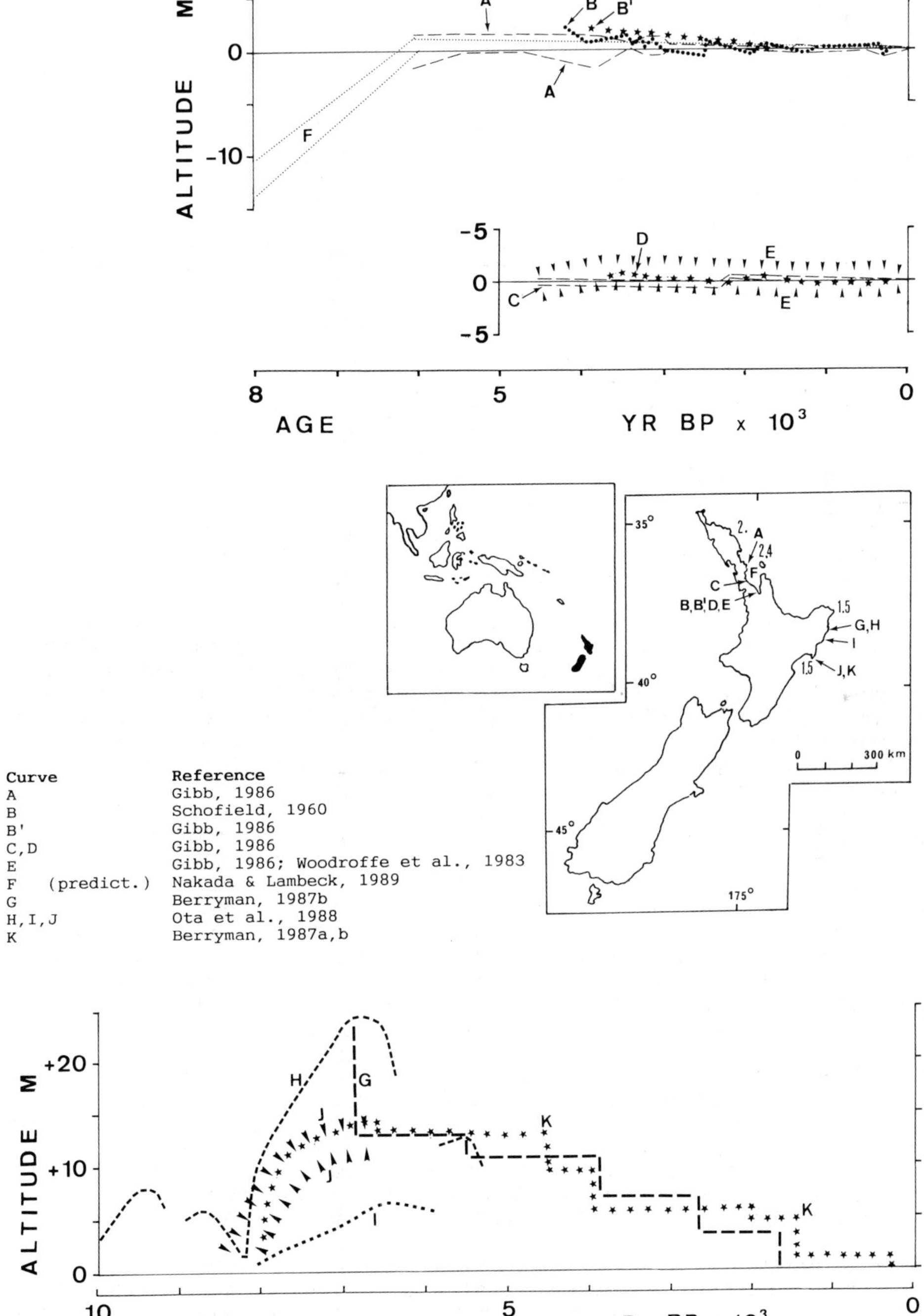

Curve	Reference
A	Gibb, 1986
B	Schofield, 1960
B'	Gibb, 1986
C,D	Gibb, 1986
E	Gibb, 1986; Woodroffe et al., 1983
F (predict.)	Nakada & Lambeck, 1989
G	Berryman, 1987b
H,I,J	Ota et al., 1988
K	Berryman, 1987a,b

Holocene emergence since 6000 yr BP attains over 8 m, only gradual movements seem to have happened (curve **B** by Ota, 1987b). At Owahanga (curve **C**, by Ota et al., 1988) marine deposits have been emerged since 8000 yr BP, though no sea-level information is available for the last 7000 years.

Near Oterei, where Singh (1973) correlated seven uplifted beach ridges with peaks of the Mörner (1969) curve, Ota et al. (1988) provided sea-level band **D**, showing that gradual emergence occurred after 9000 yr BP and that the 5000 yr BP shoreline reaches almost +10 m. Kumenga, in the lower Wairarapa Valley, on the south coast of North Island, lies in a tilted area, between the tectonic uplift to the east (see below) and the downdrop at Lake Wararapa to the west. Here Gibb (1986) deduced from five dated shell beds a very slight downdrop trend (0.05 mm/yr), though vertical uncertainty ranges do not seem to permit any clear conclusion (band **E**).

Six elevated beach ridges found up to 30 m above sea level near Cape Turakirae, a few kilometres west of the Wairarapa Fault, were interpreted by Wellman (1967) as remnants of six successive uplifts which occurred during the past 6500 years (curve **F**). The last uplift movement took place during a severe earthquake in A.D. 1855 and was sudden. Wellman inferred that the other uplifts happened during earlier earthquakes and were also sudden. Wellman (1967) estimated that the next uplift will take place in 500 years time and will be at least 1.5 m.

In the Pauatahanui inlet, Gibb (1986) obtained relative sea-level band **G**, suggesting slight emergence since 6000 yr BP, which he considers to have resulted from an uplift movement at the rate of 0.3 mm/yr. In the lower Wairau Valley, on the northeast coast of South Island, where Pickrill (1976) adopted the composite sea-level band **H**, derived from various areas of New Zealand, the sea-level trends predicted by Peltier's (1988) rheological model are shown by band **I**, and those using the ICE-3G model (Peltier, 1991) by curve **J**. Though **J** is closer to field data than **I**, the emergence it predicts in the mid-Holocene is excessive.

In the Christchurch area, Suggate (1968) dated over thirty samples of shells, peat and wood, collected from various wells and excavations. He obtained a sea-level curve (**A**, **Plate 47**) showing a gradual rise, from c. -22 m to zero, between 9400 and 5000 yr BP. More recently, Gibb (1986) assembled 16 new radiocarbon dates and proposed a sea-level band (**B**) which suggests local subsidence at the rate of 0.1 mm/yr. Predictions by the Nakada and Lambeck (1989) model (band **C**) are close to field data, whereas those by the Peltier (1988) model (band **D**) overestimates systematically former sea-level heights. In the Blueskin Bay estuary area, which is considered tectonically stable by Gibb (1986), ten dated sea-level indicators suggest no appreciable deviation from the present sea level during the last 6000 years (band **E**).

In the lower graph of **Plate 47** are assembled three composite sea-level curves proposed for New Zealand. Wellman (1962) was well aware that the changes suggested by his curve **F**, which mixes data from various areas of North Island and indicates that sea level stood a few metres higher than now some 4000 years ago, is "doubtless partly tectonic". The rapid Holocene rise of curve **G** by Cullen (1967) is based on the date of two samples collected from the floor of Foveaux Strait and on very few other data reported

Plate 46: NEW ZEALAND II

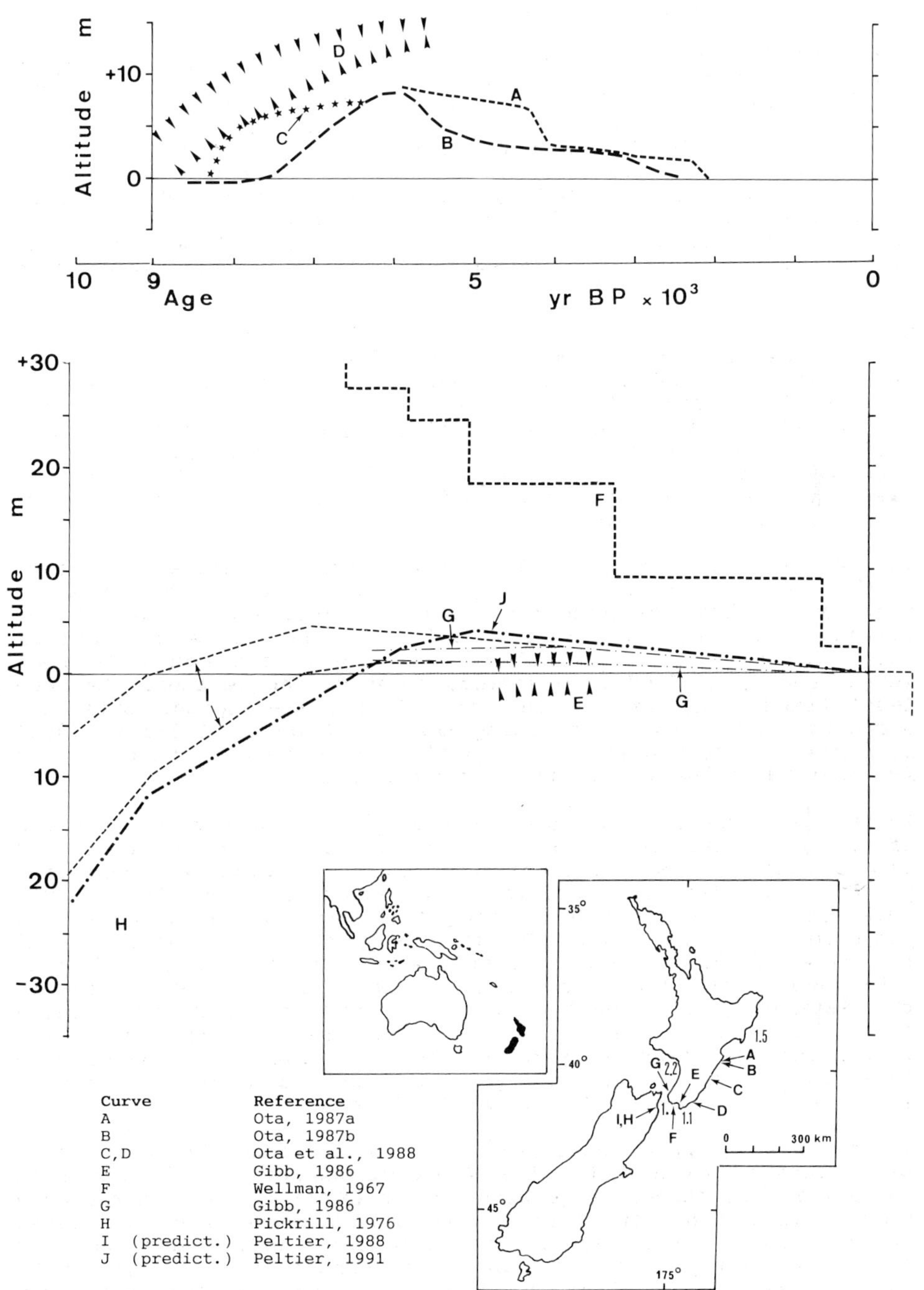

Plate **47: NEW ZEALAND III**

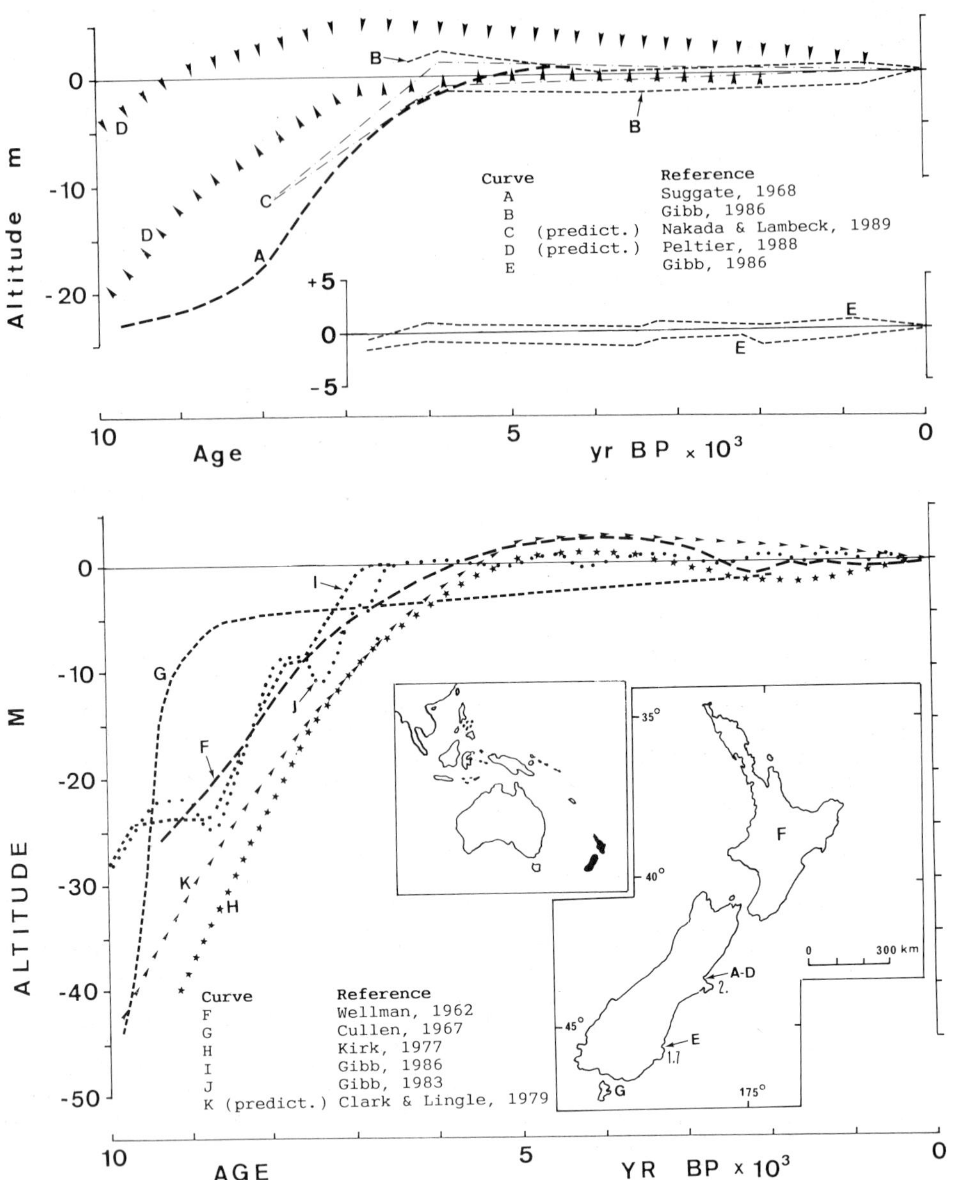

from elsewhere by other authors. Curve **H** by Kirk (1977) results from a "best fit by polynomial regression to New Zealand-wide data". Indeed, no great confidence can be placed in such composite curves in a tectonically active region like New Zealand.

Curve **I** was obtained by Gibb (1986) by assuming that the Weiti River estuary (locality A, Plate 45) and the Blueskin Bay (locality E, Plate 47) are stable areas and therefore provide eustatic sea-level data. In addition, data of curves B', C, D, E (Plate 45), E and G (Plate 46) were corrected for tectonic effects by deducting the rates of vertical movement estimated at each place, using the above two localities as zero-datum sites, and included in curve **I**, which is considered by Gibb (1986) to be representative of the regional eustatic pattern. According to **I**, the postglacial transgression was punctuated by at least two stillstands (at c. -24 m from 9200 to 8400 yr BP, and at c. -9 m from 7500 to 7200 yr BP) and terminated c. 6500 yr BP. Since that time, sea level would not have deviated from the present situation by more than a few decimetres. In a previous paper, however, Gibb (1983) inferred from about the same data the occurrence of two 3-m sea-level drops and proposed even a third oscillation around 7000 yr BP (**J**), but no trace of these fluctuations remains in the 1986 publication, nor are the reasons for this change in interpretation explained. Thus, the fact that several rates of local vertical movements estimated by Gibb (1983) were subsequently modified without explanation by Gibb (1986), and that the zero-datum used for the estimation of tectonic rates was arbitrary, leaves the impression that effective vertical uncertainty ranges may be slightly larger than those inferred for curve **I** by Gibb (1986). Finally, curve **K** corresponds to predictions of the rheological model by Clark and Lingle (1979) for "New Zealand".

AUSTRALIA (Plates 48-50)

Although the Australian continent is often quoted as tectonically stable in the literature, Holocene tectonism has been debated for various coastal areas. In western Australia, tectonic movements have been demonstrated in the Pleistocene and earthquakes occur episodically. In other areas, when differential relative sea-level changes are unquestionable during the late Holocene, at least vertical movements of isostatic origin are obvious.

In western Australia, the occurrence of emerged Holocene shorelines has been known since the pioneer work by Fairbridge (1948) and Teichert (1950), who described superimposed notches and benches at Point Peron, Rottnest, etc. These data were interpreted initially as resulting from "step by step" movements and from "spasmodic catastrophic changes of perhaps geotectonic origin" (Fairbridge, 1948, p. 64). Later, however, Fairbridge (1961) incorporated the same data in his "eustatic" curve. The tectonic-emergence interpretation was however supported by Pirazzoli (1976b, p. 303) and taken up again in Rottnest Island by Playford and Leech (1977), who suggested abrupt submergence in three steps, between c. 6500 and 5500 yr BP, followed by a 3-m sudden emergence around 5000 yr BP (curve **A, Plate 48**).

Searle and Woods (1986) found however that evidence for higher sea level using raised beach/storm deposits cited by Fairbridge (1961) and Playford and Leech (1977) was inconclusive and deduced from dated sequences of swash zone sediments, collected in pits located some 30 km south of Perth, MSL curve **B**, which suggests that sea level reached an elevation of at least 2.5 m 6400 yr BP and then declined gradually, without any evidence of fluctuations, reaching the present level shortly after 1000 yr BP. More recently, Playford (1988) published 15 radiocarbon dates of shell beds supporting his previous interpretation and even suggested that a

Plate 48: AUSTRALIA I (West & South)

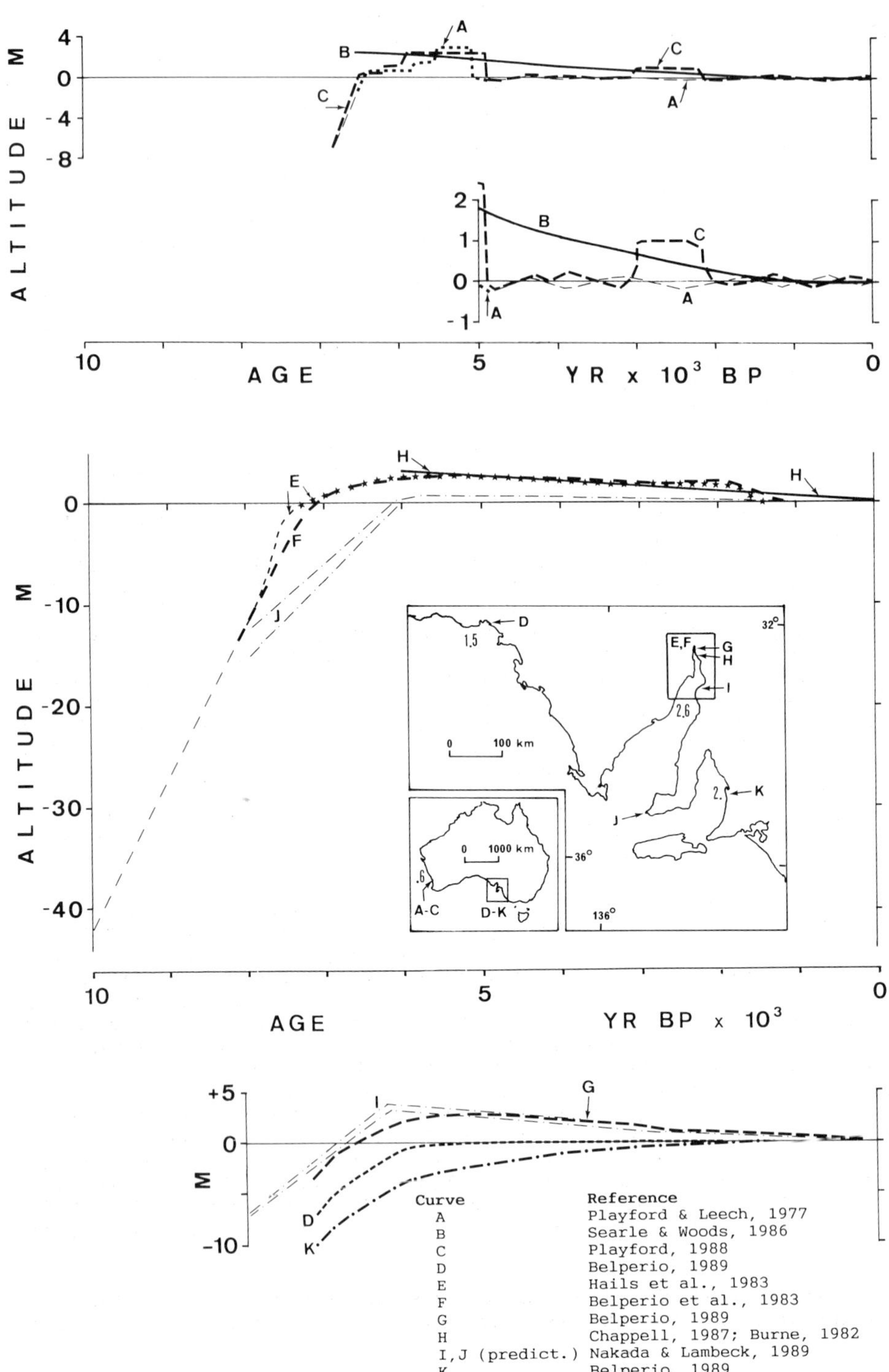

new emergence period would have occurred between 3100 and 2200 yr BP (curve **C**).

The area near the Spencer and St. Vincent Gulfs, on the south coast of Australia, provides good examples of differential Holocene emergence. Near Ceduna, well outside the Gulfs area, MSL curve **D** by Belperio (1989) suggests an intermediate trend: a rapid sea-level rise until c. 6000 yr BP, followed by standstill near the present level. In the upper Spencer Gulf, where low-energy coastal and nearshore sediments were extensively studied, using stratigraphic, vibrocoring and dating techniques, sea level was, according to Hails et al. (1983), at least 2.5 m above the present one from 6000 until 1700 years ago (curve **E**) and the subsequent fall to present level was interpreted as resulting from regional tectonic uplift. Sea-level curve **F** by Belperio et al. (1983, 1984) is identical to **E** since 7000 yr BP, but extends further in the past, suggesting a 45-m rise from 10,000 to 7000 yr BP. Both curves **E** and **F** suggest a minor sea-level fall between about 5700 and 2000 yr BP, followed by a rapid fall of nearly 2 m between about 2000 and 1500 yr BP. This rapid fall was attributed by Hails et al. (1983) to a "regional tectonic uplift".

Chappell (1987) regarded however the final part of curve **E** as unproven and preferred to consider a linear falling trend (**H**), deduced from data by Burne (1982), as the simplest approximation. This suggestion seems to have been accepted, since MSL curve **G** produced subsequently by Belperio (1989) for Port Augusta showed an almost uniform sea-level fall at a rate of 0.5 mm/yr over the past 6000 years. Near Port Adelaide, on the other hand, according to Belperio (1989), the coastal sediments record a relative sea-level rise of the same magnitude over the past 6000 years (curve **K**), which are ascribed to "geological subsidence, unknown isostatic effects and anthropogenic effects". Anthropogenic effects may have been active only over the past 150 years however, when large areas of former tidal wetlands have been drained and reclaimed. Nakada and Lambeck (1989) consider that much of the regional variability in the Spencer Gulf area can be explained in terms of hydroisostatic effects, as shown by the sea-level bands they predicted for Port Pirie (**I**) and Cape Spencer (**J**), which fit well field data from the upper Spencer Gulf (e.g. **G**) and the open coastline (e.g. **D**) respectively.

Several sea-level curves have been proposed for southeast Australia. Band **A** (**Plate 49**) by Thom and Chappell (1975) is based on a compilation of over two dozen data points reported from a wide area covering the coastal areas of Victoria and New South Wales. It suggests a broad regional sea-level rise from c. -25 m, between 9000 and 10,000 yr BP, and the present sea level, reached by 6000 yr BP. A decline in the rate of transgression is suggested between 9000 and 8000 yr BP, but was not confirmed by subsequent studies. A revision and correction of the same data enabled Thom and Roy (1983) to propose a slightly different sea-level curve (**B**). Dates of new samples recovered from inner continental shelf areas, back barrier areas and drowned river valleys in New South Wales were also considerd by Thom and Roy (1983) (band **C**). Finally, by mixing data of **B** and **C**, Thom and Roy (1983, 1985) obtained a broader regional sea-level band (**D**) for southeast Australia.

In order to clarify previous discussions about the existence or absence of Holocene emergence in southeast Australia, Donner and

Jungner (1981) studied tidal flat and beach ridge deposits at two sites: Seaspray (Victoria) and Cullendula Creek (New South Wales). Near Seaspray, though the new data by Donner and Jungner (1981) gave little evidence as to the position of sea level, they were all consistent with a sea level higher than at present, according to earlier results (curve **E**). The Cullendula Creek data (curve **F**) showed that MSL had reached a level close to its present position before 3000 yr BP and stayed near this level at least until 1900 yr BP and perhaps until now. Possible changes of MSL, according to Donner and Jungner (1981), cannot have exceeded here an amplitude of one metre during the last 5000 years. Nevertheless, Flood and Frankel (1989) reported intertidal organisms exposed on coastal cliffs in the northern New South Wales which suggest that in that area sea level was slightly more than 1 m higher about 3400 yr BP and may have maintained that level until some time after 1800 yr BP.

In the lower graph of **Plate 49** envelopes have been compiled of sea-level change predicted near Moruya (New South Wales) by Nakiboglu et al.(1983) (band **G**), Peltier (1988) (band **H**) and Nakada and Lambeck (1989) (band **I**).

Near Townsville on the Great Barrier Reef, where, according to Hopley (1978) Holocene emergence would have been up to 4.9 m, Belperio (1979) studied coastal sedimentation processes and facies and proposed for the paleointertidal zone band **A** (**Plate 50**), which does not suggest any significant deviation of MSL since 6000 yr BP, when the present MSL was reached. In the same area the predictions by Peltier's (1988) rheological model are represented by sea-level band **B**. In the Palm Group, Chappell et al. (1982) dated sequences of coral microatolls across reef flats. As unmoated coral forms are closely controlled, in their vertical growth, by the MLWST level, band **C** by Chappell et al.(1982) shows changes in the MLWST level, which has fallen about 1 m during the last 5500 years. These results are consistent with sea-level predictions in Halifax Bay by Nakada and Lambeck (1989) (band **D**).

On the basis of 45 radiocarbon dates of coral microatolls on fringing reefs between 14° and 20° S, Chappell (1983b) and Chappell et al. (1983) concluded that sea level fell smoothly from +1 m at 6000 yr BP to its present position, without any evidence for secondary oscillations; confidence limits of this MLWST level trend (**E**) were estimated at ±0.2 m.

Hopley (1982, 1983) subdivided the Great Barrier Reef area between King Island and Bowen into three sets based on regional differences in shelf and reef morphology and on reef top ages. In a first region, approximately between King Island and Innisfail where the shelf is shallow and narrow (c. 50 km) and the continental slope is very steep, many reefs are well developed, several reef top ages are in excess of 5500 yr BP and emergent corals occur on the inner shelf; here field data are consistent with sea-level band **F**. South of Innisfail the continental shelf widens to over 125 km. Here two regions were distinguished by Hopley (1982, 1983). In the inner and mid-shelf zone, between Innisfail and Bowen, reefs are larger and more irregular, there are very few reef islands and reef top ages are usually older than 4000 yr BP. The sea-level data reported from here were summarized by band **G**, which is very similar to **F**. The outer shelf zone south of Cairns, on the other hand, is characterized by younger reef top ages, smaller reefs and numerous submerged shoals. For this third

Plate 49: AUSTRALIA II (Southeast)

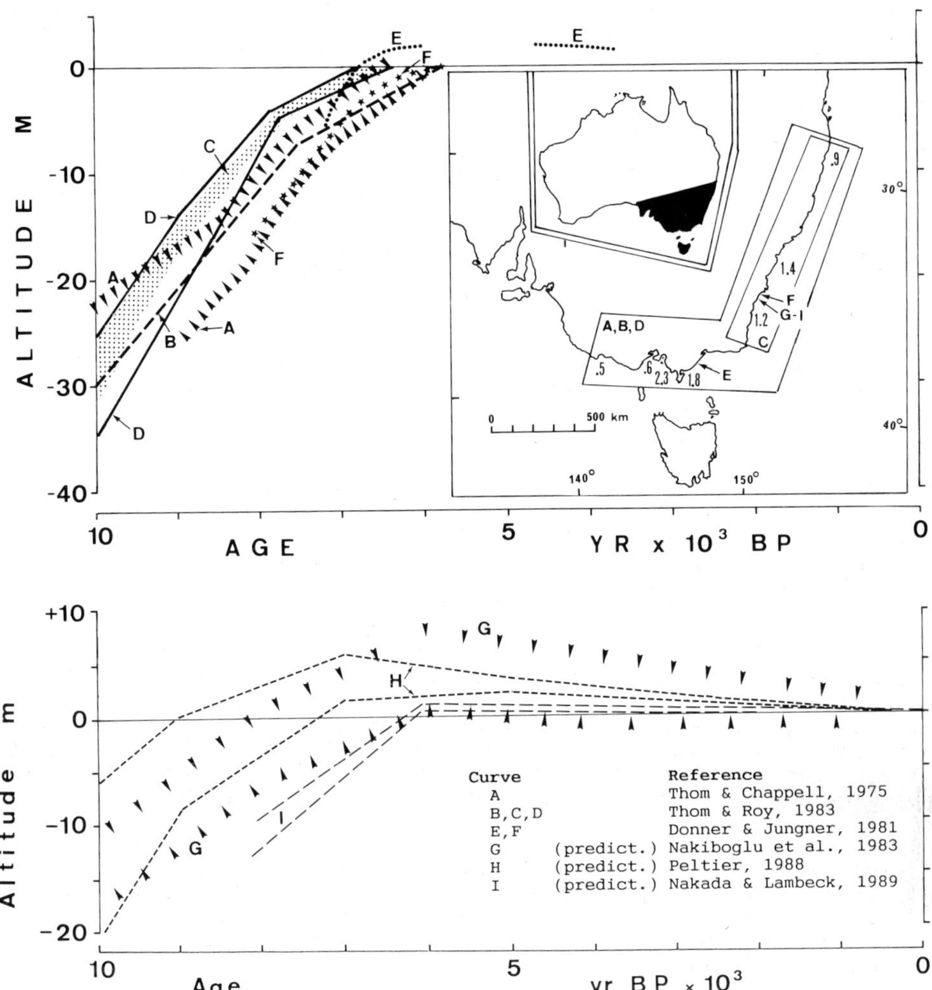

region, Hopley (1982, 1983) proposed sea-level band **H**. The difference between the inner and outer parts of the shelf gives evidence of seaward tilting (Fig. 2-29), which is ascribed by Hopley (1982, 1983) and Chappell et al.(1982) to hydro-isostatic warping. Nakada and Lambeck (1989) developed this idea in their rheological model, distinguishing in the Great Barrier Reef four parallel zones, each characterized by a typical sea-level curve (Fig. 2-30).

In the Gulf of Carpentaria, on the north coast of Australia, the coastal chenier plain near Karumba was surveyed by Rhodes (1980). Boxes **I** in the lowest graph of **Plate 50** represent the age/height

Plate 50: AUSTRALIA III (Northeast)

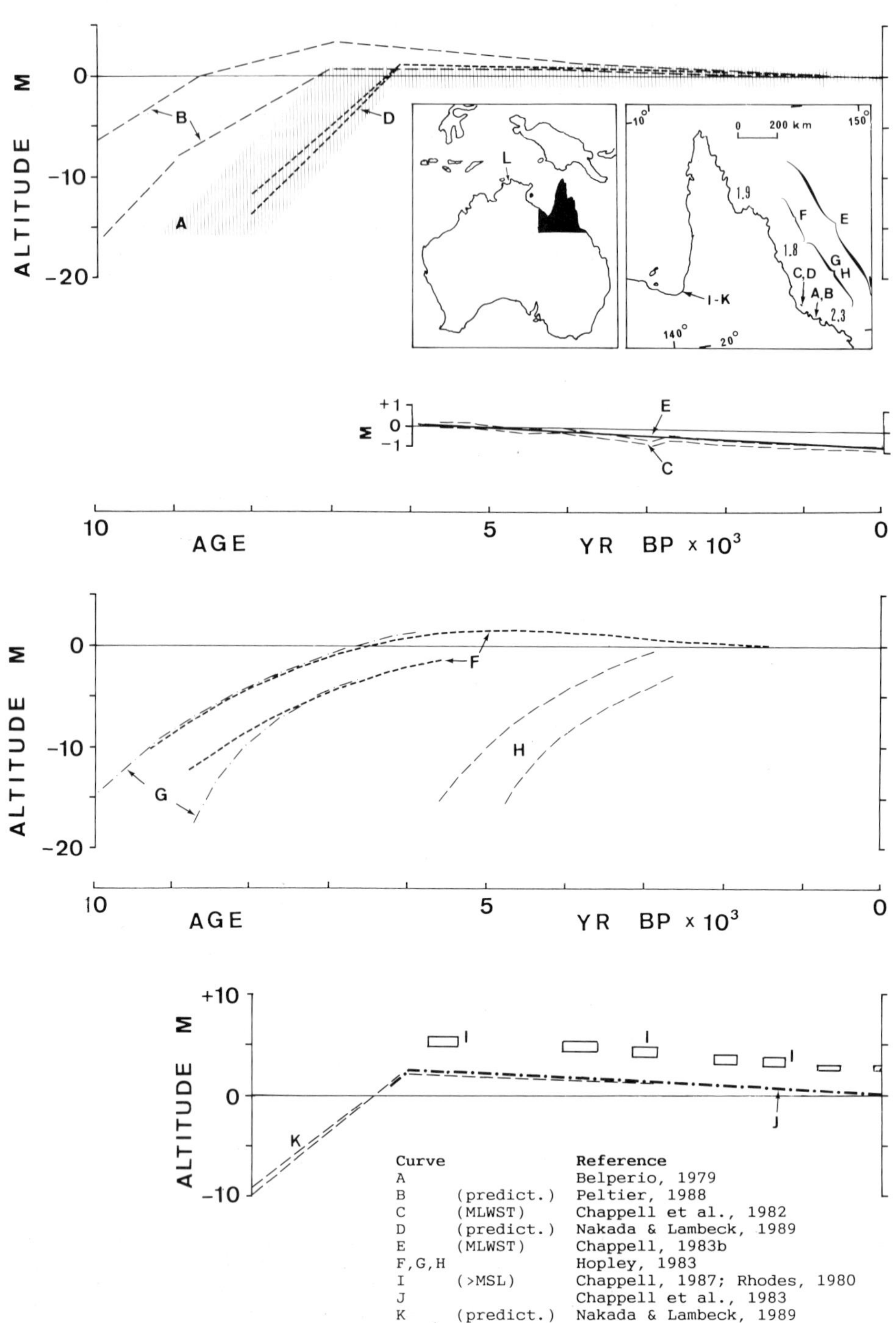

Curve		Reference
A		Belperio, 1979
B	(predict.)	Peltier, 1988
C	(MLWST)	Chappell et al., 1982
D	(predict.)	Nakada & Lambeck, 1989
E	(MLWST)	Chappell, 1983b
F,G,H		Hopley, 1983
I	(>MSL)	Chappell, 1987; Rhodes, 1980
J		Chappell et al., 1983
K	(predict.)	Nakada & Lambeck, 1989

plot of dated and modern chenier ridge bases. As contemporary counterparts range from 2.7 to 3.2 m above the local datum, a smooth fall in the relative sea level of over 2 m in the last 6000 years is indicated (**J**), which is in agreement with the predictions by Nakada and Lambeck (1989) (band **K**). In the Van Diemen Gulf, Northern Territory, the sea-level trend reported by Woodroffe et al. (1987) is however different from that described in the Carpentaria Gulf or in the inner shelf of the Great Barrier Reef; here MSL does not seem to have reached its present level until after 5800 yr BP and no evidence for falling sea level was found for the last 5000 years. This suggests that the Van Diemen Gulf is subsiding.

Recent relative sea-level trends in Australia and various Pacific islands shown by tide-gauge records are summarized in Fig. 2-31. Very few series are long enough to make an estimation of secular trends possible.

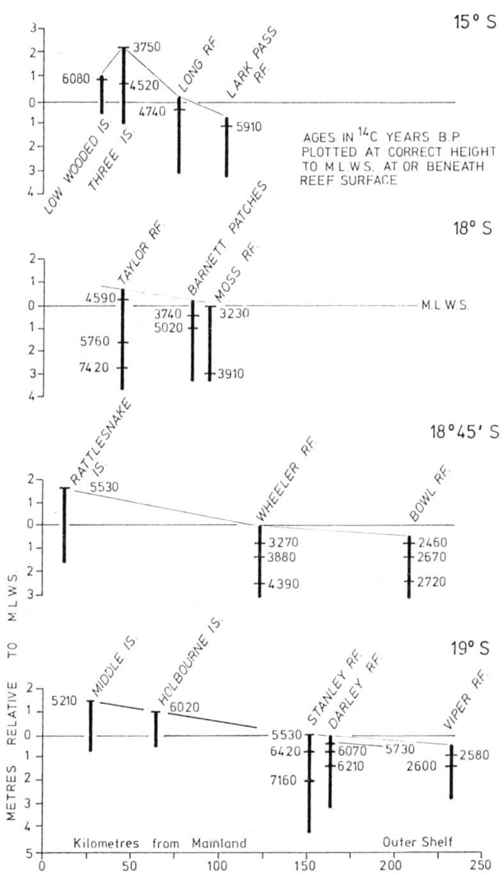

Fig. 2.29. Maximum height of *in situ* corals across the shelf of the Great Barrier Reef at about 15°00', 18°00', 18°45' and 19°00' S, clearly showing decline in height seawards. Radiocarbon ages of the reef caps are shown where available (from Hopley, 1982).

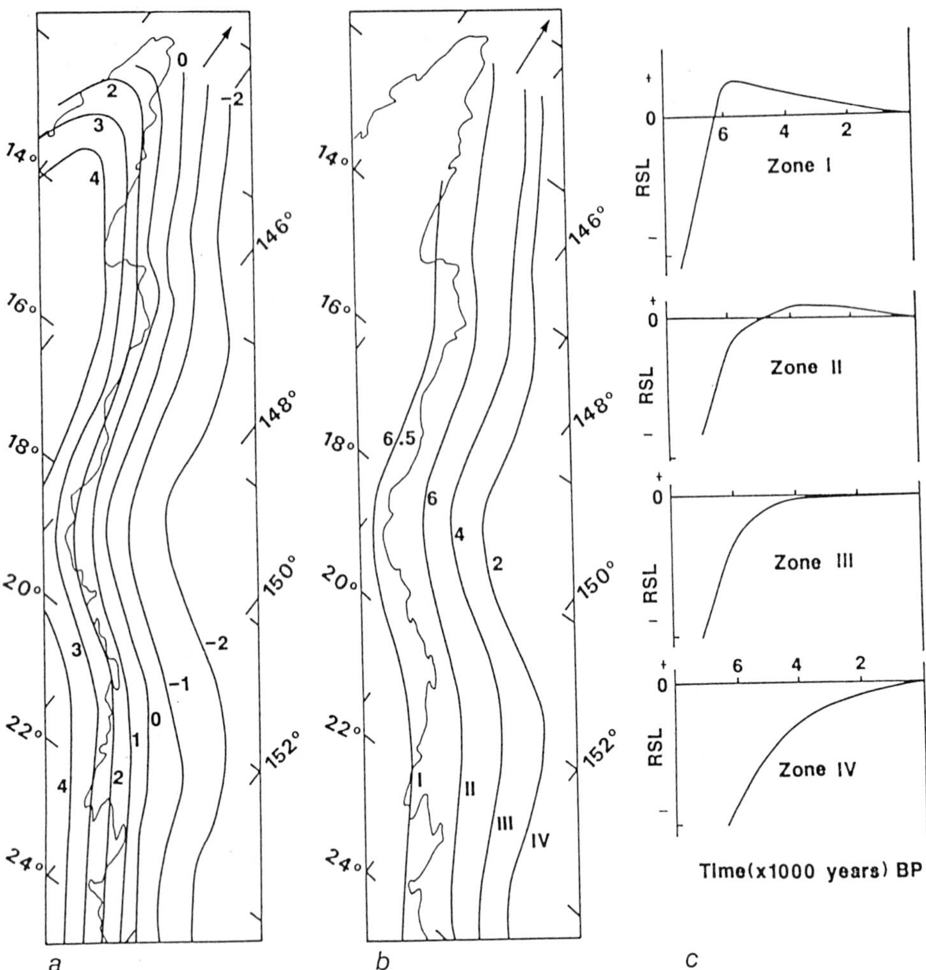

Fig. 2-30. **a)** Predicted Holocene sea levels (in m) at 6000 yr BP; **b)** the times at which sea level first reached its present value (in 1000 yr) along the margin and Great Barrier Reef of northern Queensland; **c)** characteristic relative sea-level curves in the four zones indicated by b). (From Nakada and Lambeck, 1989).

THE WEST COASTS OF THE AMERICAS (Plates 51-54)

The Aleutian Islands and the southern coast of Alaska are located near active plate boundaries and were capped by thick glacial ice until 12 000-10 000 years ago. This makes it difficult to decipher locally the exact relationship between tectonic, eustatic, glacio-isostatic, hydro-isostatic and volcanic factors, which control relative sea-level changes in combinations which vary from one area to the other, giving quite dissimilar sea-level histories

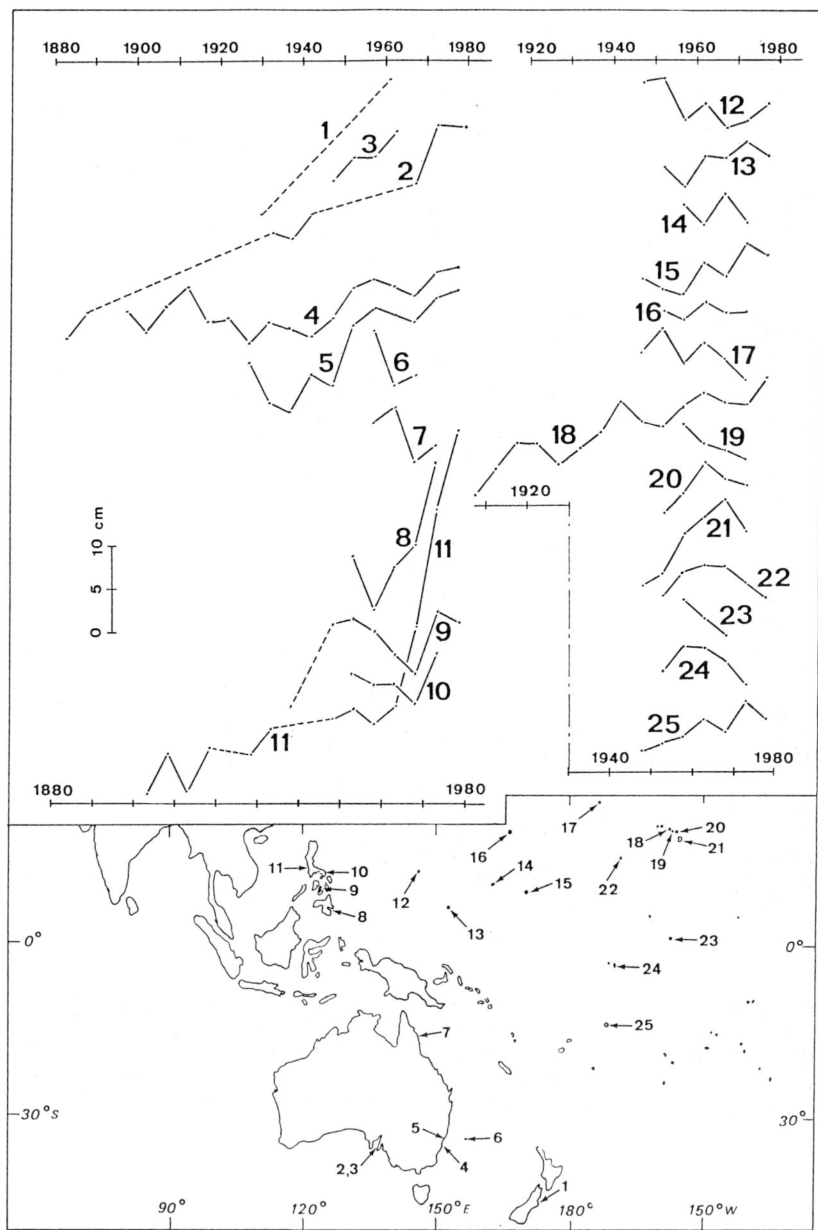

Fig. 2-31. Average 5-yr MSL variations in Australia and the Pacific as recorded by tide-gauge stations. 1) Port Litteltone; 2) Adelaide (Inner); 3) Port Adelaide (Outer); 4) Sydney; 5) Newcastle; 6) Lord Howe Island; 7) Cairns; 8) Davao; 9) Cebu; 10) Legaspi; 11) Manila; 12) Guam; 13) Truk; 14) Eniwetok; 15) Kwajalein; 16) Wake; 17) Midway; 18) Honolulu; 19) Mokuloloe; 20) Maui; 21) Hilo; 22) Johnston; 23) Christmas; 24) Canton; 25) Pago Pago. (From Pirazzoli, 1986b).

at sites less than 100 km apart. As the resultant displacements are in some cases quite great, the usual vertical scales have been modified (enlarged) in Plates 51 and 53. In Plate 51, in particular, altitudes are expressed in a logarithmic scale in the lower graph.

In Umnak Island, Aleutians, Black (1974, 1982) reported evidence of sea level at 2-3 m above the present level about 8250 yr BP and inferred from beach deposits, now above the reach of normal storm waves, that it remained high until about 3000 yr BP, when it gradually dropped to its present position (curve **A**, **Plate 51**).

In south-central Alaska, the great 1964 earthquake was accompanied by vertical tectonic deformation in two zones, which are together 700 to 800 km long and from 150 to 300 km wide. Uplift was as much as 11.5 m on land and 15 m on the sea floor which includes the earthquake focal region; subsidence in the adjacent zones was 1 m on average and 2.2 m at a maximum (Plafker and Rubin, 1967). Geological evidence along the Gulf of Alaska revealed complex Holocene histories in relative sea level, where areas of post-glacial emergence or submergence coincide approximately with areas of significant earthquake-related uplift and subsidence (Savage and Plafker, 1991). Some of these histories were summarized by Plafker and Rubin (1967). In Kukak Bay, where no vertical change occurred in 1964, the sea level of 1000-1500 yr BP was the same as today (curve **B**); in Yukon Island (**C**) and Girwood (**D**), which subsided in 1964, the relative sea-level rise since c. 3000 yr BP is almost 5 m; finally in Middleton Island (E) and Katalla (G), which were uplifted in 1964, Holocene pre-1964 emergence was c. 41.5 m after about 4500 yr BP (curve **E**) and 55 m after 7650 yr BP (curve **G**) respectively. Curve **F** represents the long-term uplift trend in Middleton Island (10 mm/yr) deduced by Kaizuka et al. (1973) from data by Plafker and Rubin (1967).

In the above areas, seismo-tectonic vertical movements include a submergence of youthful mountains recently unloaded of ice, the pulsating emergence of the continental shelf and Gulf of Alaska coasts, and local isostatic subsidences at sites of rapid Holocene sedimentation. Steplike flights of Holocene marine terraces at a number of localities along the coast suggested that the long-term vertical movements occurred, at least in part, as a series of upward pulses that were separated by intervals of stability or even gradual subsidence. Plafker and Rubin (1967) observed that gradual tectonic submergence had prevailed during at least the past 1000 years over much of the zone that was uplifted in 1964 and interpreted this submergence as direct evidence for a significant downward-directed component of regional strain preceding the earthquake.

In southeastern Alaska, only reconnaissance sea-level data seem available. Mobley (1988) summarized previous data and reported new radiocarbon-dated raised marine deposits and archaeological sites closely related to sea level from Heceta (full circles **H**) and the Prince of Wales (empty circles **I**) Islands, showing that emerged marine deposits have occurred since at least 9500 yr BP.

In British Columbia over 400 radiocarbon dates of Holocene shorelines are available, allowing regional variability to be controlled and providing relatively precise models of sea-level fluctuations, which seem to have been dominantly isostatic in nature. At the climax of the last glaciation period (Late

Plate 51: ALEUTIANS - ALASKA - BRIT. COLUMBIA I

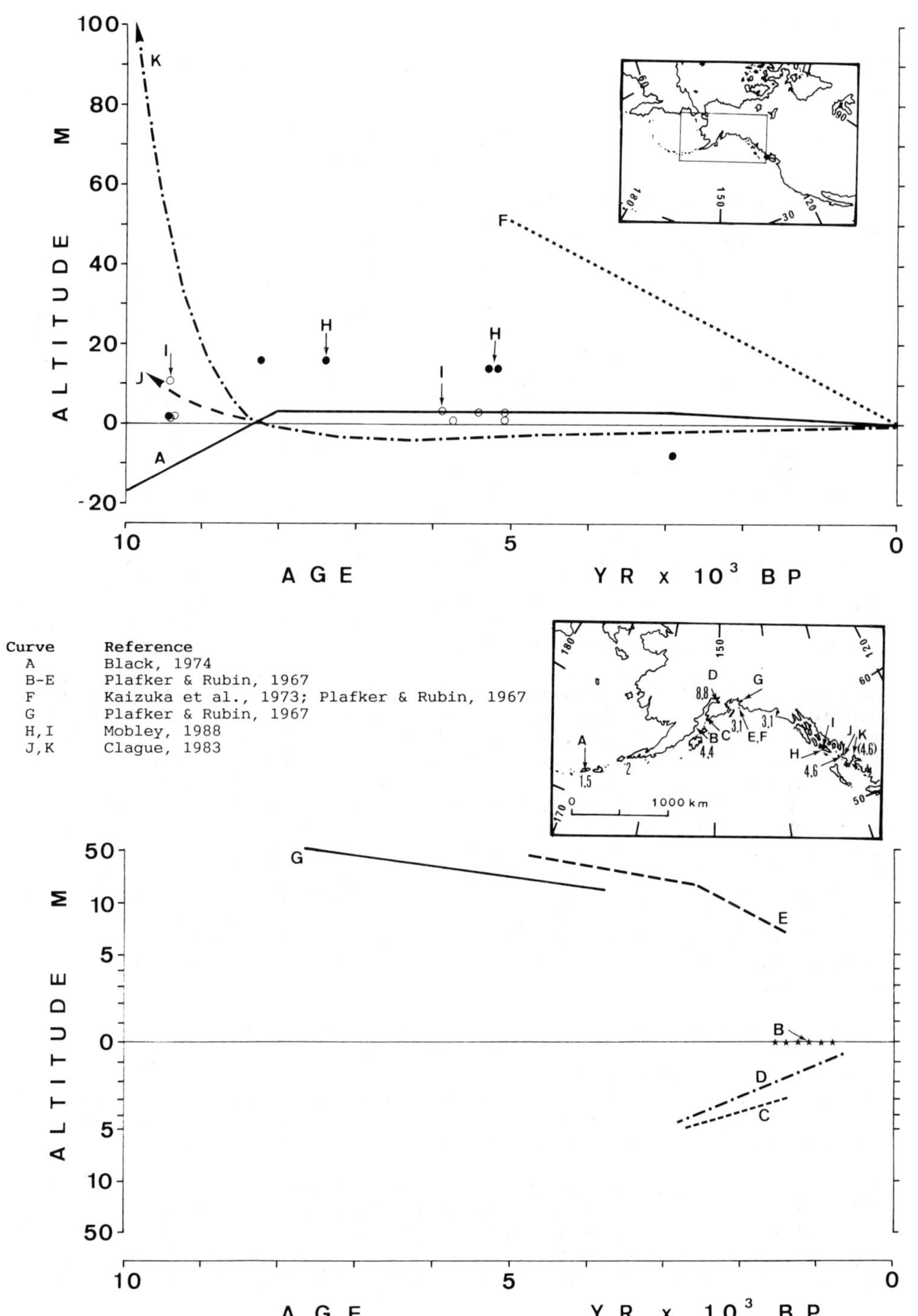

Wisconsin), ice thickness was estimated by Clague (1983) to have been sufficiently thick (c. 2000 m) over the interior of British Columbia for one or more ice domes to exist. At that time eastern Vancouver Island is estimated to have been covered by about 1000-1600 m of ice, whereas parts of the Queen Charlotte Islands and western Vancouver Island probably remained ice-free. Decay of the ice sheet was relatively rapid and accompanied by rapid isostatic adjustments which deformed the crust in a complex, non-uniform fashion. Relic shorelines are reported on the mainland coast up to about 200 m above present sea level.

At Kitimat, where the sea was at its upper limit when the area was first deglaciated, about 11,000 years ago, most of the isostatic uplift occurred, according to Clague (1983), between 10,500 and 9000 yr BP and submergence prevailed over the last 8000 years (curve **K**). Near Prince Rupert, slighter marine regression prevailed during the late Pleistocene and early Holocene (curve **J**), whereas later marine deposits suggest that sea level was within a few metres of its present position.

The pattern of sea-level change on the Queen Charlotte Islands contrasts sharply with that of the adjacent mainland coast. In northeastern Graham Island, for example, transgression culminated about 7500-8000 years ago, when shorelines were about 15 m higher than at present, and was followed by a fall of the sea in relation to land (curve **A**, **Plate 52**, by Clague, 1983). A slightly larger uplift trend (2.5 mm/yr) was reported by Hebda and Mathewes (1986) from Anthony Island for the last 4000 years (curve **B**).

For the area of islands and outer mainland centered on Bella Bella and Namu, postglacial emergence was 17 m after 10,200 yr BP according to Andrews and Retherford (1978); the rate of emergence was about 6 mm/yr around 9000 yr BP and present sea level was attained between 8000 and 7000 yr BP. Relative sea level continued to fall until the late Holocene, when a rise began, which is still continuing today (curve **C**). Curve **D** by Clague et al. (1982) for the Bella Bella-Bella Coola area suggests similar trends, which may result from migration of a collapsing proglacial forebulge, though with differences in level of several metres in relation to curve **C** during the early Holocene.

In the Fraser lowland area, near the boundary between Canada and the U.S.A., stratigraphical data provided by many borings are available. Relative sea-level curves proposed by Mathews et al. (1970) (curve **E**), Armstrong (1976) (curve **F**) and Clague et al. (1982) (curve **G**) differ however significantly from each other. On the east coast of Vancouver Island, sea-level curves **H** by Clague et al. (1982) and **I** by Mathews et al. (1970) suggest the occurrence of a sea-level fall until 8000 to 7000 yr BP, followed by a slight continuous rise continuing up to the present time. On the west coast of Vancouver Island, in contrast, Clague et al. (1982) reported a relative sea-level fall occurring since at least 4000 yr BP (curve **J**), with a pattern similar to that of the Queen Charlotte Islands.

Sea-level curves **K** by Mathews et al. (1970) for the Victoria area and **L** by Clague et al. (1982) for Victoria-Gulf Islands show similar patterns, with a transition from sea-level fall to rise between 8000 and 7000 yr BP, which contrasts sharply with the early Holocene continuous rise (curve **M**) deduced by Linden and Schurer (1988) from seismic profiling and radiocarbon dates of

Plate 52: BRITISH COLUMBIA II - WEST U.S.A. I

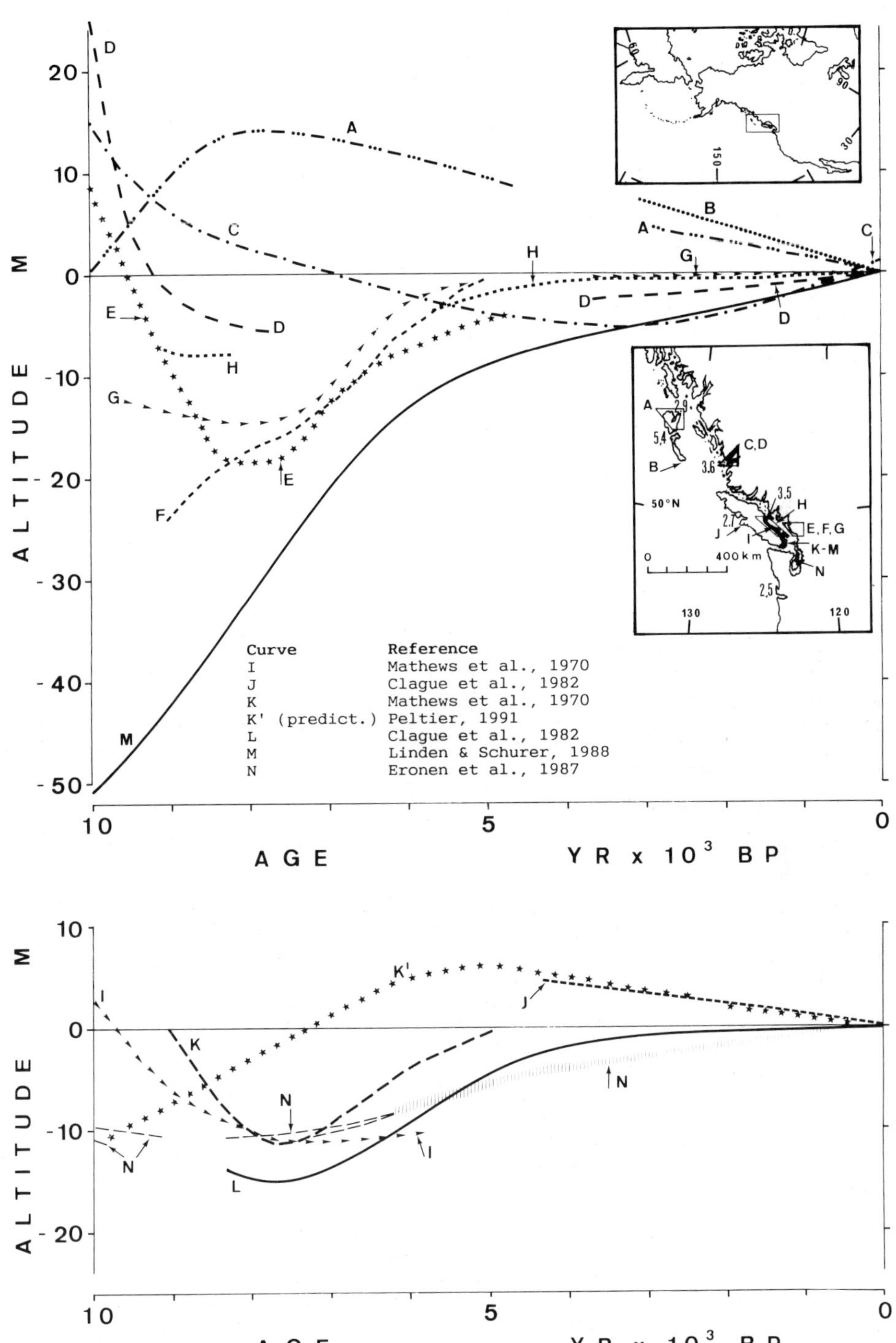

sediments cored from the sea floor of the Juan de Fuca Strait. Curve **K'**, predicted by the Peltier (1991) model for the Victoria area, does not follow the Holocene field data from this region very closely. In the Puget lowland, some 40 km northwest of Seattle, in the U.S.A., Eronen et al. (1987) deduced from analysis of core sediments, relative sea-level band **N**, which confirms a rise during the past 6000 years. A look at the various curves of Plate 52 suggests that forebulge effects may have occurred in this area.

In **Plate 53** generalized patterns of sea-level change on the British Columbia coast have been compiled which were summarized by Clague (1989) and Luternauer et al. (1989) with envelopes depicting a range of sea-level positions, which reflect regional and local differences in ice load as well as uncertainties in locating former shorelines: **A** for the Queen Charlotte Islands, **B** for Queen Charlotte Sound, **C** for mainland fiord heads, and **D** for mainland and eastern Vancouver Island.

Very few Holocene sea-level data are available along the Coastal Range of Oregon and California. In Alsea Bay, Oregon, Peterson et al. (1984) proposed sea-level band **A** (**Plate 54**), deduced from analysis of six cores, under the assumption that a tectonic uplift rate of 1 mm/yr occurred during the last 9000 years. Band **A**, which shows very close correspondence with the local sediment depth-age curve during the last 5000 years, suggests that the sea level continued to rise, though at a rate declining after 6000 years BP, in spite of assumed uplift movements.

In the vicinity of San Francisco Bay, Atwater et al. (1977) deduced Holocene sea-level changes from the elevation and radiocarbon age of plant remains from 13 core samples, obtaining curve **B**, which was given with an uncertainty range of ±2 to ±4 m in the elevations. The amount of tectonic and possibly isostatic subsidence was estimated here by Atwater et al. (1977) at 0.8 ± 0.7 mm/yr in relation to the Micronesia curve proposed by Bloom (1970a,b) (see curve I, Plate 42). As the Micronesia curve considered is probably already affected by some compaction or subsidence effects, the real subsidence trend near San Francisco may be greater than estimated by Atwater et al. (1977). At the Santa Monica shelf, relative sea-level curve **C** by Nardin at al. (1981) was obtained from high resolution seismic stratigraphic analysis, sedimentological data and six radiocarbon dates of shell material from basal transgressive deposits and lagoonal muds.

No Holocene sea-level curve from the Pacific coasts of Mexico, Central America, Colombia, Ecuador and Peru is available to the present writer. In Chile, Kaizuka et al. (1973) reported from Mocha Island, near the outer edge of the continental shelf, at 90 km from the axis of the tectonically active Chile Trench, a sequence of raised terrace surfaces. The lower of these terraces is subdivided by small cliffs into several sublevels, the highest of which reaches +33 m. Radiocarbon dates of beach deposits showed that the lower terrace was formed during the Holocene by uplift movements occurring intermittently. The average rate of uplift (5.5 mm/yr) is indicated by curve **D**.

In Tierra del Fuego, Porter et al. (1984) dated by radiocarbon marine sediments from five coastal sites along the Strait of Magellan and Beagle Channel, obtaining relative sea-level curve **E**, which suggests that a 3.5 m emergence occurred since 5000 yr BP.

According to Rabassa et al. (1987), however, emergence in the Beagle Channel area, where the marine environment was fully established by 7900 yr BP, was 8 m since approximately 5500 yr BP. Mörner (1989) argued that the sea-level data available in Tierra del Fuego had been affected by local seismo-tectonic uplift displacements and could not be combined, therefore, into a meaningful sea-level graph. According to Mörner (1989) the maximum Holocene sea level would only have reached the elevation of 0.5 to 1.0 m above the present sea level (curve **F**) outside areas of local seismotectonic uplift.

As a matter of fact, evidence of Holocene emergence is scanty along the east coast of Tierra del Fuego, which remained ice-free during the last glaciation: a relative sea-level fall of 1.8 m was

Plate **53: BRITISH COLUMBIA III**

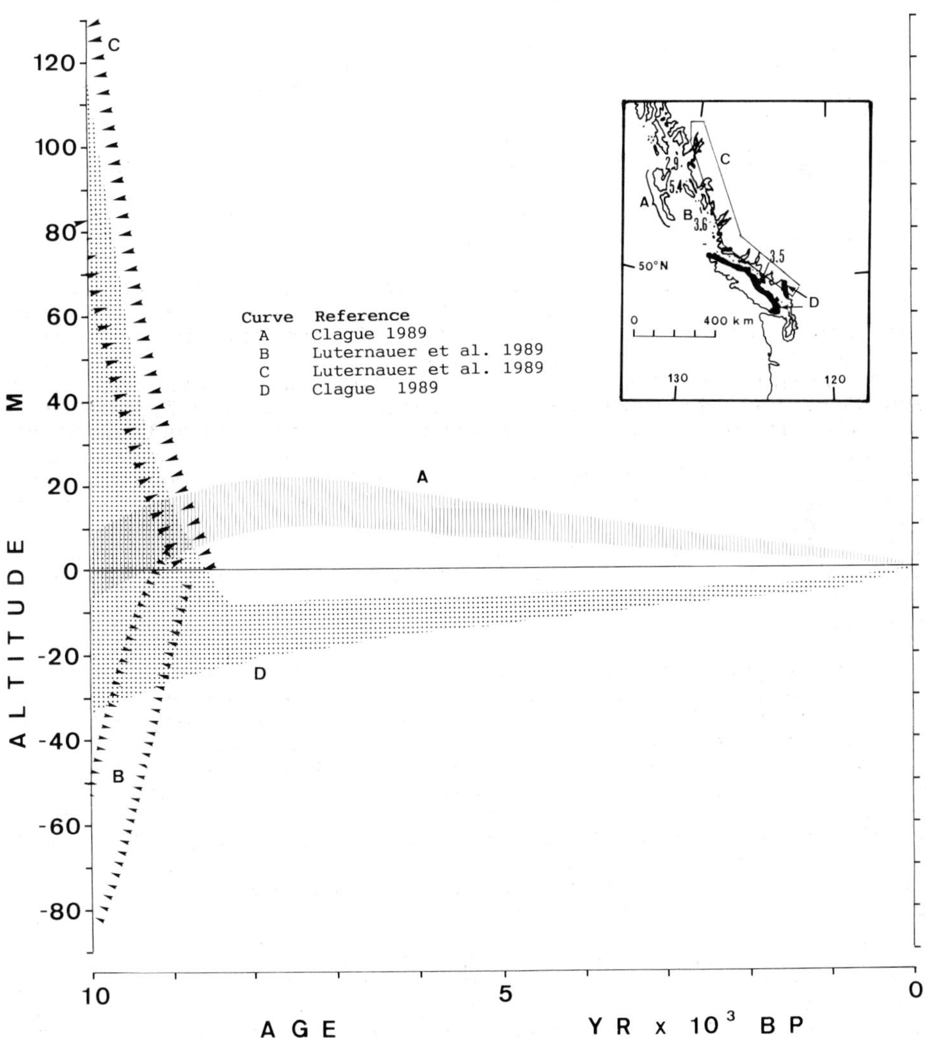

Plate 54: WEST U.S.A. II - CHILE - TIERRA DEL FUEGO

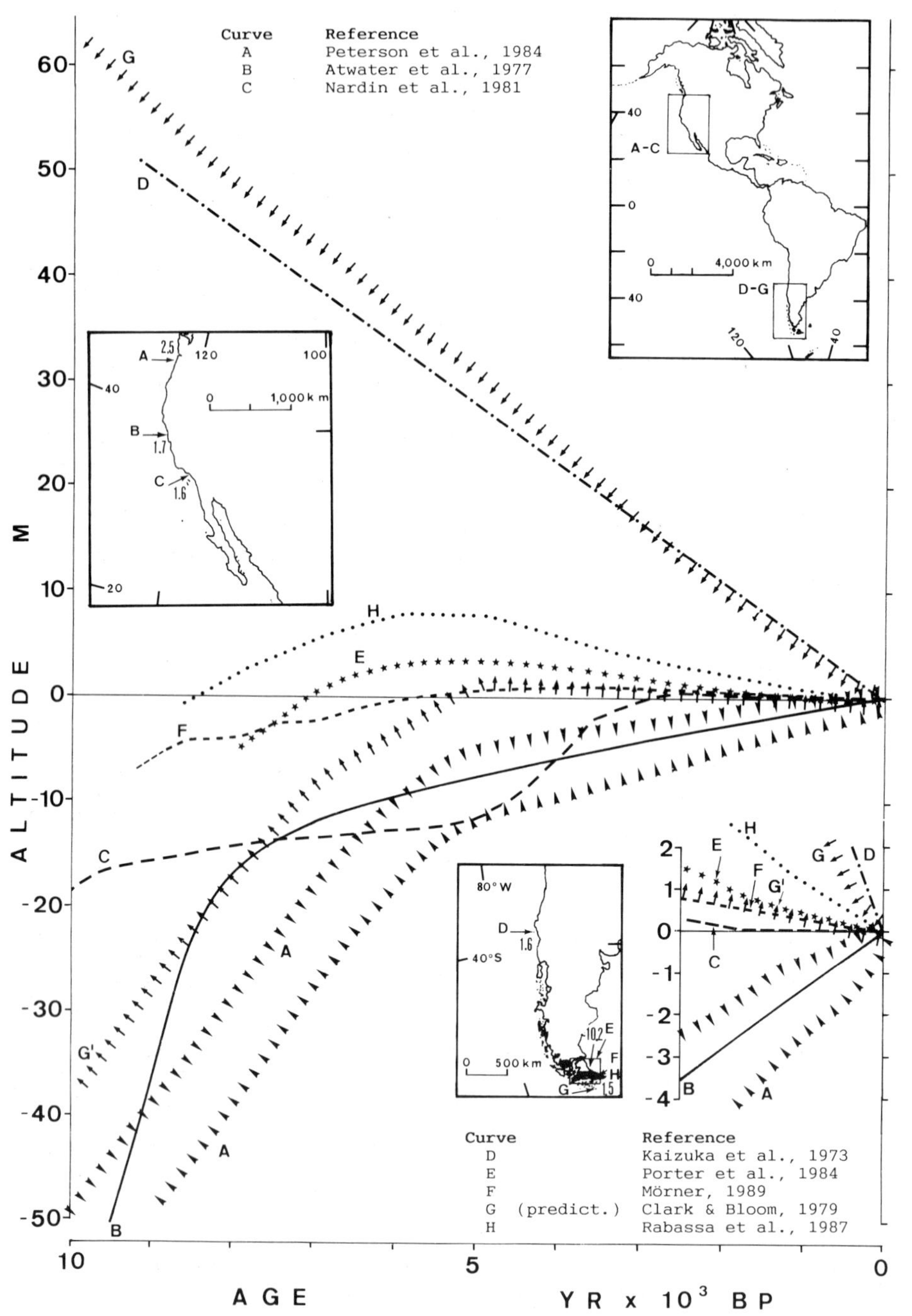

deduced by Bujalesky and González Bonorino (1990) from the supratidal plain elevation in San Sebastián Bay, whereas at La Misión, north of Rio Grande, the elevation of tidal flat deposits does not exceed, according to Mörner (in press), +0.9 m above the present "high tide level" (the local tidal range being about 10 m), suggesting emergence of the same order. In the Beagle Channel, on the other hand, where the tidal range is only about 1 m and elevated shorelines are frequent, reaching a maximum at +12 m (Rabassa et al., 1990), emergence may be tectonic in origin, since glacio-isostatic processes seem to have been completed in the early Holocene. The sea-level changes predicted at Cape Horn by Clark and Bloom (1979) vary from curve **G** (strong emergence resulting from the melting of a large Patagonian ice sheet) to curve **G'** (if no Patagonian ice sheet had melted). Divergent trends shown by curves **A** to **G'** for the last 2500 years are shown at an enlarged scale in the inset on Plate 54.

Relative sea-level changes indicated by tide-gauge records along the west coasts of the Americas (Fig. 2-32) show irregular trends in the Aleutian Islands and strong uplift movements in the Gulf of Alaska, at least from Unalaska (4) to Sitka (8). In the Canadian Straits relative vertical stability is suggested. More to the south, the relative sea-level rise recorded at Seattle (1.9 mm/yr) contrasts with the apparent sea-level fall (1.3 mm/yr) in the nearby Neah Bay (13). On the west coast of the U.S.A., where the shoreline is parallel to the San Andreas Fault, tide gauge records indicate slight emergence in Astoria (16) and Crescent City (17) and slight submergence at different rates further south. In Central and South America the records available are not lengthy enough, with the exception of Balboa (33) to estimate long-term tendencies.

ANTARCTICA (Plate 55)

Very few data on Holocene sea-level changes are reported from the Antarctic continent. Denton et al. (1975) and Stuiver et al. (1976) studied delta-like features in the McMurdo Sound, providing evidence of emergence exceeding 8 m since 5400 yr BP. However these dates are uncorrected for deficiency of ^{14}C in Antarctic marine waters and are thus too old by 850 to 1400 years. Undated raised beaches up to over +30 m were found at several localities by Nichols (1968) and Denton et al. (1975).

The Antarctic ice sheet is estimated to have contributed to the post-glacial global rise in sea level with an equivalent of 30 ± 6 m (Fig. 1-27), the main cause of melting being the sea-level rise caused by the disintegration of the large late-glacial ice sheets in the Northern Hemisphere. Lingle and Clark (1979) used a numerical model to estimate the relative sea-level changes caused by re-distribution of ice and water loads on the Earth's surface, and of rock within the Earth's mantle, caused by the vanishing of the Northern Hemisphere ice sheets and the consequent Antarctic ice sheet retreating to its present position. Their predictions, which unfortunately are poorly constrained by sea-level data, concern some key areas in the Weddel Sea, Pine Island Bay and the Ross Sea.

In the Weddell Sea area, two predicted relative sea-level curves were provided. Near the edge of the continental shelf, pronounced

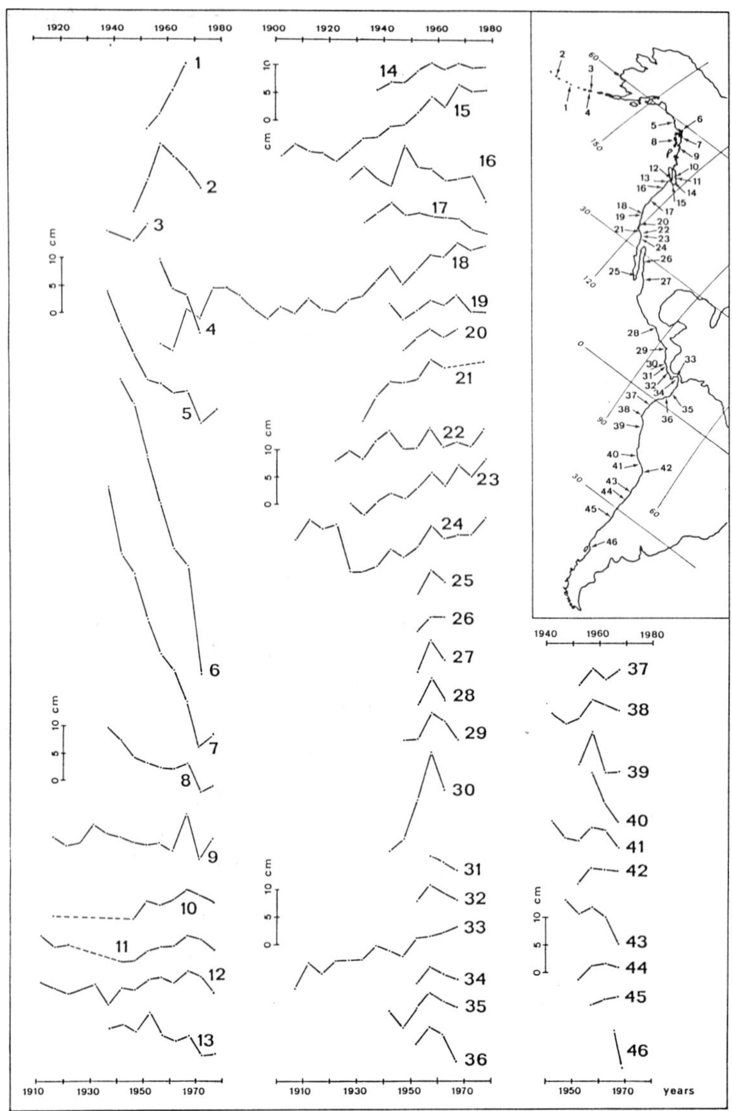

Fig. 2-32. Average 5-yr MSL variations on the west coasts of the Americas as recorded by tide-gauge stations. 1) Massacre Bay; 2) Sweeper Cove; 3) Dutch Harbor; 4) Unalaska; 5) Yakutak; 6) Skagway; 7) Juneau; 8) Sitka; 9) Ketchikan; 10) Point Atkinson; 11) Vancouver; 12) Victoria; 13) Neah Bay; 14) Friday Harbor; 15) Seattle; 16) Astoria; 17) Crescent City; 18) San Francisco; 19) Alameda; 20) Avila; 21) Santa Monica; 22) Los Angeles; 23) La Jolla; 24) San Diego; 25) La Paz; 26) Guaymas; 27) Mazatlan; 28) Salina Cruz; 29) La Union; 30) Puntarenas; 31) Quepos; 32) Puerto Armuelles; 33) Balboa; 34) Naos Island; 35) Buenaventura; 36) Tumaco; 37) La Libertad; 38) Talara; 39) Chimbote; 40) San Juan; 41) Matarani; 42) Arica; 43) Antofagasta; 44) Caldera; 45) Valparaiso; 46) Puerto Montt. (From Pirazzoli, 1986b).

Plate 55: ANTARCTICA

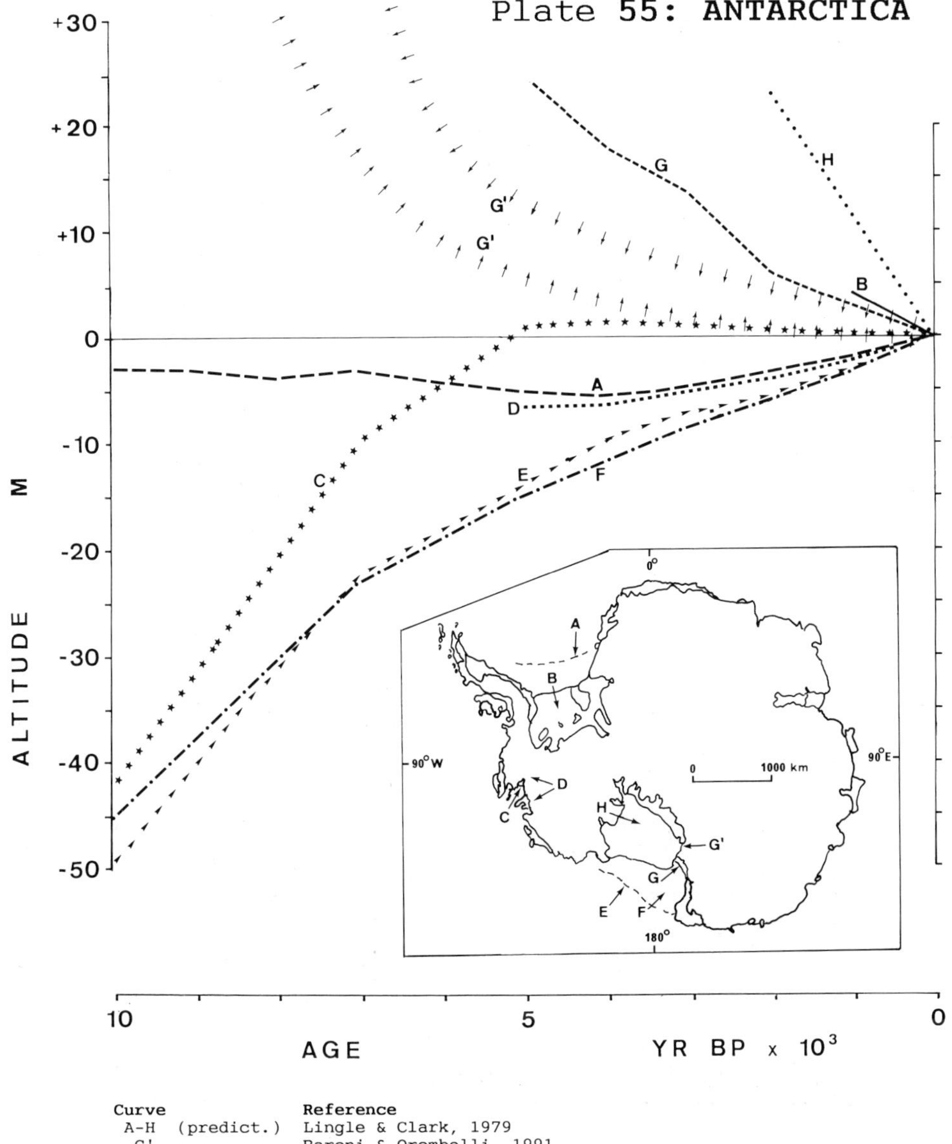

Curve	Reference
A-H (predict.)	Lingle & Clark, 1979
G'	Baroni & Orombelli, 1991

emergence due to Antarctic ice sheet retreat was expected to have been largely counteracted by submergence due to Northern Hemisphere deglaciation, so that relative sea level in the early Holocene was only a few metres lower than at present (curve **A**, **Plate 55**). In this area relative sea level was predicted to have remained within 6 m of the present sea level since 11,000 yr BP, with net submergence of about 5.5 m since 4000 yr BP. In the southern Weddell Sea, in contrast, near the present grounding line of the Ronne Ice Shelf, about 4.3 m of net emergence were predicted during the past 1000 years (curve **B**).

North of Pine Island Bay, in the Southern Ocean near Lat. 61°S, continuous submergence was expected to have been caused by addition of water to the ocean, combined with the collapse of a pro-glacial fore bulge (curve **C**). Nevertheless net emergence of about 1.2 m was predicted since 5000 yr BP due to completion of northern deglaciation. In Pine Island Bay, on the other hand, near the present grounding lines of Pine Island and Thwaites Glaciers, grounded ice was assumed to exist until 6000 yr BP; here net submergence of 5 m was predicted since 5000 yr BP by the Lingle and Clark (1979) model (curve **D**).

In the Ross Sea region, relative sea-level changes were predicted in four areas: near the edge of the continental shelf (**E**) and at Pennell Bank (**F**) net submergence was expected to have been predominant; in McMurdo Sound (**G**) 25-m emergence was predicted since 5000 yr BP; finally in the centre of the present Ross Ice Shelf 23 m of emergence was predicted to have occurred during the past 2000 years (**H**).

Only one sea-level band deduced from field data (**G'**) is available to test the above predictions. It was obtained by Baroni and Orombelli (1991) in Terra Nova Bay (Victoria Land) using over 40 radiocarbon dates of samples collected from raised beaches, ranging from c. 7500 yr BP to the present after correction for the reservoir effect. The upper part of band **G'** is constrained by the local marine limit (situated at about +30 m) and by maximum elevations and minimum ages for sea level deduced from guano deposits and penguin remains collected from abandoned nesting sites on raised beach deposits. The lower boundary of **G'** is delimited by minimum elevations and maximum or contemporary ages for sea level deduced from *in situ* marine shells, mostly unbroken and articulated. According to Baroni and Orombelli (1991) the real sea-level curve is probably close to the lower limit of band **G'** and deglaciation of Terra Nova Bay was accomplished before 7000 yr BP. The latter point is consistent with the Antarctic deglaciation history proposed by Peltier (1988), who placed the ice-sheet melting between 10,000 and 6000 yr BP, but challenges the conclusions by Nakada and Lambeck (1989) concerning ongoing Antarctic melting into late Holocene time.

THE EAST COASTS OF SOUTH AMERICA (Plate 56)

Auer (1951) investigated peat bogs in Tierra del Fuego and Patagonia. By using three thin layers of volcanic ashes as a means of correlation, he attempted comparisons with the Baltic region and proposed an early relative sea-level curve (**A**) for the area from Tierra del Fuego to Bahia Blanca. Urien (1970) interpreted radiocarbon-dated shell layers from the continental shelf of Argentina as former shorelines and deduced sea-level curve **B**, with wide fluctuations which seem however rather unlikely. Clark and Bloom (1979) predicted for a locality at 60° S latitude on the Argentina shelf sea-level band **C**. According to their earth model with a strong lithosphere, the upper part of band **C** corresponds to the assumption that a large Patagonian ice sheet existed during glacial times, whereas the lower part of band **C** corresponds to a small ice sheet or no ice sheet. Anyway, with either a large or small Patagonian ice sheet, submergence would be the dominant effect according to Clark and Bloom (1979), because a collapsing proglacial forebulge would cause subsidence at this locality.

Plate 56: ARGENTINA - BRAZIL

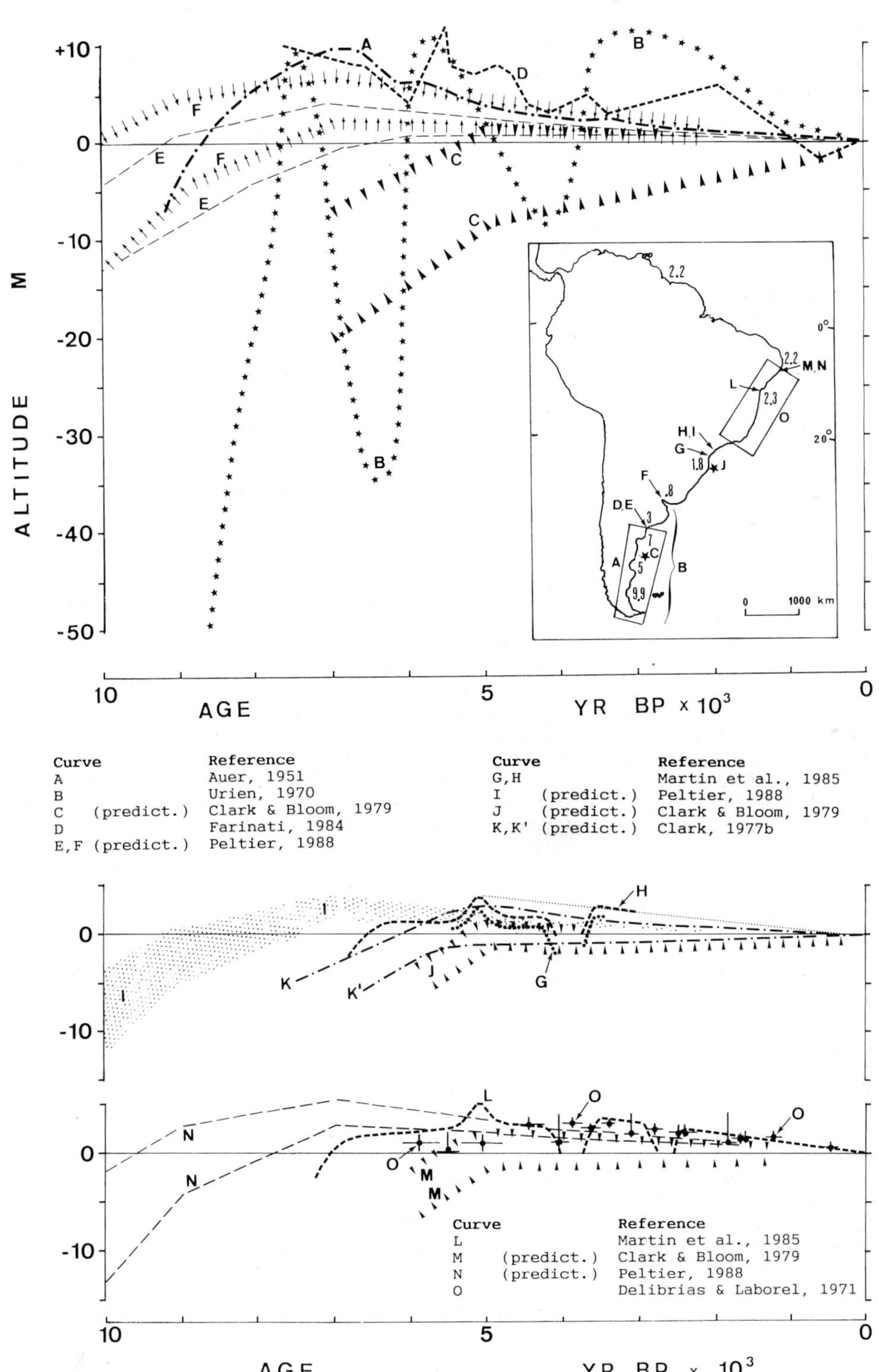

At Bahia Blanca, Farinati (1984) proposed relative sea-level curve **D**, constructed by using 16 radiocarbon dates of shell deposits left in beach ridges by the regression of the Holocene sea. As altitudes were deduced by Farinati from topographical maps, vertical uncertainty ranges are of the same order of the interval between contour lines (5 m) in the maps. This, and the fact that the local spring tidal range is over 3 m, suggest that no great confidence can be placed on the fluctuations suggested by curve **D**. Nevertheless, most of the field data from Bahia Blanca are above the **E** sea-level band predicted in the same area by Peltier (1988). More to the N, at Rio Plata, sea-level predictions by Peltier (band **F**) suggest the occurrence of emergence since as early as 9000 ± 1000 yr BP. This seems unlikely, however. In the Mar del Plata area earliest evidence of emergence is no older than 6200 yr BP (Fasano et al., 1984; Schnack et al., 1982) (Fig. 2-33).

In Brazil, systematic investigations carried out along 2000 km of coastline, supported by many radiocarbon dates (over 250 dates are listed by Martin et al., 1979-1980) made possible the construction of relative sea-level curves in eight coastal sectors (Fig. 2-35). Three of these curves have been reproduced in **Plate 56** at the usual scale of the Atlas, for comparison with other data available: curve **G** for the Paranaguá sector, curve **H** for the Cananeia-Iguape sector and curve **L** for the Salvador sector. It can be noted that common trends are a sea-level maximum occurring slightly earlier than 5000 yr BP (Fig. 2-34), followed by a gradual sea-level fall continuing to the present time. Superimposed on this long-term gradual drop, two rapid oscillations, with sea level falling below zero near 4000 and 2800 yr BP, were advocated by Martin et al., which seem poorly established however. The relatively lower position of curve **G** in relation to nearby curve **H**, was ascribed by Martin et al. (1985) to deformation of the present geoid surface.
Band **I**, which corresponds to the changes in relative sea-level predicted by Peltier (1988) at Iguape, fits almost satisfactorily field data for the last 6000 years, but tends to overestimate early Holocene sea-level positions. Band **J** was predicted by Clark and Bloom (1979) for the continental shelf near Santa Catarina. Curve **K** was predicted by Clark (1977) for a "continental" locality of "coastal Brazil" and curve **K'** for a locality 100 km E on the continental shelf. At Recife sea-level bands **M** and **N** were predicted by Clark and Bloom (1979) and Peltier (1988) respectively. Sea-level index points **O** refer to data obtained by Delibrias and Laborel (1971) along the coast between 24° S and Recife. Some of these points, for which uncertainty ranges are specified for time and elevation, clearly do not fit the sea-level fluctuations claimed by Martin et al. (1985).

WEST INDIES - BERMUDA (Plate 57)

Fairbanks (1989) employed *Acropora palmata* corals, drilled from coral reefs off the south coast of the Barbados, to provide interesting indications of trends in sea-level change from 17,000 to 7000 yr BP. After correction for an estimated local uplift trend of 0.34 mm/yr, Fairbanks mixed his sea-level data with similar data reported by Lighty et al. (1982) from various areas of the tropical western Atlantic (see curve **D**, below). The Holocene part of the "Barbados sea-level curve" proposed by Fairbanks (1989) is reproduced in **Plate 57** (curve **A**). It shows a

Plate 58: GUIANA - CENTRAL AMERICA

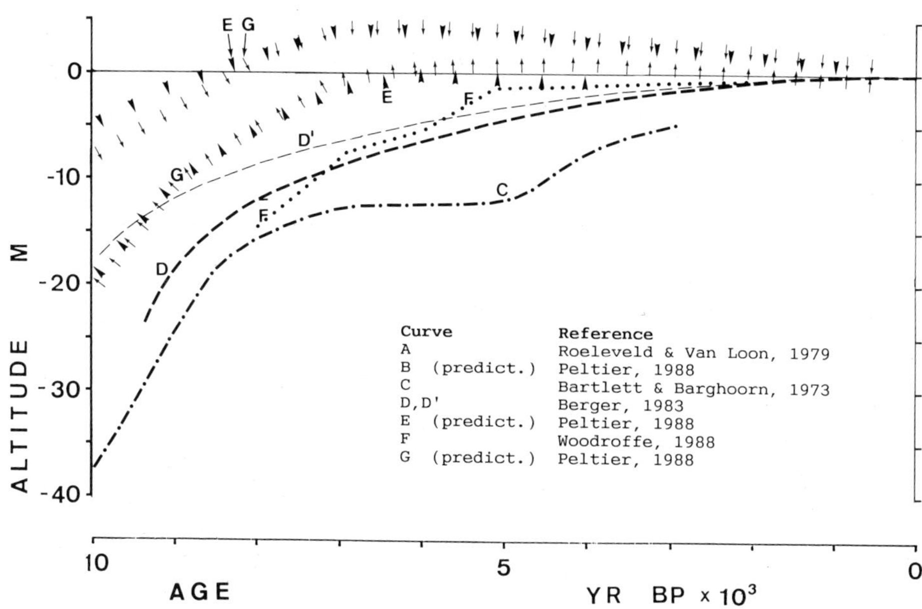

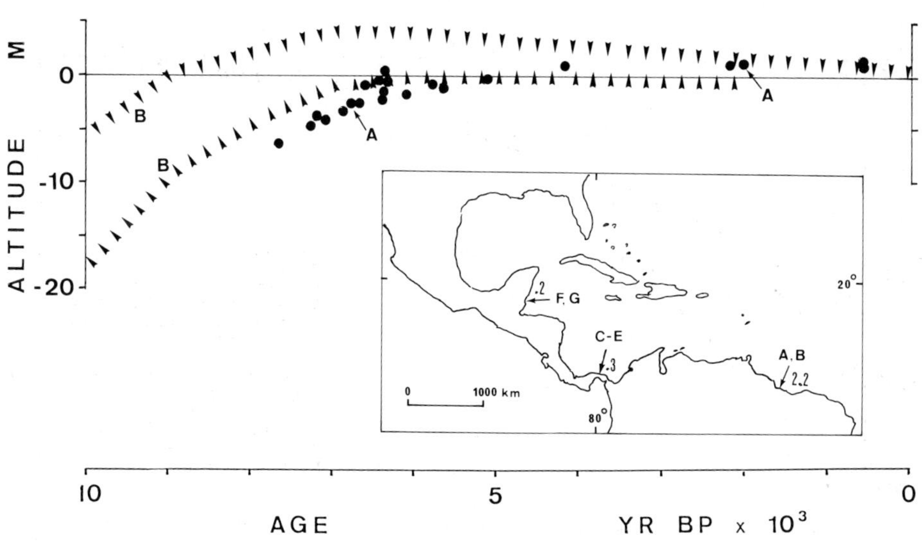

GUIANA - CENTRAL AMERICA (Plate 58)

In the coastal plain of Surinam, Roeleveld and Van Loon (1979) dated 19 samples of peat layers (**A** filled circles in **Plate 58**) mostly formed in brackish coastal *Rhizophora* swamps near MSL. Data fit into the picture of a sea-level rise in this area until about 6000 yr BP. Roeleveld and Van Loon (1979) concluded that sea-level remained essentially at its present position from 6000 yr BP onward. However the surface elevation at the sampling localities could not be established, and this may constitute a significant source of error in vertical estimations. Band **B** predicted by Peltier (1988) for Surinam tends to overestimate the elevations of early Holocene sea levels.

In the Gatun basin, Panama, a mangrove pollen study of deep core sediments by Bartlett and Barghoorn (1973), supported by fifteen radiocarbon dates of peat samples, provided a chronology of rise in the relative sea level from 11,300 yr BP to the present, showing a very rapid early postglacial rate of rise, and a pronounced slowing after about 7300 yr BP (curve **C**), until about 2000 years ago, when the present level was reached. However, as peat layers had undoubtedly experienced some autocompaction, as well as compaction caused by overlying clays and gravels deposited by the meandering Trinidad River, **C** must be considered as a minimum MSL curve.

In the same area and probably using the same data, Berger (1983) proposed a new smooth curve (**D**), which is all the time positioned above curve **C**, and added a calibrated curve (**D'**) for comparison of the effect of tree-ring calibrated and non calibrated radiocarbon dating. The time scale of curve **D'**, which illustrates that earlier Holocene dates become progressively older when calibration is applied, should be read in sideral yr BP and not in radiocarbon yr BP. Relative sea-level band **E** predicted on the east coast of Panama by Peltier (1988) is systematically higher than field data.

In Belize, Woodroffe (1988) proposed relative sea-level curve **F**, obtained from a compilation of radiocarbon dates on mangrove sediments provided by various authors. Belize field data are again systematically below the sea-level predictions of the Peltier (1988) model (band **G**).

THE GULF OF MEXICO AND THE ATLANTIC SHELF OF THE UNITED STATES OF AMERICA (Plate 59)

Most Holocene sea-level curves of the continental shelf off the south and east coasts of the United States date from the 1960s, a period when "as never before, geologists turned to the sea to learn more about the geological history of the Earth" (Emery and Milliman, 1970).

Emery and Garrison (1967) have summarized, with least-squares best-fit straight lines, radiocarbon ages of the period 17,000 to 7000 yr BP for salt-marsh peat and shallow water (0 to 4 m) shells from the Texas continental shelf. The Holocene part of their sea-level line is represented in **Plate 59** by **A**.

The Sabine-High Island area, located off the Texas coast, between the Galveston Lagoon to the west and the Mississippi River delta

Plate 59: - GULF OF MEXICO - ATLANTIC SHELF OF THE U.S.A.

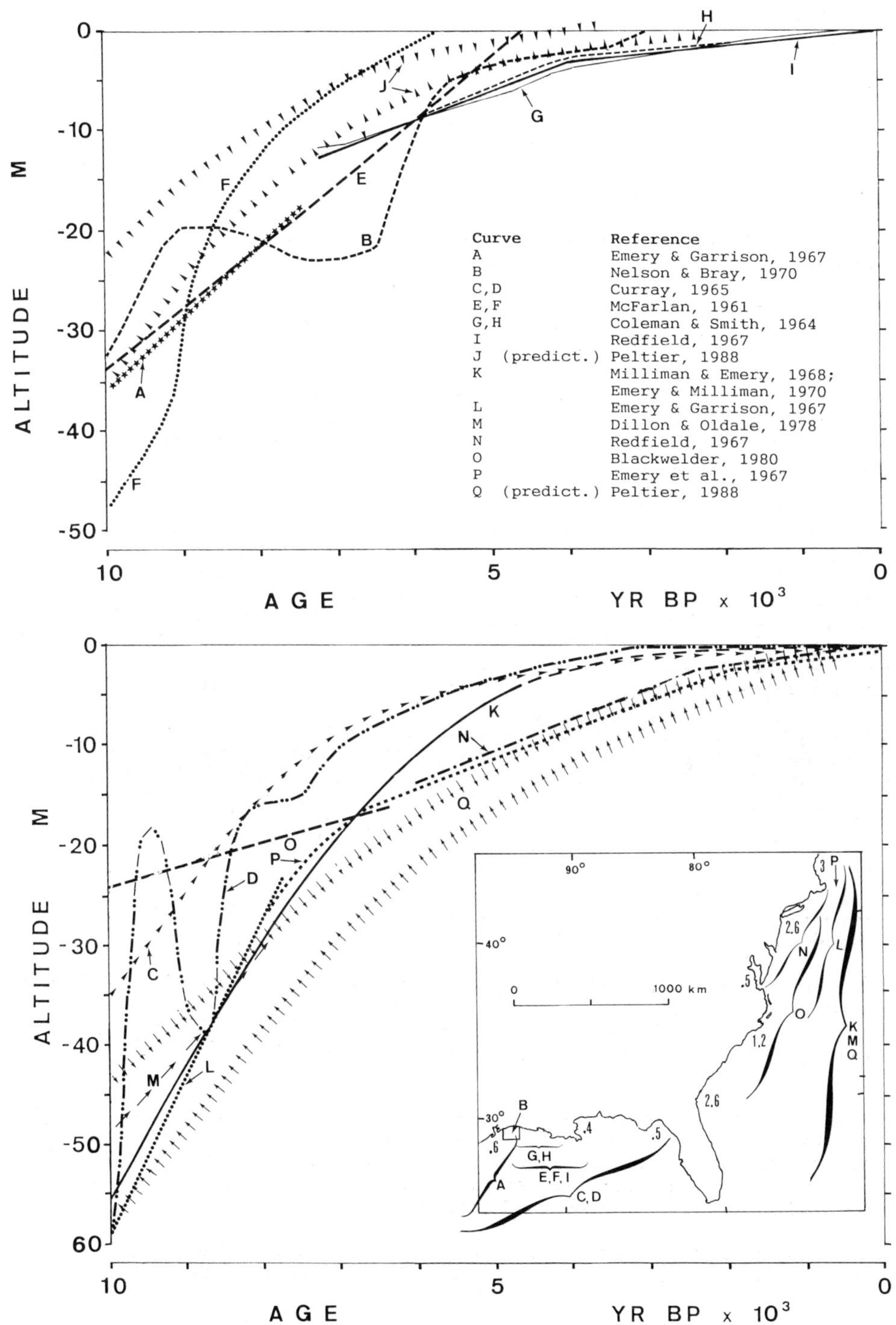

to the east, began to be flooded by the rising sea in 10,200 yr BP. Nelson and Bray (1970) used more than one hundred radiocarbon dates of samples cored or dredged from this area to reconstruct the local Holocene environment history. Relative sea-level curve **B** is mostly based on one dozen samples of wood and peat and on the assumptions that the peat and wood developed in a sea-level swamp, that there had been no compaction beneath the peat since it was formed and that no regional tilting or tectonism had occurred. The wide fluctuations shown by curve **B** may be a consequence of the assumptions made.

Curray (1965) updated his previously published sea-level graphs and provided two curves for the continental shelf of the "Gulf of Mexico": curve **C** shows approximate means of the data compiled, whereas curve **D** updates previous curves by Curray (1960, 1961).

In southern and offshore Louisiana, McFarlan (1961) dated by ^{14}C some 120 samples from both the surface and the subsurface, the elevation of which was subsequently corrected in order to deduct structural movements. Two sea-level curves were proposed: curve **E**, which includes all corrected data and shows a linear sea-level rise until about 5000 yr BP, and curve **F**, based on fewer points provided by "hingeline" samples (which have required no structural adjustment). As it is not the aim of **E** and **F** to describe relative sea-level histories, but rather to attempt an estimation of eustatic changes in sea level, these curves are of limited usefulness in local applications.

In south-central Louisiana, Coleman and Smith (1964) dated by ^{14}C 13 samples of marsh peat now buried at depths ranging from 1.2 to 12.2 m and obtained two relative sea-level curves: **G**, connecting points of the greatest depth for each known date, and **H**, connecting points of the least depth. Curves **G** and **H** define a submergence band which includes regional subsidence and peat compaction, since most of the samples dated were collected at the surface of peat layers. Coleman and Smith (1964) estimated the rate of subsidence in the study area at about 0.7 mm/yr. Using the same data, Redfield (1967) proposed curve **I** for "Louisiana", which often crosses the middle part of the band defined by **G** and **H**. Band **J** corresponds to relative sea-level changes predicted by the Peltier (1988) model in the "NW Gulf of Mexico".

Milliman and Emery (1968) and Emery and Milliman (1970) proposed for the past 36,000 years a well-known sea-level curve for the whole continental shelf of the eastern U.S.A., based on radiocarbon dates of various dredged or cored materials. The Holocene part of this curve is shown as curve **K**. Emery and Garrison (1967) used 16 radiocarbon dates of sediments from a 1300-km wide area of the Atlantic continental shelf of the U.S.A., from Cape Hatteras to Gulf of Maine, to obtain a least-squares straight line (curve **L**) which was assumed to correspond to sea-level positions. Dillon and Oldale (1978) found evidence of differential subsidence in part of the continental shelf south of Long Island. Accordingly, they corrected curves **L** and **K** and obtained for the "U.S. East coast" curve **M**, that they considered as eustatic. Redfield (1967) assembled dated sea-level indicators from northern Virginia to Cape Cod, to construct relative sea-level curve **N**. He deduced, from a comparison with similar data from Bermuda, southern Florida, North Carolina and Louisiana which were assumed to be stable regions, that coastal subsidence movements of about 4 m occurred on the northeastern coast of the

U.S. between 4000 and 2000 yr BP and has continued at a rate of about 0.3 mm/yr since then.

Blackwelder (1980) mixed 28 radiocarbon-dated samples of salt-marsh peat and oyster beds cored from a 1000-km wide continental shelf area, extending between South Carolina and New York City, and proposed a sea-level rise line between 7000 and 12,000 yr BP, the Holocene part of which is represented as curve **O**. Though vertical errors of samples were estimated by Blackwelder to be about ±2 m, it appears that many samples deviate from curve **O** by several metres. In particular, samples from off South Carolina plot below and generally within 5 m of samples from off Delaware, the mouth of Chesapeake Bay, and the Cape Fear arch area of North Carolina. In spite of the very wide area studied, Blackwelder (1980) claimed that the locations of the 12,000-yr to 9000-yr old shorelines did not undergo major differential vertical deformation. This conclusion is not consistent however with early Holocene sea-level curves provided by other authors (see Plates 60-62).

In Cape Cod Bay, Emery et al. (1967) proposed sea-level curve **P**, based on several radiocarbon-dated samples of peat and shells. Lastly, the predictions made by the Peltier (1988) model for the "Cont. Shelf USA", which are summarized by relative sea-level band **Q**, appear at a lower level than most field observations.

Continental shelf data are, in most cases, essential to reconstruct past sea levels which were much lower than the present one. However, the approach which consists in mixing shelf data coming from very wide areas, in order to construct a regional sea-level curve, may have been acceptable in the 1960s, but is known today to be inaccurate or misleading, since it does not take into account glacio-isostatic and tectonic effects, which may change with the area investigated, and hydro-isostatic effects, which probably vary on the same shelf transect perpendicular to the coast, depending on the distance from the present-day shoreline and the water depth.

THE ATLANTIC COAST OF THE UNITED STATES OF AMERICA: INTRODUCTION

On the southernmost eastern coast of Florida, south of Miami, carbonate reefs protect wide lagoons. Between Miami and Chesapeake Bay there are many small estuaries; here wide lagoons and barrier islands fringe most of the coast. From Chesapeake Bay to Long Island the coast becomes deeply embayed, with the mainland coastal plain fringed by long sand barriers, mostly separated from the mainland by lagoons. The latter are partly filled, and large tracts of marsh exist where the fill has been almost complete. There are also three long estuaries, the origin of which is a former river valley (Chesapeake Bay, Delaware Bay), combined in New York Harbor with a partially filled fiord (Shepard and Wanless, 1971). North of Nantucket Island, Massachusetts, continental glaciers extended beyond the present shoreline. The outermost margin of the ice traversed the islands located just south of Cape Cod (Nantucket, Martha's Vineyard, Block Island, Long Island), but west of New York Harbor it curved inland away from the coast (Fig. 2-36).

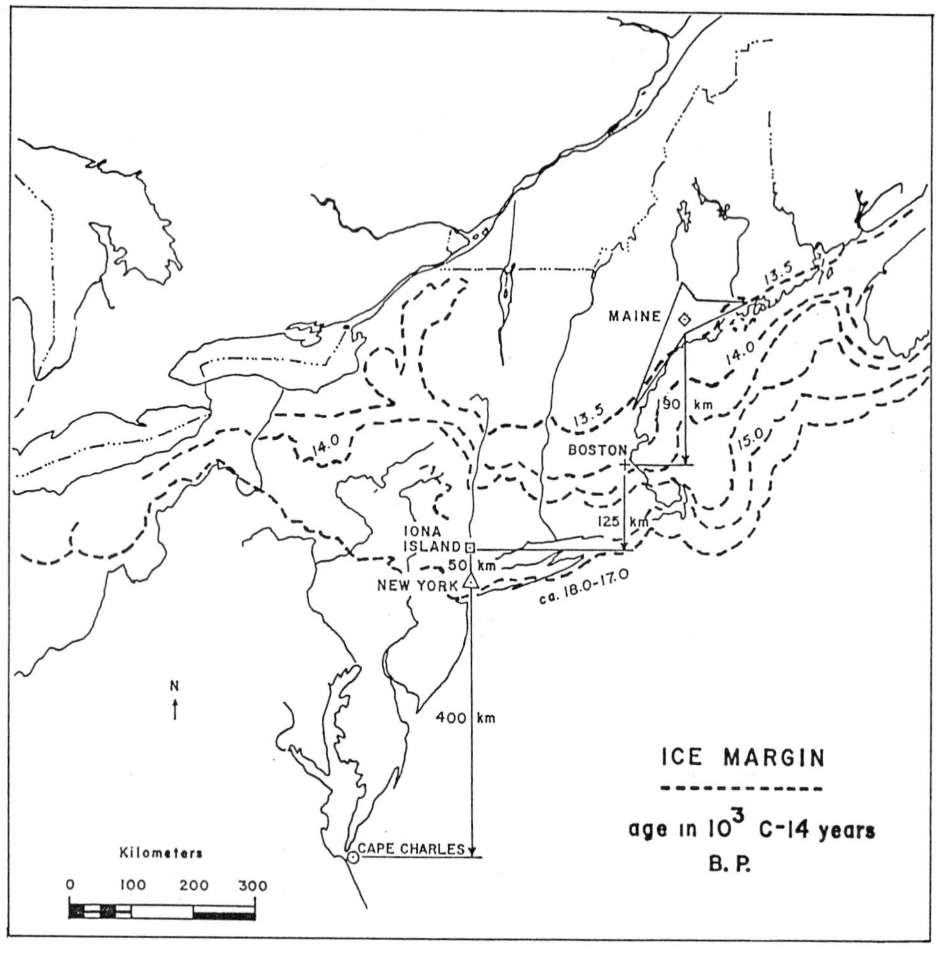

Fig. 2-36. Location map of stations mentioned in Fig. 2-38 and their relationship with late glacial ice margin positions (from Newman et al., 1971, adapted).

The fact that submergence curves of various authors diverge with increasing age along the Atlantic coast of the U.S.A. has been interpreted in various ways. For the northernmost coastal area, the occurrence of glacio-isostatic rebound is obvious, though it can hardly account alone for all the observed patterns. Developing an hypothesis suggested by Daly (1934) (see Introduction), Bloom (1967, p. 1491) concluded that "the postglacial submergence histories of five eastern United States coastal sites support the hypothesis that the load of the water added to the continental margins by the postglacial rise of sea level has been sufficient to deform coastal areas isostatically in proportion to the average depth of water in the vicinity". Another idea discussed by Daly (1934) was that of a "viscous bulge" which would have formed peripherally to an ice cap and would have moved like a wave towards the centre of the ice cap during deglaciation (Fig. 2-37).

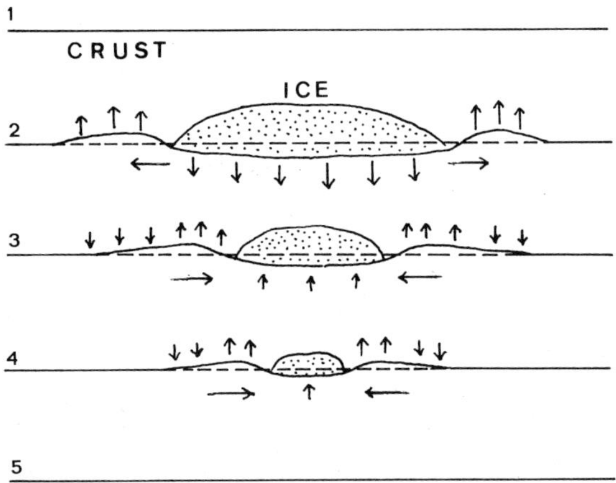

Fig. 2-37. Sections illustrating the "viscous-bulge" hypothesis according to Daly (1934, Fig. 69, adapted).

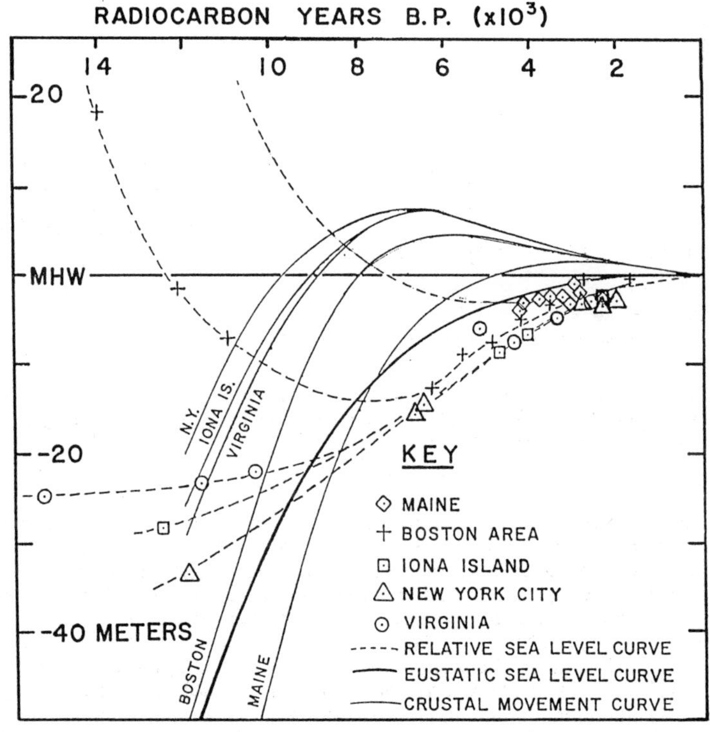

Fig. 2-38. Apparent sea-level curves, the Shepard and Curray (1967) "eustatic" sea level curve, and crustal movement curves for five stations (located in Fig. 2-36) along the northeastern Atlantic coast of the U.S.A. Note displacement of wave crests with time (from Newman et al., 1971, adapted).

This hypothesis was applied by Newman et al. (1971) to the data for the northeast coast of North America, suggesting possible dimensions and dynamics for a peripheral bulge (Fig. 2-38). Since the late 1970s, several generations of global isostatic models have developed, among others, the above glacio- and hydro-isostatic hypotheses, to obtain predictions of past sea-level changes which in several cases are fairly close to existing field data.

FLORIDA - CAROLINA (Plate 60)

In the Everglades region, southern Florida, Scholl (1964) and Scholl and Stuiver (1967) reported a relative sea-level history for the last 4400 years based on over 30 radiocarbon dates of mangrove swamp deposits or marine and brackish-water sediments overlying freshwater calcitic muds (curve **A**, **Plate 60**). Almost the same data were summarized by Redfield (1967) with a straight line (**B**). Two years later, Scholl et al. (1969) corrected the earlier part of curve **A**, extending it to 5500 yr BP (curve **C**). According to them, the sea-level rise, which would include 1 to 2 m of isostatic subsidence, occurred at the rate of 0.83 mm/yr until 3500 years ago and of 0.35 mm/yr after that. A least-square polynomial curve very similar to **C** was found by Gornitz and Seeber (1990), using data from South Florida available in the sea-level data bank described by Pardi and Newman (1987).

Robbin (1984) integrated previous data with newly acquired radiocarbon dates of *in situ* mangrove peats and subaerially-formed soilstone crusts (caliche), to present a relative sea-level curve for the upper Florida Keys and adjacent Florida reef tract (curve **D**). According to D.F. McNeill (pers. comm., August 1988), the part of curve **D** older than 8000 yr BP, was poorly constrained; it has been revised by Lidz and Shinn (1991). At Daytona Beach, relative sea-level changes predicted by the Peltier (1988) model are summarized by band **E**, which is slightly above least-square polynomial curve **E'** obtained by Gornitz and Seeber (1990) from "North Florida" (27.4°-31.6°N, 80.0°-82.0°W) field data.

In South Carolina, where Peltier (1988) predicted sea-level band **F** for Port Royal, very oscillating sea-level histories were proposed at a much higher elevation by Moslow and Colquhoun (1981) (curve **G**) and Colquhoun and Brooks (1987) (curve **H**) in estuary areas near Charleston. These oscillations were obtained mainly by comparing archaeological data with radiocarbon-dated basal peats (Brooks et al., 1979) and interpreted as corresponding to sea-level changes. However such oscillations, which are of the same order as the local tidal range, may also be related, at least in part, to water table variations in the estuary areas investigated, reflecting periods of drier or wetter climate conditions.

At Southport, North Carolina, a linear sea-level trend since 4000 yr BP (curve **I**, almost coincident with curve **K**) was deduced by Redfield (1967) from four peat dates. In the same place, sea-level predictions by the Peltier model (1988) are summarized by band **J**, which deviates appreciably from **I**. Sea-level curve **K** was proposed by Moslow and Heron (1981), using a number of unspecified radiocarbon dates obtained from cored samples collected within the Cape Lookout area or compiled from two unpublished Ph.D. theses

Plate 60: FLORIDA - CAROLINA

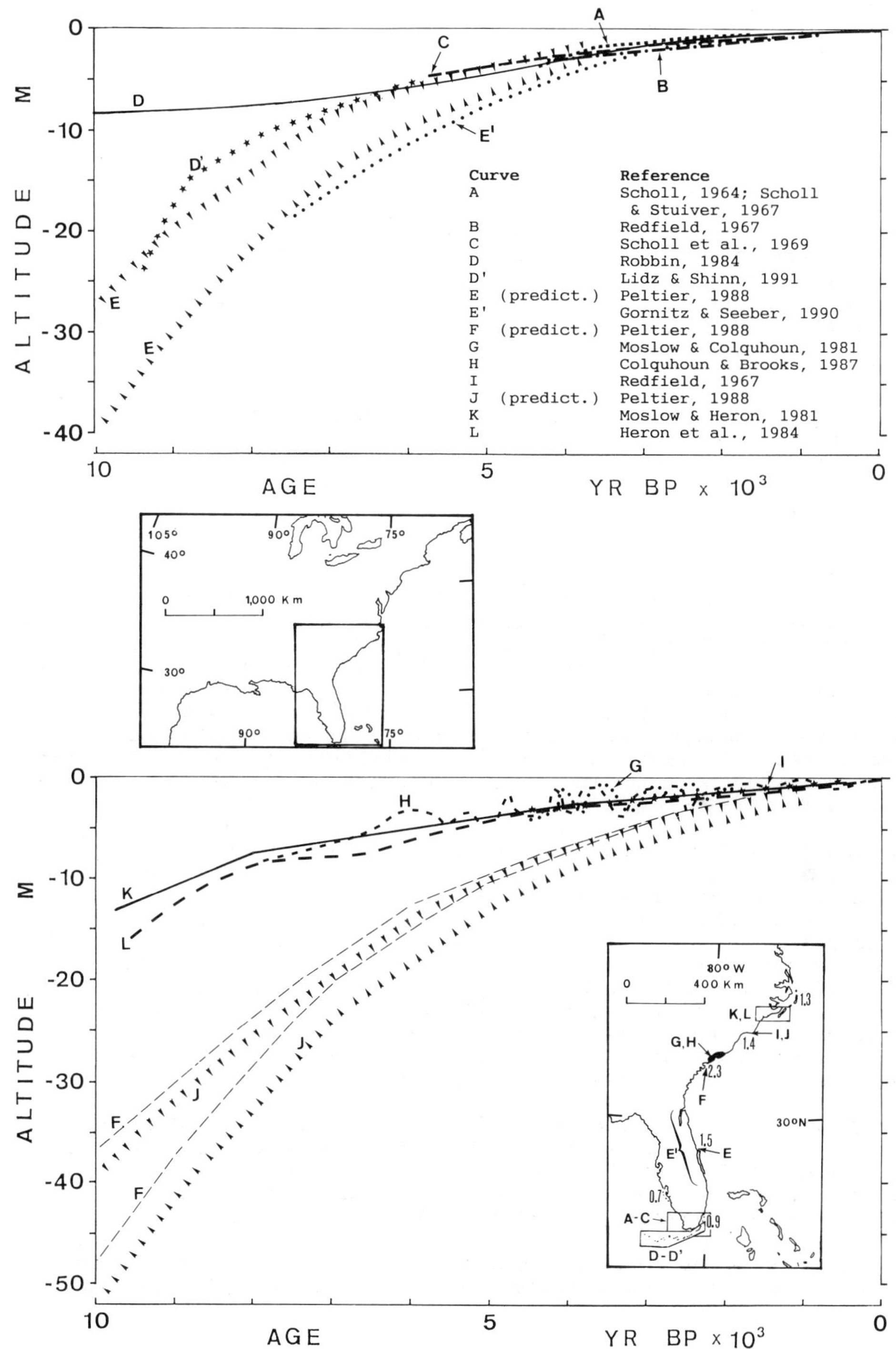

(Berelson, 1979; Steele, 1980). Later, Heron et al. (1984) published curve **L** for the same area.

VIRGINIA - DELAWARE (Plate 61)

On the Virginia coasts, Newman and Rusnak (1965) and Newman and Munsart (1968) produced in the Wachapreague area submergence curve **A** (**Plate 61**), based on radiocarbon dates of four basal peat samples. The oldest of these dates suggests a relative sea level position about one metre higher around 5100 yr BP than around 4350 yr BP, which is interpreted as a consequence of crustal warping (Newman and Rusnak, 1965) and/or of an eustatic sea-level fluctuation (Newman and Munsart, 1968).

Curve **B**, ascribed by Newman et al. (1971) to Cape Charles, results in fact from a compilation of earlier data from a wider area, published by Harrison et al. (1965) and Newman and Munsart (1968). It can be compared with relative sea-level band **C**, predicted at Cape Charles by the Peltier (1988) model, which appears however lower than all field data. Other compilations led Newman et al. (1980) to propose curve **D** for the west shore of Chesapeake Bay and curve **E** for the Chesapeake Bay entrance; curves **D** and **E** are in fact simply line segments connecting data points and this may explain their irregular trends. Gornitz and Seeber (1990), refining data previously assembled by the late W.S. Newman, proposed that the sea level had risen, following linear trends, at rates of 1.2 mm/yr since 6000 yr BP in South Chesapeake Bay and of 1.81 mm/yr since 4000 yr BP in North Chesapeake Bay.

Finkelstein and Ferland (1987) suggested that the dates of the two deeper samples used by Newman and Munsart (1968), both obtained from small samples of humic acid, may be inaccurate and added some new data points indicating that along the eastern shore of Virginia a gradual sea-level rise occurred for approximately the last 4600 yr BP (curve **F**). Van de Plassche (1990) added to curve **F** the approximate trend for the mid-Holocene relative sea-level rise that he deduced from Finkelstein and Ferland (1987) (shaded band **G**) and compared it with his own curve (**H**), which, almost coincident with the later part of curve **B**, is situated slightly below **G**. As most sea-level curves are very near to each other in the Virginia coastal area, **A**, **B**, **D**, **E**, **F** and **H** are shown at an enlarged vertical scale in the inset for the period 4500 to 2000 yr BP.

In the coastal Delaware area, Kraft (1971) used various radiocarbon-dated samples, encountered by drill or borings, to obtain relative sea-level curve **I**, which is related to mean low sea level, includes compaction and subsidence effects and covers all the Holocene period. Belknap and Kraft (1977) used 88 dates of samples collected with various methods (for details see Kraft, 1976; Kraft and John, 1976) to construct relative sea-level curves at various time scales. Curve **J** corresponds, as usually in this Atlas, to the 5570-yr half-life of radiocarbon. Two possibilities are shown by **J** prior to 7000 yr BP, indicating the uncertainty of the curve for that period of time.

Curve **K**, by Newman et al. (1980), consists of line segments connecting data points compiled from Kraft (1976). Using almost the same data but after having applied various corrections, Preuss

Plate 61: VIRGINIA - DELAWARE

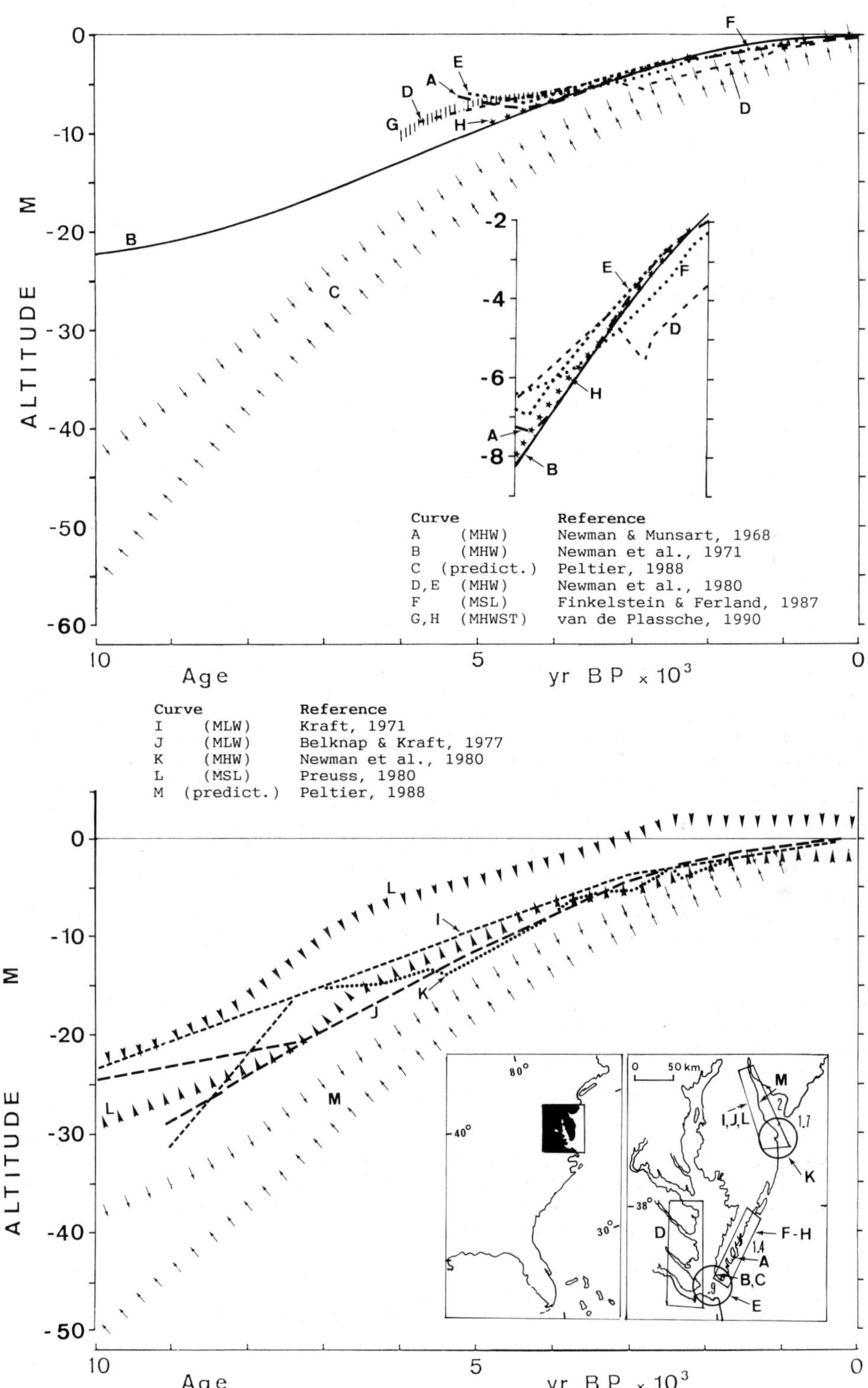

(1980) produced MSL band **L**. Lastly, Peltier (1988) predicted the relative sea-level band **M** at Bowers.

NEW JERSEY - NEW YORK - CONNECTICUT (Plate 62)

Several sea-level curves are available for the vicinity of Atlantic City (New Jersey). Stuiver and Daddario (1963) used five radiocarbon dates of basal peats, cored from the floor of the lagoon between the Brigantine City Barrier and the mainland, to obtain MHW curve **A** (**Plate 62**), which shows a submergence of 3 mm/yr between 6000 and 2600 yr BP, but of only 1.2 to 1.4 mm/yr for the past 2600 years. Curve **B** by Bloom (1967) is based on the same data as curve **A**, but radiocarbon ages were corrected to take into account fluctuations of atmospheric ^{14}C. The time scale of curve **B** is therefore expressed in calendar years BP and not in radiocarbon years BP, as most curves in this Atlas. The main effect of this correction is to decrease the submergence slope prior to 2200 years ago. Curve **C** by Newman et al. (1980) consists simply of line segments drawn from data point to data point for the same data set as curve **A** and is consequently very near to **A**. Still in the Brigantine area, curve **D** corresponds to the predictions by the Nakada and Lambeck (1987) model and band **E** to the predictions by the Peltier (1988) model. Both curve **D** and band **E** appear clearly below the relative sea-level positions indicated by field data in this area, where, according to Gornitz and Seeberg (1990), a linear 1.87 mm/yr sea-level rise has occurred since almost 7000 yr BP.

For New York City, Newman et al. (1971) proposed relative sea-level curve **F**, based on data obtained by Newman (1966). For the same area, Newman et al. (1980) published curve **G**, obtained by joining available data points to each other with straight segments, whereas Gornitz and Seeberg (1990), using Newman's sea-level data, suggested a 2.17 mm/yr relative sea-level rise during the last 7000 yr BP (curve **G'**). For Great South Bay, on the south coast of Long Island, Rampino (1979) obtained relative sea-level curve **H** from new radiocarbon dates of basal peat or organic silt clay samples and other data compiled from the literature. Curve **H** suggests a drop in the submergence rate around 3000 yr BP.

In Iona Island (Hudson River Estuary, some 45 km north of New York City), Newman et al. (1971) proposed relative sea-level curve **I**, which seems to be based on only a few data points provided by Newman et al. (1969). Curve **J** for Hudson River by Newman et al. (1980) consists of straight segments connecting 13 data points, including those later than 6000 yr BP from Iona Island.

In New Haven Harbor (Connecticut), a relative sea-level rise of 10.5 to 12 m during the last 5900 years was inferred by Upson et al. (1964) from pollen analyses and one radiocarbon date of cored plant remains. Bloom and Stuiver (1963) used seven radiocarbon-dated samples from coastal Connecticut, most of them collected from the Hammock River tidal marsh near Clinton, to obtain submergence curve **K** which, as most sea-level curves from this region, does not show any evidence of pauses or reversals in the continuous submergent trend for the last 7000 years. The same data plus two additional data points enabled Newman et al. (1980) to produce curve **L** for the same area. Van de Plassche et al. (1989) brought new evidence showing a significantly faster submergence

Plate 62: NEW JERSEY - NEW YORK - CONNECTICUT

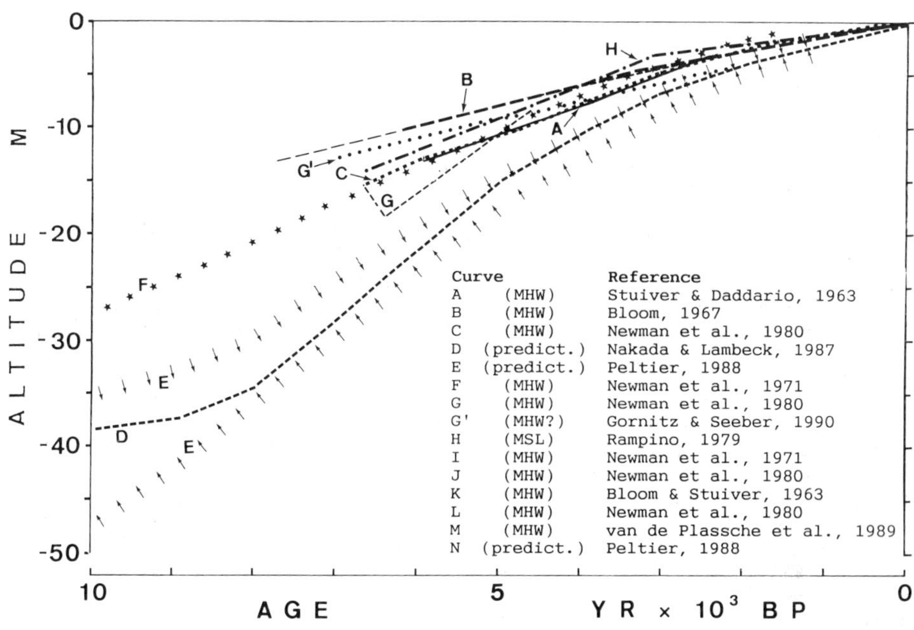

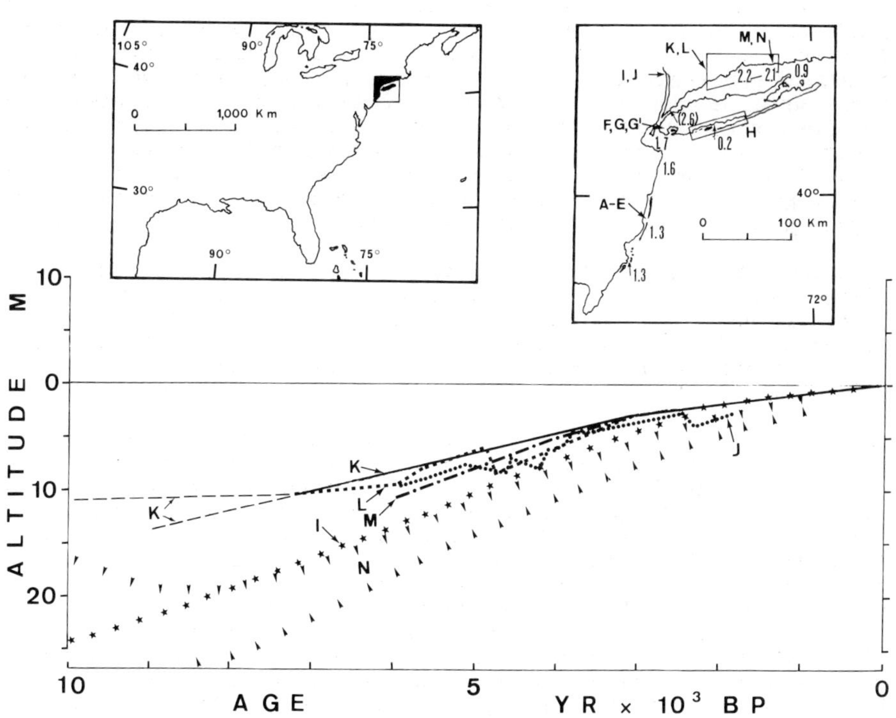

between 7000 and 3000 yr BP in this area (curve **M**) than previously believed; the difference in gradient between curves **K** and **M** was explained by contamination of two of the earlier samples with older residual peat. Lastly, sea-level band **N**, predicted by Peltier (1988) in the Clinton area, is most often below field data.

MASSACHUSETTS (Plate 63)

In southeastern Massachusetts, Oldale and O'Hara (1980) used ten radiocarbon dates of shells and freshwater peat, obtained from vibracores, to draw sea-level curve **A** (**Plate 63**), which suggests that the rate of relative sea-level rise dropped gradually, from 17 mm/yr around 10,000 yr BP to about 3 mm/yr between 6000 and 2000 yr BP.

In the Barnstable marshes area, Redfield and Rubin (1962) cored basal peat samples from five sites and inferred from vertical accretion rates of high-marsh *Spartina* peat the relative MHW curve **B**, which is supported by over 20 radiocarbon dates. According to **B**, the rise of the relative sea level has been continuous since at least 3700 yr BP, though an abrupt change in the rate of rise (from 3.0 mm/yr to 1.0 mm/yr) is suggested around 2100 yr BP. MHW band **C** was obtained by Redfield and Rubin (1962) by plotting sea-level data from the coast between Boston and New York. Curve **D** was deduced by Bloom (1967) from the Redfield and Rubin (1962) Barnstable data, after calibration of the radiocarbon dates; its time scale should be read therefore in calendar years and not in radiocarbon years BP as the other curves. Curve **E** results from a compilation of data for "Cape Cod" by Newman et al. (1980), which includes in fact data from Eastham, Barnstable and Nantucket.

In the Boston area, relative sea-level curve **F**, constructed by Kaye and Barghoorn (1964) using eight radiocarbon dates of organic remains, the elevations of which were corrected for the compaction of peat, shows a very limited rise (0.6 m) for the last 3000 years, following a more rapid rise (about 3 mm/yr) during the middle Holocene. Some late Pleistocene dates by Kaye and Barghoorn (1964) suggest that the Holocene sea-level rise may have been preceded by a relative sea-level drop between 14,000 and 10,000 yr BP. Curve **G** proposed by Newman et al. (1971) for the Boston area was obtained in fact by mixing data by Kaye and Barghoorn (1964) from Barnstable and by McIntyre and Morgan (1964) from Plum Island, and also by drawing a relative sea-level curve across points which correspond to freshwater samples. Early Holocene trends shown by curve **G** are therefore somewhat arbitrary. Nevertheless they are in qualitative agreement with sea-level data compiled for Boston by Quinlan and Beaumont (1981) (band **H**) and with the predictions of the model by Quinlan and Beaumont (1982) (curve **I**). Sea-level predictions for Boston by Nakada and Lambeck (1987) (curve **J**) and by Peltier (1988) (band **K**) also show a change from a relative sea-level drop to a rise during the early Holocene. Lastly curve **K'**, predicted by the Peltier (1991) model, does not seem to improve the fit to field data very much prior to 6000 yr BP.

In the Plum Island River marshes, detailed floral and faunal studies of sediment and peat samples recovered from six borings and eight radiocarbon dates enabled McIntire and Morgan (1964) to

Plate 63: MASSACHUSETTS

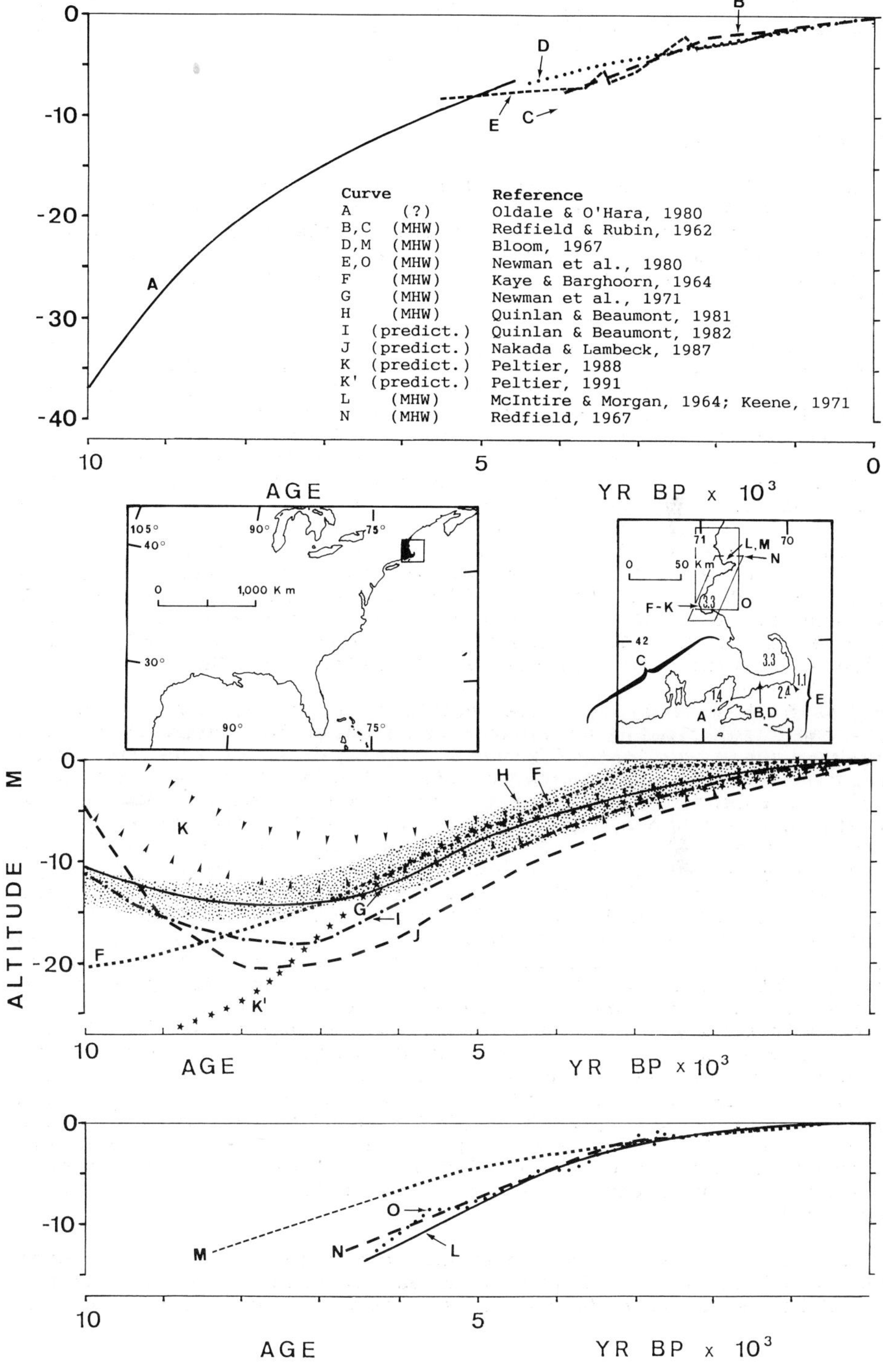

define some former MWH positions, which were subsequently used by Keene (1971) to construct submergence curve **L**. According to McIntire and Morgan (1964), reversal of local movement from land uplift towards subsidence occurred somewhere between 10,500 and 6300 yr BP and the relative sea-level rise which took place after 3000 yr BP, at a rate of about 1 mm/yr, is ascribed essentially to crustal downwarping. Curve **M** by Bloom (1967) is the same as **L**, but with a time scale expressed in calendar years instead of radiocarbon years BP. Curve **N** was constructed by Redfield (1967) for the "eastern coast of Massachusetts" by adding to the sea-level data from Plum Island six peat dates obtained from the Neponset River area. Lastly curve **O**, proposed for "northeastern Massachusetts" by Newman et al. (1980), is made of straight segments joining up 17 data points from Plum Island, Neponset and the Boston area already published by various authors.

NEW HAMPSHIRE - MAINE (Plate 64)

On the 25 km-long New Hampshire coast, submergence curve **A** (**Plate 64**) was proposed by Keene (1971) using four peat dates from the base of salt marshes near Hampton (rectangles **A** indicate uncertainty margins for the samples dated). These data suggest that, as in most of New England, the relative sea level has risen since 7000 yr BP, though at a rate decreasing gradually since about 4000 yr BP. At Odiorne Point, submergence curve **A'** was constructed by Harrison and Lyon (1963) on the basis of four radiocarbon dates of stumps from a drowned forest. At the Isles of Shoals, some 10 km off the New Hampshire coast, sea-level band **B** predicted by the Peltier (1988) model suggests the occurrence of a relative sea-level fall prior to 7000 yr BP.

Studies on late Quaternary sea-level changes in coastal Maine were reviewed by Belknap et al. (1987). According to these authors, levelling measurements, tide-gauge records and other evidence "suggest that the coast is being warped downward to the east, possibly due to non-glacially induced tectonics". Bloom (1963) proposed a postglacial crustal rebound curve which, superimposed on the Shepard (1960) "eustatic" curve, would "account for all of the known late-glacial and postglacial relative fluctuations of land and sea level in southwestern Maine"; for the late Holocene, most field data consist of submerged stumps. Bloom's (1963) data were transformed by Belknap et al. (1987) into relative sea-level curve **C**, which suggests a 50-m relative sea-level drop between 10,000 and 6000 yr BP (or a 80-m drop since the deposition of three shell samples dated about 12,000 yr BP), followed by a gradual rise during the past 6000 years. Newman et al. (1971), however, deduced from about the same data relative sea-level curve **D**, which shows a twice smaller early Holocene emergence for "Maine". Curiously, curve **E** for "SW Maine" by Newman et al. (1980), which is based again on the same data assembled by Bloom (1963), does not show any Holocene emergence.

Schnitker (1974), on the other hand, studying sonic subbottom profiles in the nearshore area of the Gulf of Maine, off the Kennebec River, suggested that a 65-m submergence could have occurred since about 9000 yr BP (curve **F**, transformed by Belknap et al., 1987). The 65-m lowstand was inferred from undated apparent subaerial erosion of Pleistocene sediments at that level. With new data, Belknap et al. (1987) suggested a similar 65-m

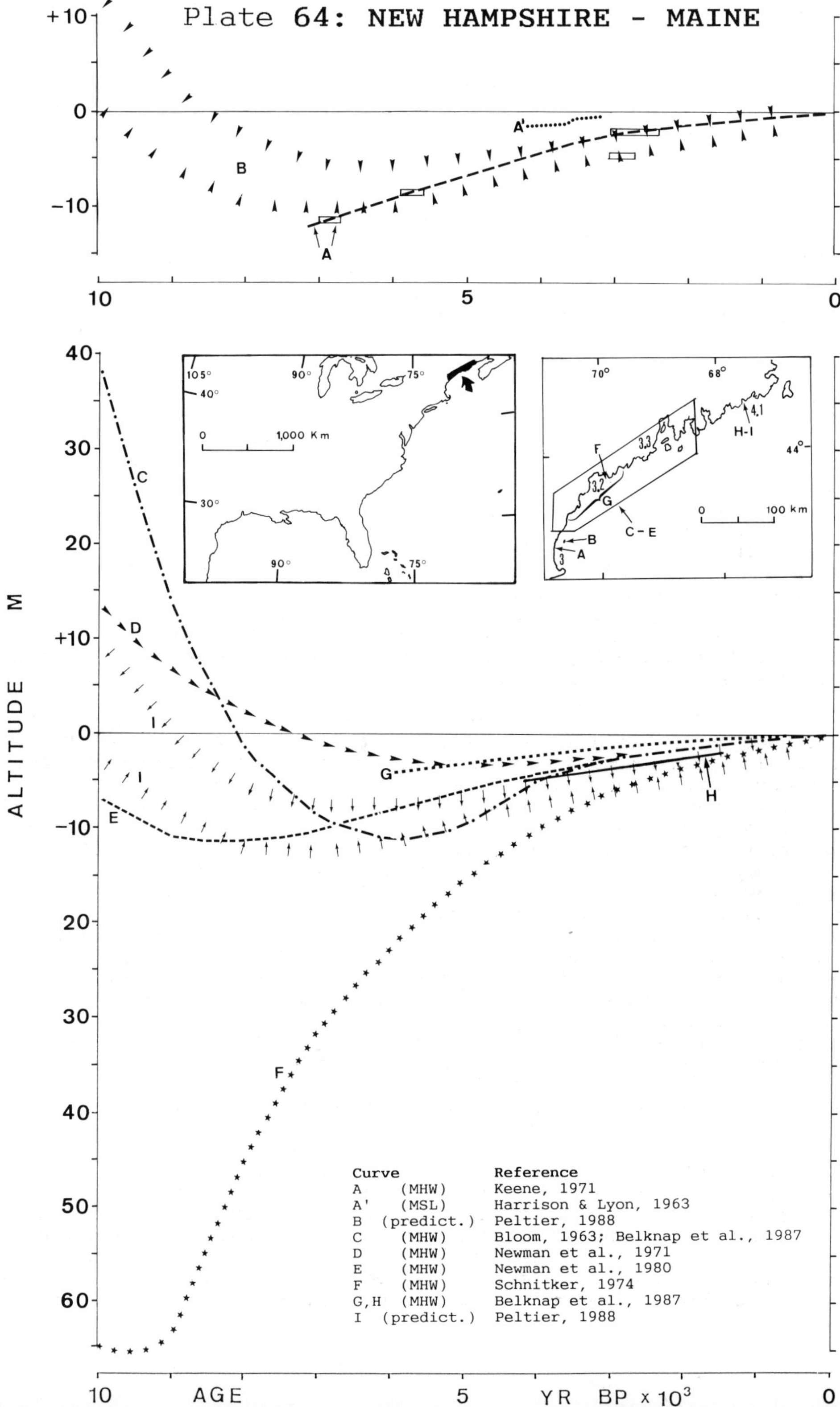
Plate 64: NEW HAMPSHIRE - MAINE
+10
0
-10
A'
B
A
10
5
0
40
30
20
+10
0
-10
20
30
40
50
60
ALTITUDE M
C
D
I
G
E
H
F
10
AGE
5
YR BP x 10³
0
105°
90°
75°
40°
0
1,000 Km
30°
90°
75°
70°
68°
4.1
H-I
3.3
F
3.2
G
44°
0
100 km
C-E
B
A
3
Curve
Reference
A (MHW) Keene, 1971
A' (MSL) Harrison & Lyon, 1963
B (predict.) Peltier, 1988
C (MHW) Bloom, 1963; Belknap et al., 1987
D (MHW) Newman et al., 1971
E (MHW) Newman et al., 1980
F (MHW) Schnitker, 1974
G,H (MHW) Belknap et al., 1987
I (predict.) Peltier, 1988

lowering. Though contrasting trends in nearby areas, such as those shown by curves **C** and **F**, are not impossible (e.g., see Plate 53), such a strong gradient of crustal movements would require confirmation. If it is correct however, the interpretation of "non-glacially induced tectonics" can hardly be the only explanation for differences in elevation of several dozen metres in a few thousand years in a relatively low seismic area such as coastal Maine.

For the Wells-Popham area, sea-level trend **G** is considered possible by Belknap et al. (1987). Between 4000 and 1500 yr BP, a sea-level rise trend of 1.2 mm/yr was calculated at Addison (Belknap et al., 1987) (curve **H**). In the same area, Peltier (1988) predicted sea-level band **I**.

DEVELOPMENT AND DEGLACIATION OF THE LAURENTIDE ICE SHEET

The extent of ice, the variation in ice thickness and the timing of glaciation and deglaciation of the Laurentide Ice Sheet is currently a subject of intense debate. According to Dyke et al. (1989), the last ice sheet grew from plateau-based centres in Baffin Island, Keewatin, Quebec-Labrador, and possibly northern Ontario. Subsequently northeastern Ontario and parts of northwestern Ontario would have been invaded by the Labrador sector of the ice sheet during buildup, followed by a bipartition of the sector to produce the Hudson and Labrador centres. During deglaciation, earlier retreat, mostly in the interval 14,000 to 10,000 yr BP, probably melted 50% of the volume of northern hemisphere ice sheets and made the interior plains of the St. Lawrence lowlands ice-free. A second stage of deglaciation occurred between 10,000 and 8000 yr BP.

An approach which is often used (e.g. Peltier, 1976; Clark, 1980a; Quinlan and Beaumont, 1982; Quinlan, 1985) is to deduce glacial history from the pattern of glacial isostatic rebound. Some of these models, of global applicability, are commented in the Introduction. Among the studies limited to the North American continent, Quinlan and Beaumont (1981) compared postglacial sea-level responses at six sites between Boston (Massachusetts) and northwestern Newfoundland Island with those expected on wastage of "maximum" versus "minimum" ice loads which correspond to ice shelf coverages proposed by Flint (1971) and Grant (1977) respectively. When the ice sheet retreats, the forebulge migration (Daly, 1934, Newman et al., 1971) will play an important role in determining the local sea-level history. Quinlan and Beaumont (1981) predicted the existence of four types of sea-level response, found in zones within and beyond the ice border (Fig. 2-39, the ice sheet producing the bulge retreats from right to left in the direction of the arrow): in zone A, sea level has been falling since deglaciation; in zone B an early fall is followed by a smaller rise: many emerged shorelines will occur here; in zone C an early fall is followed by a greater rise: no raised marine features will be found here; in zone D there will only be a relative sea-level rise.

In a subsequent paper, Quinlan and Beaumont (1982) refined their model by deducing ice distribution from geological evidence of late Wisconsinian ice limits and showed several stages in the deglaciation history. In the proposed reconstruction, little or no grounded ice is shown in the southern Gulf of St. Lawrence, and a thin ice layer, often less than 1 km thick, appears over much of the Canadian Atlantic area. As noted by Brookes et al. (1985), over the Maritime Provinces and the Gulf of St. Lawrence the ice limits were closer to the earlier (Quinlan and Beaumont, 1981) "minimum" model, while over Newfoundland, although the ice extent was reduced, thickness was closer to the "maximum" model used earlier. The major differences lay in the placement of a thick dome of ice north of the Gulf of St. Lawrence.

Later, Quinlan (1985) used the Quinlan and Beaumont (1982) approach to analyse the relative sea-level record in the region around Baffin Island (see Plates 70 and 72-75), with two principal objectives: first, to derive an ice reconstruction consistent with the known relative sea-level record; and second, to use this reconstruction to estimate the local sea-level trends for sites where direct sea-level indicators are either absent or ambiguous in their implications.

The highest level reached by the Late Pleistocene/early Holocene seas - called the marine limit - serves as a fundamental stratigraphic marker in studies of deglaciation. In areas within the maximum limit of the former ice sheets, the marine limit was formed successively as the ice retreated. This limit is therefore a strongly diachronous feature ranging in age from c. 14,000 to 5000 yr BP for the last deglaciation and between over 70,000 yr BP to the present for the highest raised marine deposits in Canada. The area of the highest Holocene rebound observed is still eastern Hudson Bay where the marine limit measured has been close to +300 m (Andrews, 1989) (Fig. 2-40).

Though relatively well represented in this Atlas, the sea-level curves concerning the Canadian Shield reproduced here are far from

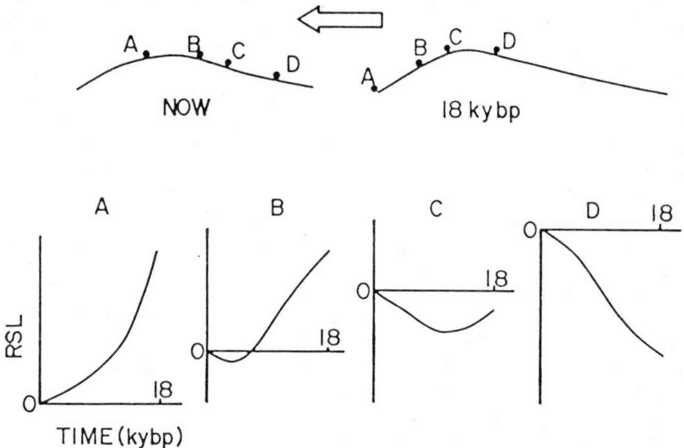

Fig. 2-39. Above: Schematic representation of a peripheral bulge at two points in time, 18,000 yr BP and now. The bulge migrates in the direction of the arrow, affecting the relative sea level at sites A, B, C and D. Below: The relative sea-level history at each of the sites A through D resulting from bulge migration (from Quinlan and Beaumont, 1981).

being exhaustive; some additional curves can be found in the work by Fulton (1989) and Andrews and Peltier (1989), which reached the present author only after the Canadian chapters and plates had already been completed.

NEW BRUNSWICK - NOVA SCOTIA (Plates 65 and 66)

Scott and Medioli (1980) applied the "Scandinavian" method in the Maritimes, which consisted in the use of sediments as sea-level indicators, at the transition between marine to freshwater in raised basins which were formerly below the present sea level. For the region comprising "northeastern Maine and southwestern New Brunswick", they obtained cores from five small lakes and provided

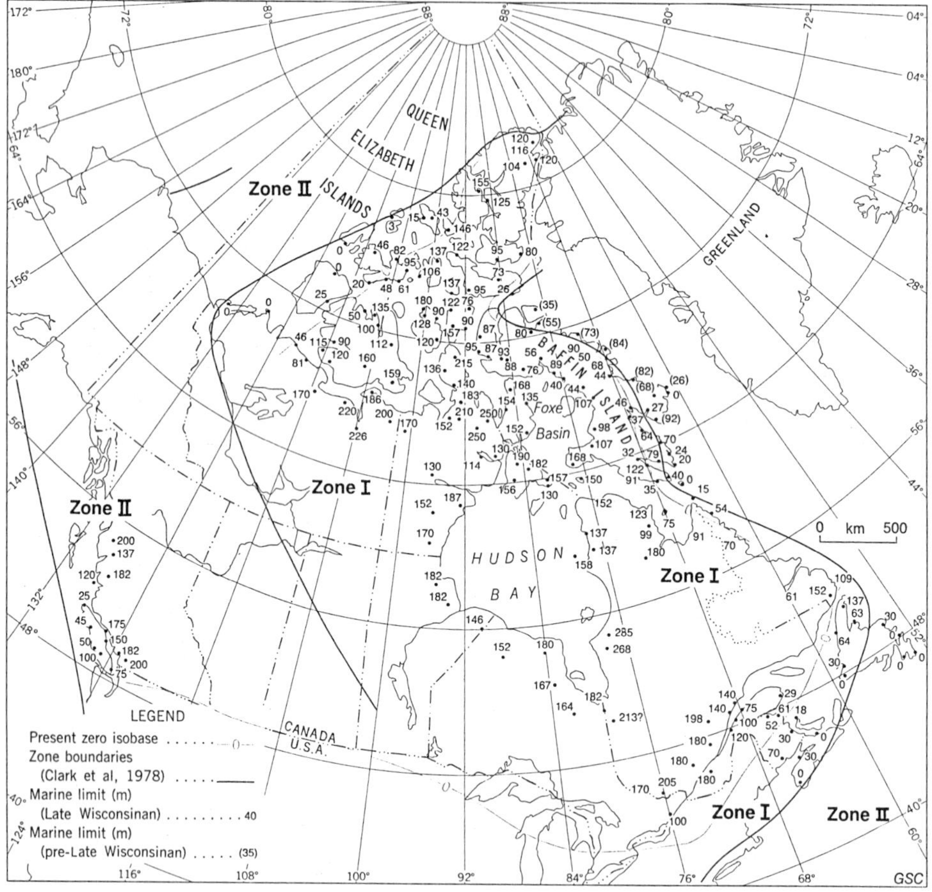

Fig. 2-40. Elevation (m) of marine limits from sites around Canada. Map also shows the boundaries of the various sea-level response zones of Clark et al. (1978) (see Fig. 1-4) and the location of the present zero. (From Andrews, 1989).

emergence curve **A** (**Plate 65**), which indicates an early Holocene emergence rate of about 5 mm/yr.

In southern New Brunswick, Grant (1980), inspired by data from Gadd (1973), proposed curve **B**, showing the occurrence of an initial postglacial emergence, until uplift waned in the early Holocene, then a prevailing submergence trend due to subsidence for the past 7000 years. Predictions by Quinlan and Beaumont (1982) for southeastern New Brunswick are summarized in relative sea-level curve **C**. In St. John, predictions by Peltier (1988) are shown by sea-level band **D**. A considerable divergence exists between curves **A**, **B**, **D**, and **C**; refinements of the existing estimations would certainly help to clarify this uncertain situation.

In a study aimed at determining the development of the tidal amplification in relation to the sea-level rise in the Bay of Fundy, Scott and Greenberg (1983), using radiocarbon-dated basal peat samples, produced relative sea-level curves **E** for Mary's Point, and **F** for Fort Beauséjour. Curve **F** is slightly different from submergence curve **G**, proposed for Fort Beauséjour in a previous study by Grant (1970). The difference between curves **F** and **G** is ascribed by Scott and Greenberg (1983) to a slumping which would have affected the section sampled by Grant (1970).

In Baie Verte, on the Northumberland Strait coast of New Brunswick, Scott et al. (1987a) obtained new cores in sites already studied by McRoberts (1968), determining former sea-level positions from faunal assemblages, which were radiocarbon-dated by six peat samples. According to their curve (**H**), the sea-level rise increased dramatically from 1 mm/yr between 5400 and 4500 yr BP to 10 mm/yr between 4500 and 4000 yr BP, then decreased between 4000 and 2000 yr BP to its present rate of 1.5 mm/yr. This complex sea-level record is explained by Scott et al. (1987a) by a locally complex ice history, with three separate ice caps thinning towards this area in glacial times.

In Nova Scotia, Harrison and Lyon (1963) dated fossil stumps from remaining drowned forests, at Fort Lawrence (**I**) and Grand Pré (**J**). Assuming that each stump was killed by rising salt water and that its radiocarbon age represented its true age at death, Harrison and Lyon (1963) constructed curves **I** and **J** respectively, suggesting a sequence of submergence and emergence at each site. At Kingsport, a relatively slow rate of sea-level rise was observed for the last 4400 yr BP by Scott and Greenberg (1983) (curve **K**). At Granville Ferry, relative sea-level curve **L** proposed by Scott and Greenberg (1983) is based on only three data points.

Belknap et al. (1987) combined the data used to plot curves **E**, **F**, **K**, **L** (Plate 65) and **A** (Plate 66), to obtain a linear regression trend (curve **M**, Plate 65) corresponding to the "Bay of Fundy" salt marsh data points prior to 2000 yr BP. Finally, Stea et al. (1987) proposed **N** as a composite relative sea-level curve for the "head of the Bay of Fundy", where Gornitz and Seeberg (1990) calculated, for the last 4000 years a sea-level rise trend of about 2 mm/yr (curve **N'**).

Near Chebogue Harbour, Scott and Greenberg (1983) determined four former HHW positions using marsh foraminiferal zones as indicators. They dated these boundaries with basal peats and obtained relative sea-level curve **A** (**Plate 66**).

Plate 65: NEW BRUNSWICK - NOVA SCOTIA I

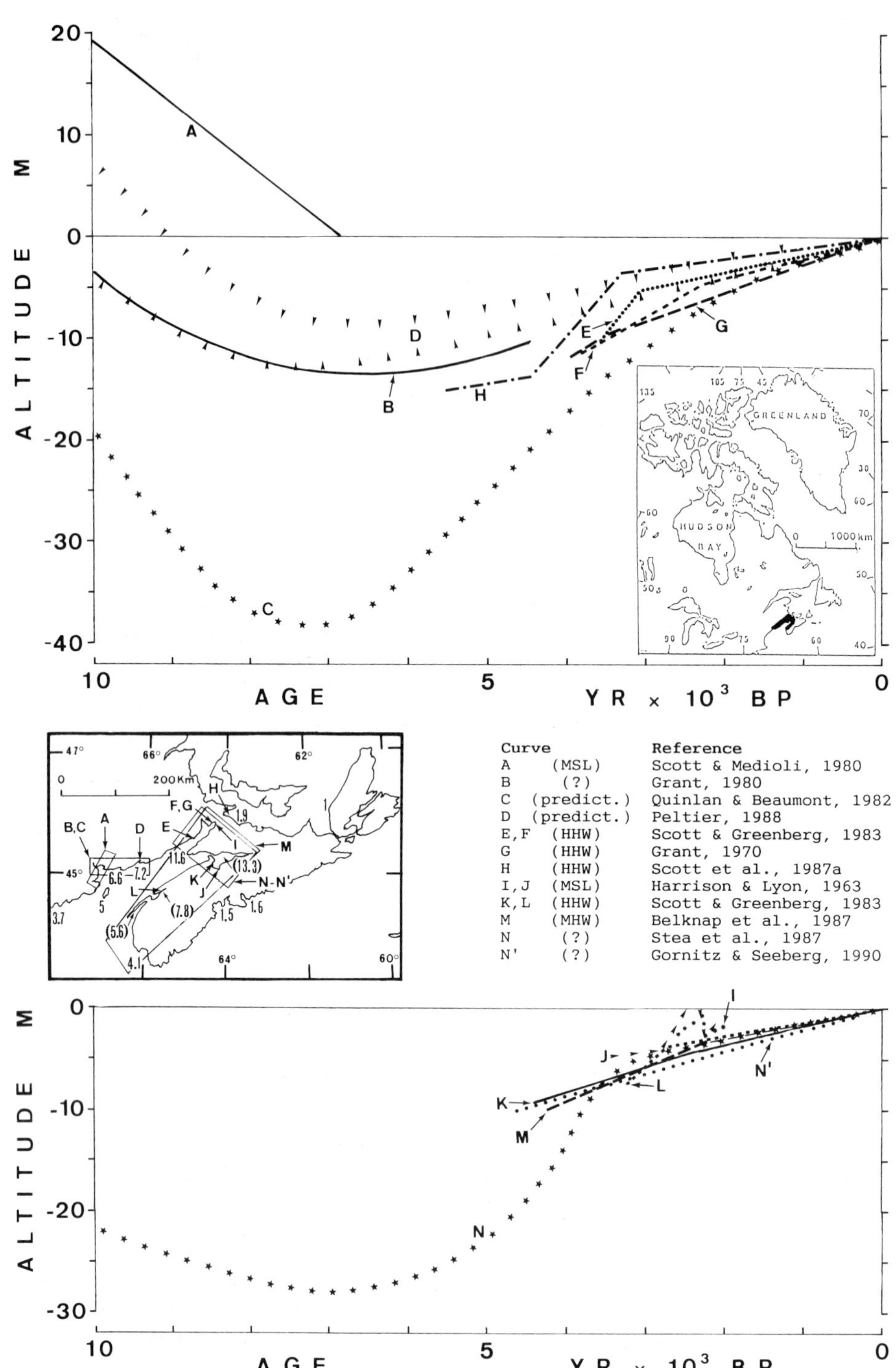

Plate 66: NOVA SCOTIA II

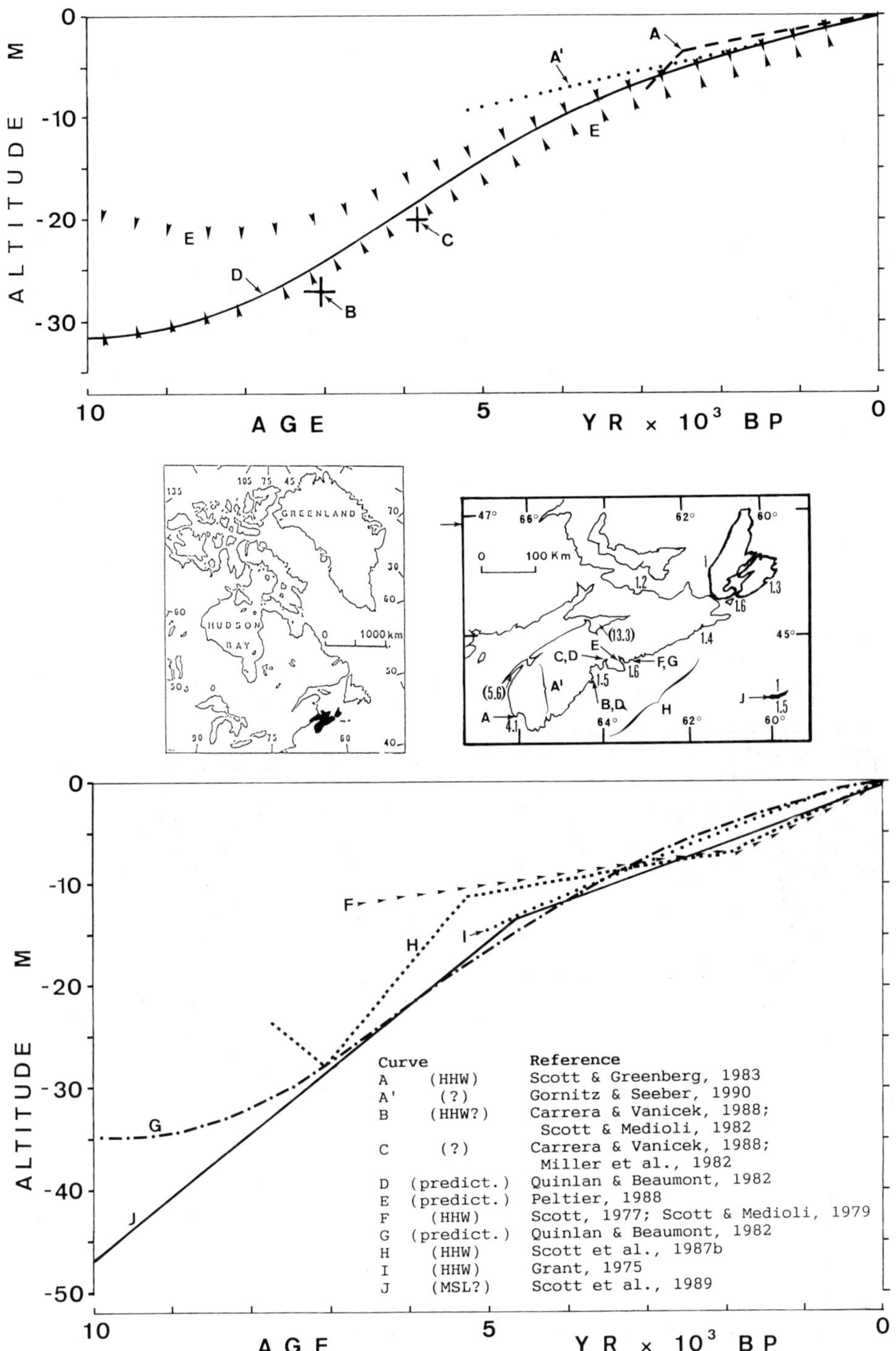

On the eastern coast of Nova Scotia, isolated sea-level data points were reproduced by Carrera and Vanicek (1988), for Lunenburg (**B**, from Scott and Medioli, 1982) and for the Bedford Basin (**C**, from Miller et al., 1982). These two data points can be compared with sea-level curve **D**, predicted by Quinlan and Beaumont (1982) for the "Lunenburg Bay-Bedford Basin" and with band **E**, predicted by Peltier (1988) for Halifax.

In the Chezzetcook Inlet area, HHW curve **F** provided by Scott (1977) and Scott and Medioli (1979) can be compared with curve **G**, predicted by Quinlan and Beaumont (1982). **H** is a composite sea-level curve for the eastern shore of Nova Scotia proposed by Scott et al. (1987b), whereas **I** is a HHW curve proposed by Grant (1975) for the "Maritimes".

At Sable Island, located on the outer Scotian shelf over 200 km from the Nova Scotian coast, curve **J** was obtained by Scott et al. (1989) from a long borehole sequence and nine radiocarbon dates of peat samples, some of which had already been published by Scott et al. (1984). The relative sea-level rise shown by **J** indicates a rate decreasing from 6.2 mm/yr, between 8100 and 4500 yr BP, to 3.0 mm/yr from 4500 yr BP to the present.

PRINCE EDWARD ISLAND (Plate 67, above)

In the eastern part of Prince Edward Island, which seems to have experienced only submergence during the last 13,000 years, Scott and Medioli (1979) determined sea-level trends for the the last 2000 years at Orwell Bay (curve **A**) and Pisquid (curve **B**), using cored marsh material described in a previous publication (Scott and Medioli, 1978). The rates of relative sea-level rise suggested are of 1.4 mm/yr at Orwell Bay and of 1.9 mm/yr at Pisquid. For comparison, tide-gauge records indicate a slightly higher submergence rate (2.9 ± 0.6 mm/yr) at the nearby Charlottetown station (Pirazzoli, 1986b) (Fig. 2-41, curve 47).

Quinlan and Beaumont (1982) summarized existing field data in eastern Prince Edward Island with relative sea-level band **C**, and their calculated trend with curve **D**, which predicts the occurrence of emergence in the late Pleistocene. The western part of Prince Edward Island, where emerged features (indicating a formerly higher sea level) have been observed, obviously experienced first an emergence in post-glacial times, followed by submergence. In the Percival River area, according to Scott and Medioli (1979), only the latest submergence trend could be confirmed, with a rate of relative sea-level rise of about 0.8 mm/yr for the last 3000 yr BP (curve **E**: uncertainty margins of two dated samples are indicated by filled boxes). Quinlan and Beaumont (1981, 1982) proposed sea-level band **F** (which is indeed quite wide, especially for the early Holocene) to summarize observed sea-level data available in western Prince Edward Island. Curve **G** corresponds to the sea-level predictions by the Quinlan and Beaumont (1982) model for the same region; for the last 2000 years, their predictions almost coincide with curve **A**. At Tignish, in the westernmost part of the island, predictions by the Peltier (1988) model are shown by band **H**, which suggests the occurrence of emergence during the early Holocene and of relative sea-level stability since about 7000 yr BP. Lastly, curve **I** was proposed by Grant (1980) for "Prince Edward Island".

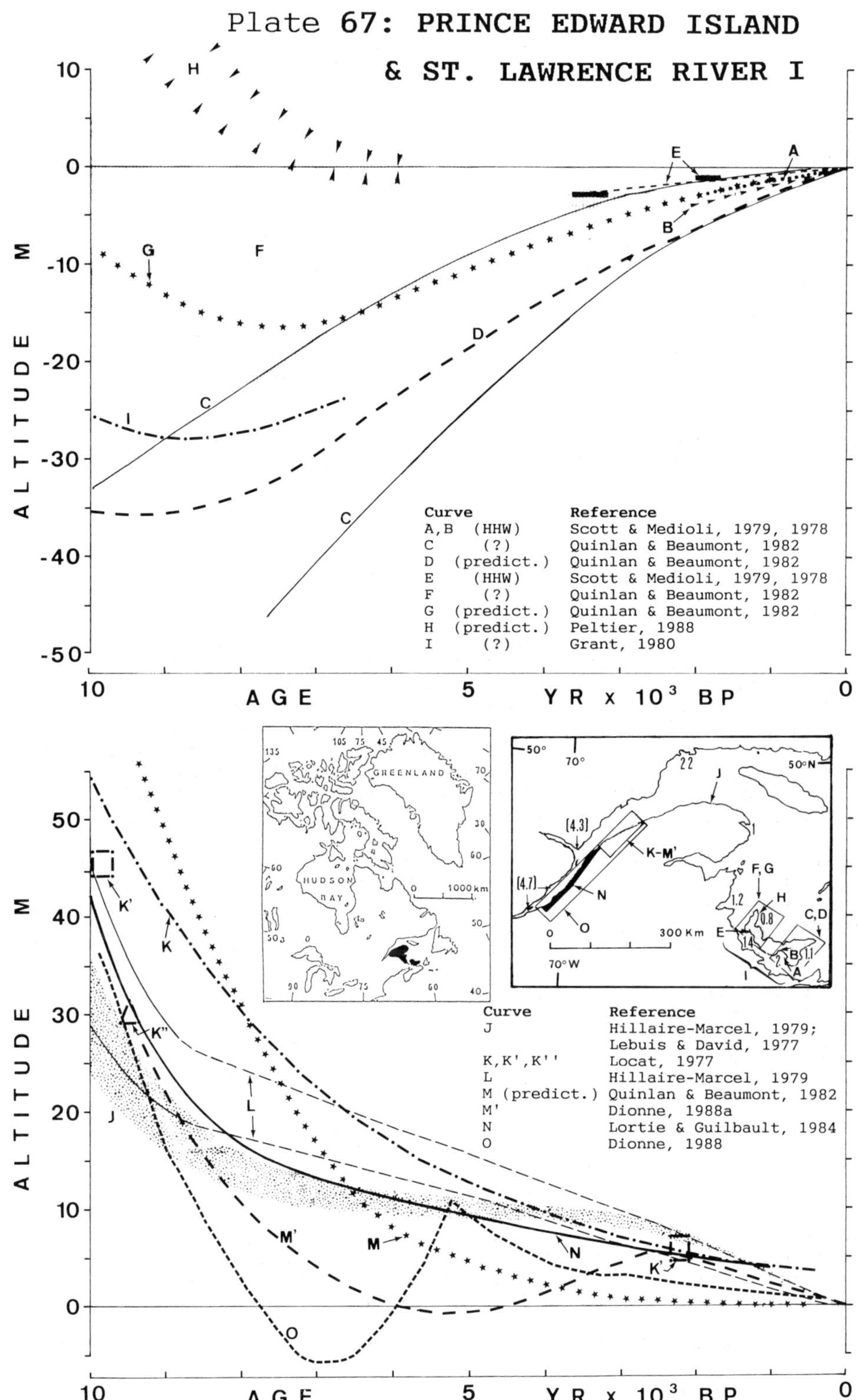
Plate 67: PRINCE EDWARD ISLAND
& ST. LAWRENCE RIVER I
ALTITUDE M
AGE YR x 10³ BP
Curve | Reference
A,B (HHW) | Scott & Medioli, 1979, 1978
C (?) | Quinlan & Beaumont, 1982
D (predict.) | Quinlan & Beaumont, 1982
E (HHW) | Scott & Medioli, 1979, 1978
F (?) | Quinlan & Beaumont, 1982
G (predict.) | Quinlan & Beaumont, 1982
H (predict.) | Peltier, 1988
I (?) | Grant, 1980
GREENLAND
HUDSON BAY
1000 km
300 Km
70°W
K-M'
F,G
C,D
Curve | Reference
J | Hillaire-Marcel, 1979; Lebuis & David, 1977
K,K',K'' | Locat, 1977
L | Hillaire-Marcel, 1979
M (predict.) | Quinlan & Beaumont, 1982
M' | Dionne, 1988a
N | Lortie & Guilbault, 1984
O | Dionne, 1988

Plate 68: ST. LAWRENCE RIVER & GULF II

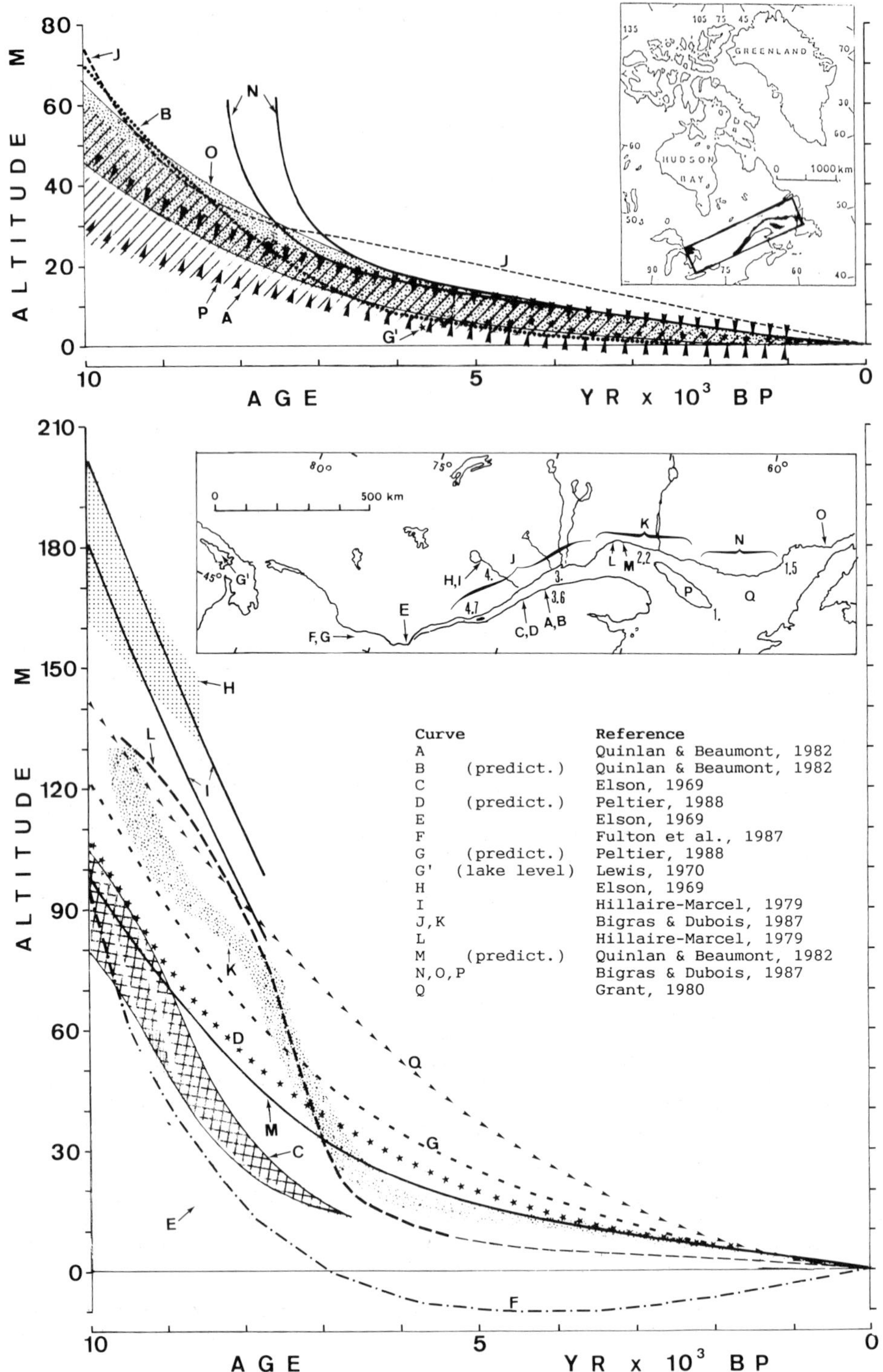

ST. LAWRENCE RIVER & GULF (Plates 67 & 68)

Near Sainte-Anne-des-Monts, in the southern St. Lawrence estuary, field data provided by Lebuis and David (1977) were assembled by Hillaire-Marcel (1979), who produced emergence band **J** (**Plate 67**, lower graph). In the Baie des Sables-Trois Pistoles area, where the marine limit (about 13,500 yr BP old) is found between +120 and +166 m, Locat (1977) used 11 radiocarbon-dated shell samples to construct emergence curve **K**, the Holocene part of which is based however on only two new data points (shown in Plate 67 by boxes **K'**), a third data point being compiled from Dionne (1972) (triangle **K''**). The difference in elevation between the dated samples **K'** and **K''** and curve **K** is due to the fact that most shell samples dated were not in living position, so that the sea level associated with them had to be determined from other features (beach ridges, terraces, paleodeltas, etc.). In the same region and using the same data, Hillaire-Marcel (1979) obtained however quite a different emergence band (**L**) and Quinlan and Beaumont (1982) predicted curve **M** (see also curves **A** and **B**, Plate 68). At Saint-Fabien-sur-Mer, Dionne (1988a) deduced from sand deposits containing *Mya* shells in growth position that the sea level was similar to the present one between 5800 and 4750 yr BP. According to him, this low sea level was followed by a transgression episode, reaching +4 to +8 m between 2500 and 2000 yr BP, then by a gradual relative sea-level drop to the present zero (curve **M'**).

In the area between Trois Pistoles and Saint-Eugène, Lortie and Guilbault (1984) deduced average emergence curve **N** from the study of diatoms and foraminifera in raised marine post-glacial sediments. Of the several samples dated, however, only one gave an age younger than 9500 yr BP and former sea-level positions had to be estimated using *ad hoc* paleobathymetric criteria; this makes curve **N** approximate.

More detailed for the Holocene period, relative sea-level curve **O** was obtained by Dionne (1988) for the Montmagny-Baie des Sables area, using stratigraphic and geomorphological data supported by 55 radiocarbon dates of stumps, logs and basal peats. Curve **O** shows a low sea level around 7000 yr BP, followed by a 20-m sea-level rise, culminating with a short high stand and, after 4400 yr BP, again coastal emergence. Though the cause of the mid-Holocene resubmergence of the St. Lawrence estuary remains unexplained, some similarities can be found between curve **O** and certain emergence curves from the Atlantic coasts of North Europe (see Plates 1, 3, 5, 7, 8).

Quinlan and Beaumont (1981, 1982) deduced from the data assembled by Locat (1977) (i.e., for the last 10,000 years, from **K'** and **K''**, see Plate 67), relative sea-level band **A** (**Plate 68**), that they ascribed at Rimouski, and compared band **A** with curve **B**, which they calculated for the same locality. For comparison, curve **B** (Plate 68) is the same as curve **M** (Plate 67), though at a different vertical scale.

At Rivière du Loup, Elson (1969) suggested that crustal uplift could be represented by band **C**, which was interpreted from data published by Lee (1962). In the same place, predictions by the Peltier (1988) model are shown by curve **D**.

At Montréal, local crustal uplift was summarized by Elson (1969) with band **E**. In the Ottawa area, where marine data are available only until shortly after 10,000 yr BP, when the Champlain Sea regressed, relative sea-level curve **F** was proposed by Fulton et al. (1987).

Curve **F**, which suggests the occurrence of a submergence episode during the late Holocene, is very different from the predictions for Ottawa by Peltier (1988), which show continuous emergence throughout the Holocene (curve **G**). Also shown in Plate 68, for comparison, are the changes in the elevation of water in the Huron Lake at its outlet, from 5500 yr BP to the present (curve **G'**), obtained by Lewis (1970): the upward movement proceeded here at an almost constant rate of 2.8 ± 0.8 mm/yr during the period considered. For southeast Lake St. John, emergence envelopes were proposed for the early Holocene by Elson (1969) (band **H**) and by Hillaire-Marcel (1979) (band **I**), both based on the data provided by Lassalle and Rondot (1967).

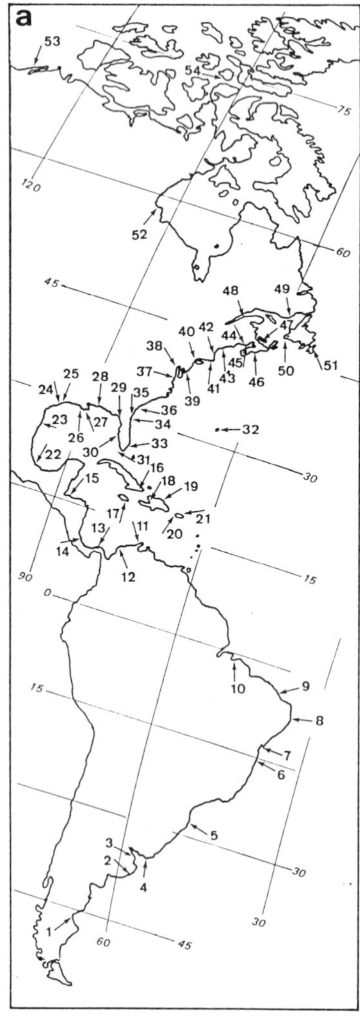

Fig. 2-41.
a) Location of tide-gauge stations on the east coasts of the Americas. **b**) Average 5-yr MSL variations on the east coasts of the Americas as recorded by tide-gauge stations:
1) Comodoro Rivadavia; 2) Mar del Plata; 3) Buenos Aires; 4) Montevideo; 5) Imbituba; 6) Cannavieiras; 7) Salvador; 8) Recife; 9) Fortaleza; 10) Belem; 11) Riohacha; 12) Cartagena; 13) Cristobal; 14) Puerto Limon; 15) Puerto Plata; 16) Guantanamo; 17) Port Royal; 18) Port-au-Prince; 19) Puerto Cortes; 20) Mayaguez; 21) San Juan; 22) Vera Cruz; 23) Port Isabel; 24) Freeport; 25) Galveston; 26) Eugene Island; 27) Bayou; 28) Pensacola; 29) Cedar Keys; 30) St. Petersburg; 31) Key West; 32) St. Georges; 33) Miami; 34) Mayport; 35) Fernandina; 36) Charleston; 37) Hampton Roads; 38) Baltimore; 39) Atlantic City; 40) New York; 41) Newport; 42) Boston; 43) Portland; 44) Eastport; 45) St. John NB; 46) Halifax; 47) Charlottetown; 48) Pointe-au-Père; 49) Harrington H.; 50) Port-aux-Basques; 51) St. John Nfld.; 52) Churchill; 53) Tuktoyaktuk; 54) Resolute.
(From Pirazzoli, 1986b).

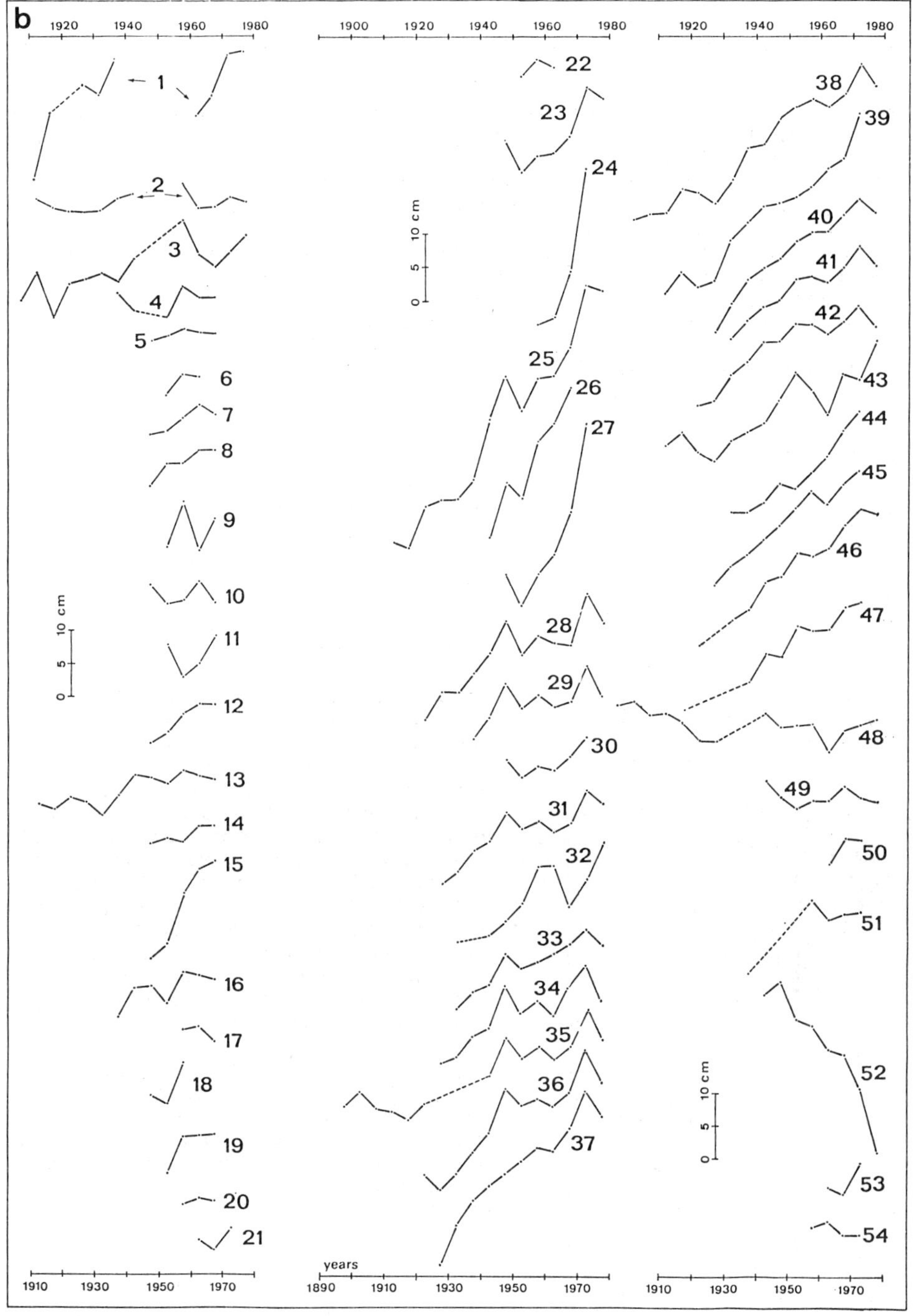
b
1920 1940 1960 1980
1900 1920 1940 1960 1980
1920 1940 1960 1980
1
2
3
4
5
6
7
8
9
10
11
12
13
14
15
16
17
18
19
20
21
22
23
24
25
26
27
28
29
30
31
32
33
34
35
36
37
38
39
40
41
42
43
44
45
46
47
48
49
50
51
52
53
54
0 5 10 cm
years
1910 1930 1950 1970
1890 1910 1930 1950 1970
1910 1930 1950 1970

Bigras and Dubois (1987) assembled and published in a standard form a detailed description of 272 radiocarbon dates, 73 of which are new, for the northern part of the Estuary and the Gulf of St. Lawrence. This material enabled them to propose five updated emergence curves for the region considered: curve **J** for the Haute Côte, band **K** for the Moyenne Côte, band **N** for the Natashquan/Harrington area, band **O** for the Blanc-Sablon area, and band **P** for Anticosti Island. On the Moyenne Côte, band **K** can be compared with curve **L**, proposed by Hillaire-Marcel (1979) for the Sept-Iles area, and with curve **M**, predicted by Quinlan and Beaumont (1982) at Moisie. Curve **Q** was proposed by Grant (1980) for the "northern Gulf of St. Lawrence".

NEWFOUNDLAND ISLAND (Plate 69)

The east coast of Newfoundland Island provides a clear example of a transition area between predominating Holocene emergence (in the north) and submergence (in the south). Emergence band **A** (**Plate 69**), which is based on a dozen radiocarbon dates of marine shells collected in life position from beach formations, was proposed by Grant (1980) for "northern Newfoundland". In the same study, relative sea-level changes for "NW Newfoundland" were summarized by curve **B**.

In "northwestern Newfoundland", Quinlan and Beaumont (1982) compared relative sea-level data (summarized by Quinlan and Beaumont, 1981) (band **C**) with the predictions of their 1982 model (curve **C'**). In the same area, band **D** predicted by Peltier (1988) appears to be slightly below the observed data. More localized curves **E**, **F**, **G**, **H** and **I**, about which no details are known to the present writer, were proposed by Grant (unpublished?).

In St. George's Bay observed relative sea-level data compiled by Quinlan and Beaumont (1981) (band **J**) are compared with the predictions of the Quinlan and Beaumont (1982) model (curve **J'**), which appear poorly constrained, especially in the late Holocene. In the same area new field data collected by Brookes et al. (1985) made possible the construction of relative sea-level curve/band **K**, which deviates considerably from Quinlan and Beaumont (1982) predictions **J'**. Again in St. George's Bay, sea-level predictions by the Peltier (1988) model (band **L**) show only emergence during the Holocene, and deviate therefore from field data.

On the east coast of Newfoundland Island, Shaw and Forbes (1990) dated a dozen samples of organic deposits collected by coring at four sites between Hamilton Sound and Bonavista Bay, showing (band **M**) that the relative sea level fell below the present datum in the late Pleistocene and that, as in the southeastern part of the island, submergence predominated during most of the Holocene: the relative sea level was still below -4 m at 5500 yr BP and reached approximately -0.7 m by 3000 yr BP.

Finally the model by Nakada and Lambeck (1987) predicts relative sea-level curve **N** in "Newfoundland". However, the part of the island for which this prediction was calculated is not specified.

Plate 69: NEWFOUNDLAND ISLAND

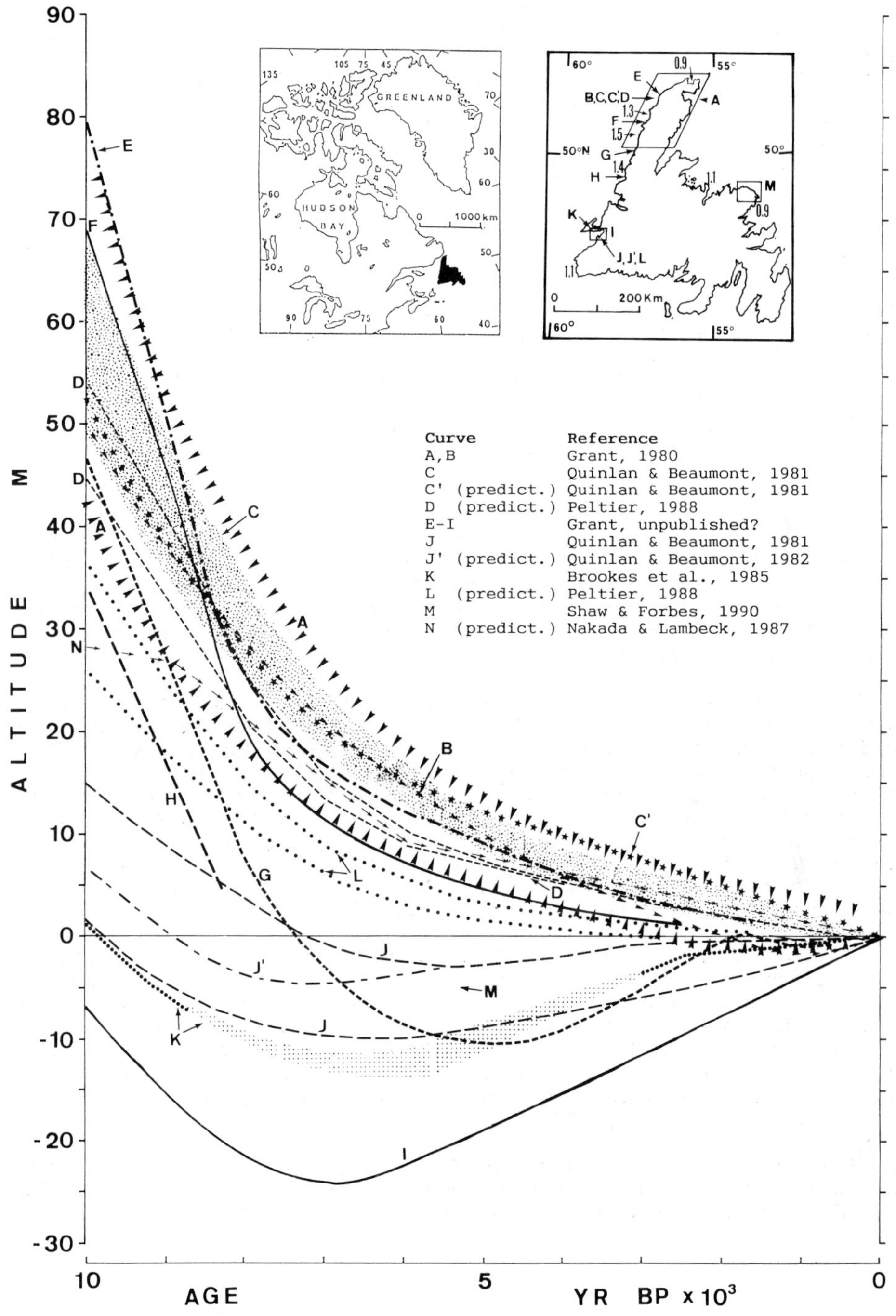

Plate 70: LABRADOR

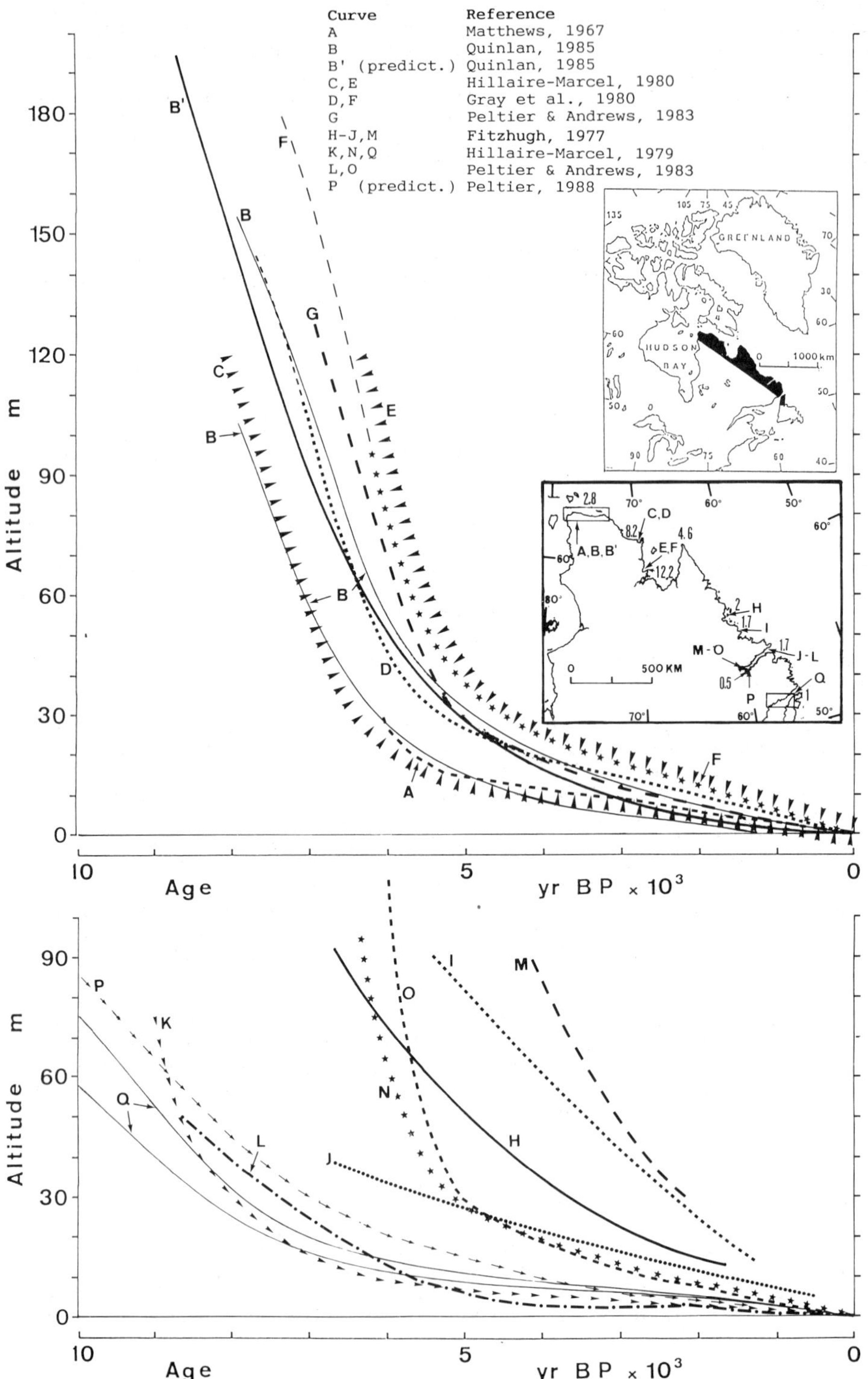

LABRADOR (N & E) (Plate 70)

In northern Ungava, curve **A** (**Plate 70**), showing a relative sea-level drop of 2.9 mm/yr for the past 5200 years, was obtained by Matthews (1967) using three ^{14}C dates of molluscs from raised marine terraces. The part of curve **A** prior to 6000 yr BP, which was constructed by deducting assumed values of a eustatic sea-level rise, in order to represent an isostatic uplift curve, is not reproduced in Plate 70. According to Quinlan (1985) field data from Ungava Peninsula fall within relative sea-level band **B**, and can be compared with curve **B'** predicted by the Quinlan (1985) model.

On the western coast of Ungava Bay a tilting of the land towards the SW appears. According to Hillaire-Marcel (1980) field data are comprised between emergence curve **C** for Koartac and Diana Bay, and curve **E** for Hopes Advance Bay. The latter curve is very similar to **F**, found in the same area by Gray et al. (1980); on the other hand curve **D** for Diana Bay-Wakeham Bay by Gray et al. (1980) shows a much steeper emergence than **C**. For Ungava Bay (site not specified), Peltier and Andrews (1983) proposed relative sea-level curve **G**, suggesting a 70-m emergence since 6000 yr BP and a 130-m emergence since 7000 yr BP, the latter value being deduced from inland data.

On the central Labrador coast a great variability in uplift rates appears. Fitzhugh (1977) used archaeological data - evidence of prehistoric occupations for the last 7000 years - and radiocarbon dates of shells, peat and whalebones to construct three emergence curves for the outer coast: **H** for Nain, **I** for Windy Tickle-Davis Inlet, and **J** for Groswater Bay. The relative steepness of curve **I** suggests that either a greater ice thickness, or a longer persistence of the ice cover occurred there.

In the outer Hamilton Inlet, the emergence history was summarized by Hillaire-Marcel (1979) with curve **K** and by Peltier and Andrews (1983) with curve **L**, which are quite different from curve **J** for the nearby Groswater Bay (located at the extreme eastern end of Hamilton Inlet). In the inner part of Hamilton Inlet there is a striking difference between curve **M** obtained by Fitzhugh (1977) in the North West River area and curves **N** and **O**, which were proposed by Hillaire-Marcel (1979) and by Peltier and Andrews (1983) respectively. Predictions by the Peltier (1988) model in the same area (Goose Bay) are shown by curve **P**. Lastly, emergence band **Q** was proposed for the Belle-Ile Strait area by Hillaire-Marcel (1979).

The variability in uplift rates on the Labrador coasts suggests different glacio-isostatic relaxation histories, in relation to the distance from significant ice masses which, according to Peltier and Andrews (1983), were probably located close to the central-east sector of the Labrador-Ungava plateau.

HUDSON BAY (Plates 71 and 72)

In the eastern part of Hudson Bay, curve **A** (**Plate 71**) was reported by Andrews (1987), without details, as an example in Arctic Canada that "the change of relative sea level after ice withdrawal has frequently been graphed as a smoothly decelerating curve". In the

Plate 71: HUDSON BAY I

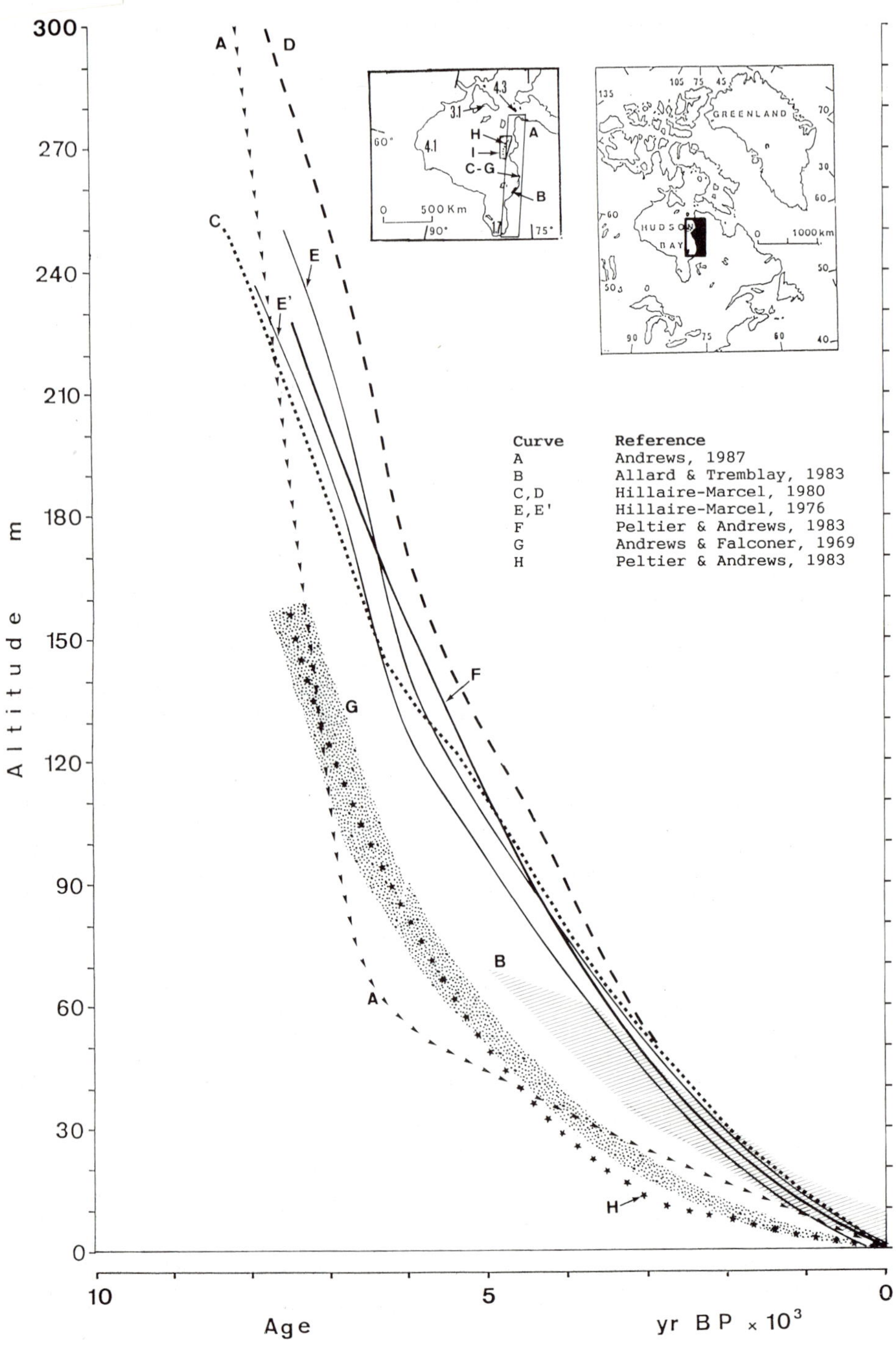

Manitounuk Islands, according to Allard and Tremblay (1983), who made a detailed study on Holocene coastal dynamics, supported by over 60 radiocarbon dates, emergence began at about 6000 yr BP. Rapid in the beginning, the uplift slowed down afterwards and the emergence rate remained nearly stationary (about 10 mm/yr) for the last 2800 years (band **B**).

In the Richmond Gulf, where a spectacular "staircase" air-photo of raised beaches showing clearly the main drops of relative sea level on the eastern side of Richmond Gulf Peninsula was provided by Hillaire-Marcel (1979, 1980), various relative sea-level curves are available. A series of strandlines located not far from the site was used by Hillaire-Marcel (1979, 1980) to establish emergence curves **C** (expressed in sideral years BP) and **D** (expressed in radiocarbon years BP), based on a dozen radiocarbon dates, which show the occurrence of four successive phases. The oldest phase corresponds to a relatively rapid emergence, during which the high rate of isostatic uplift was partially compensated by the rising eustatic sea level. Then, around 6000 yr BP a strong inflection appears, followed by an uplift acceleration beginning around 5500 yr BP. This is ascribed by Hillaire-Marcel to a composite isostatic readjustment, the Richmond Gulf area being situated between two distinct ice loads the melting of which was not synchronous. Finally, gradual decrease in isostatic uplift appears during the last few thousand years.

Using several series of beaches in the Richmond Gulf, Hillaire-Marcel (1976) defined a regional emergence band for the whole gulf area, which is delimited above by curve **E**, deduced from data in the S-SW part of the gulf, and below by curve **E'**, deduced from N-NW data, thus indicating a tilting of the basement rock towards S-SW. Relative sea-level curve **F** for Richmond Gulf by Peltier and Andrews (1983), based on data by Hillaire-Marcel and Vincent (1980), is curiously traced above curve **C** and band **E-E'** between 6300 and 5000 yr BP, suggesting plotting inaccuracies of ±15 m at the scale used.

In the Ottawa Islands, which were deglaciated between 7610 and 7250 yr BP, Andrews and Falconer (1969) proposed emergence band **G** for Gilmour Island, with the support of nine radiocarbon dates of *in situ* marine mollusc samples whose elevations agree closely with calculated uplift curves. Later, Peltier and Andrews (1983) proposed for the "Ottawa Islands" curve **H**, which seems based on the same data as band **G**, but was drawn below this band between 5000 and 2000 yr BP.

In the southeast Hudson Bay area, uplift curve **A** (**Plate 72**), proposed by Elson (1969) using data published by Lee (1962) suggests that emergence since 7000 yr BP is of the order of 250 m (if eustatic changes are deducted from curve **A**). In the same area, sea-level data points deduced by Walcott (1977) from dated *Mytilus* shell samples, represented here with error margins as boxes **B**, suggest that emergence did not exceed 200 m since 7000 yr BP and 230 m since 8000 yr BP. Nevertheless in the east coast of James Bay, according to curve **C** by Hillaire-Marcel (1979), who summarizes earlier data by other authors, emergence would be about 210 m since 7000 yr BP and almost 300 m since 8000 yr BP.

Near Cape Henrietta Maria, where no marine limit elevations and ages are available, emergence curve **D** was obtained by Webber et al. (1970) as "best" fitting, with assumptions, two radiocarbon

Plate 72: HUDSON BAY II

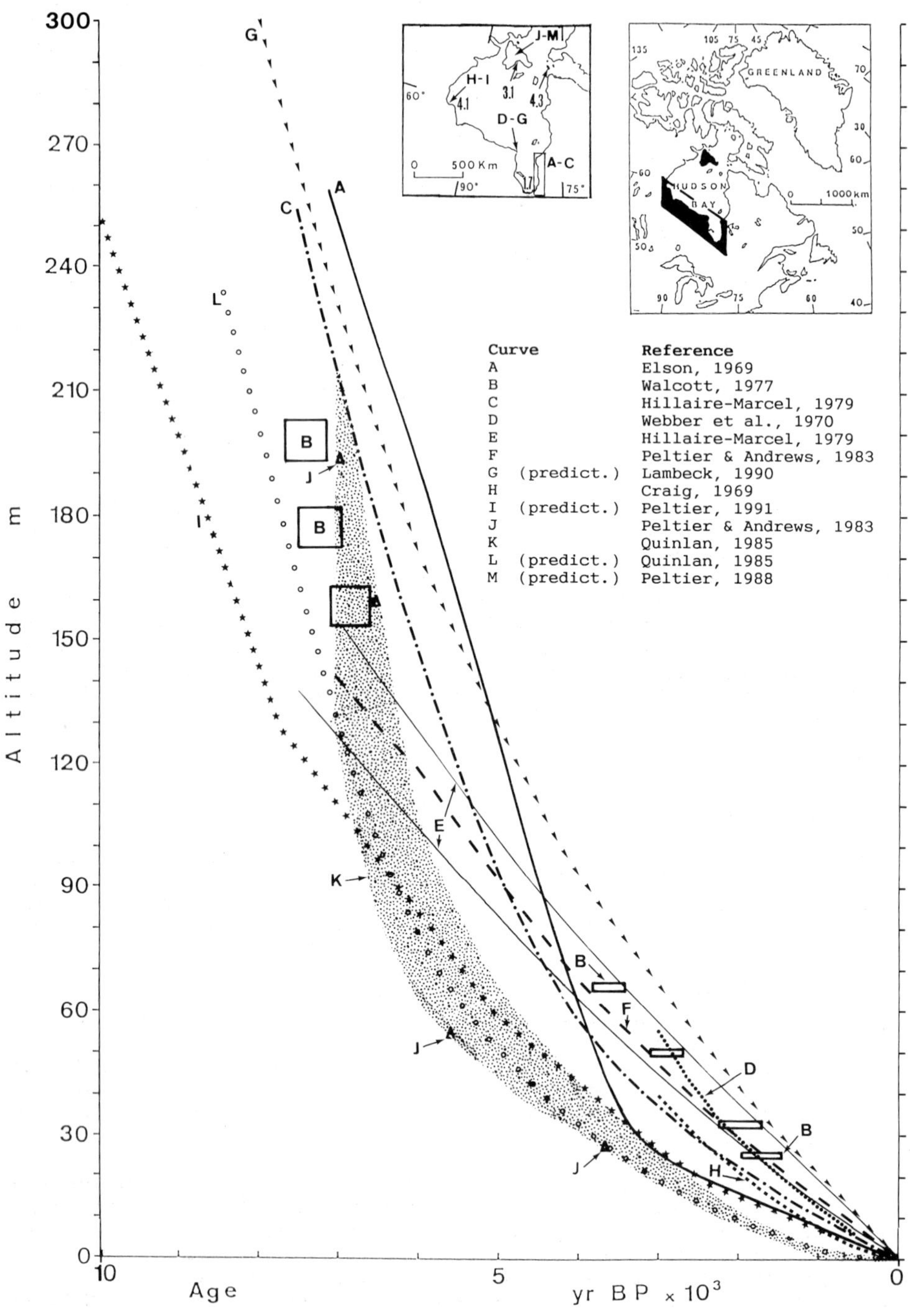

dates of *Mytilus edulis* shells from carefully levelled raised marine features (whereas four other dates of marine shells and driftwood were considered invalid). Curve **D** suggests that during the last 1000 years uplift has been about 12 mm/yr. Extrapolation of **D** to likely dates of deglaciation (8000-7000 yr BP) indicates hypothetical elevations of the marine limit above 300 m, which are the highest estimations available for eastern and arctic Canada and support the hypothesis that a center of uplift and ice-loading was situated in southeastern Hudson Bay and northern James Bay. However, according to a compilation made by Hillaire-Marcel (1979) of data by Webber et al. (1970) and Walcott (1972a), emergence since 7000 yr BP would be less than 150 m at Cape Henrietta Maria (band **E**). Relative sea-level curve **F** for the same place, proposed by Peltier and Andrews (1983) quoting the same sources as Hillaire-Marcel (1979) falls fortunately inside band **E**, while curve **G** obtained by the Lambeck (1990) rheological model at the same site predicts a much larger emergence.

In the Churchill area, where tide-gauge records prove that the present-day emergence rate is 6.6 ± 1.5 mm/yr, radiocarbon dates on marine shells from shoreline deposits indicate that, for the period 3000 to 1000 yr BP, the relative sea level was dropping at a rate of about 16 mm/yr (curve **H**, by Craig, 1969). In the same region, curve **I** corresponds to predictions by the Peltier (1991) model.

In Southampton Island, Peltier and Andrews (1983) obtained seven relative sea level data points (triangles **J**) indicating that almost 200-m emergence occurred since 7000 yr BP. Quinlan (1985) made in Southampton Island a comparison between envelope **K**, representing likely limits on the positions of observed relative sea-levels, and curve **L** predicted by his model, which shows observed emergence rates greater than expected by model computation earlier than 7000 yr BP. Finally emergence predictions in Southampton Island by the Peltier (1988) model are indicated with band **M**.

FOXE BASIN - BAFFIN ISLAND (Plates 73 and 74)

Quinlan (1985) used sea-level observations from 11 sites located in or near Baffin Island to constrain a model of ice reconstruction and predict the form of local relative sea-level histories. Four of these sites are located in the Foxe Basin area. For each of them a comparison is given in **Plate 73** between sea-level curves predicted by Quinlan (1985) and envelopes representing likely limits on the positions of observations. The sites are: Northern Melville Peninsula (predicted curve **A** and observed band **A'**), Brodeur Peninsula (predicted curve **B** and observed data-points **B'**), Baird Peninsula (predicted curve **D** and observed band **D'**) and Cape Tanfield (predicted curve **E** and observed band **E'**). Lastly, curve **C** was reported from the northeastern Foxe Basin by Ives (1964).

Andrews (1987) published a relative sea-level curve for "SE Baffin Is." showing a significant fluctuation around 8200 yr BP (curve **F**) which is based on geomorphic and stratigraphic records preserved in Meta Incognita Peninsula. Here the relative sea level would have remained at the marine limit phase briefly, then would have fallen rapidly, followed by a new transgression that would have

Plate 73: FOXE BASIN - BAFFIN ISLAND I

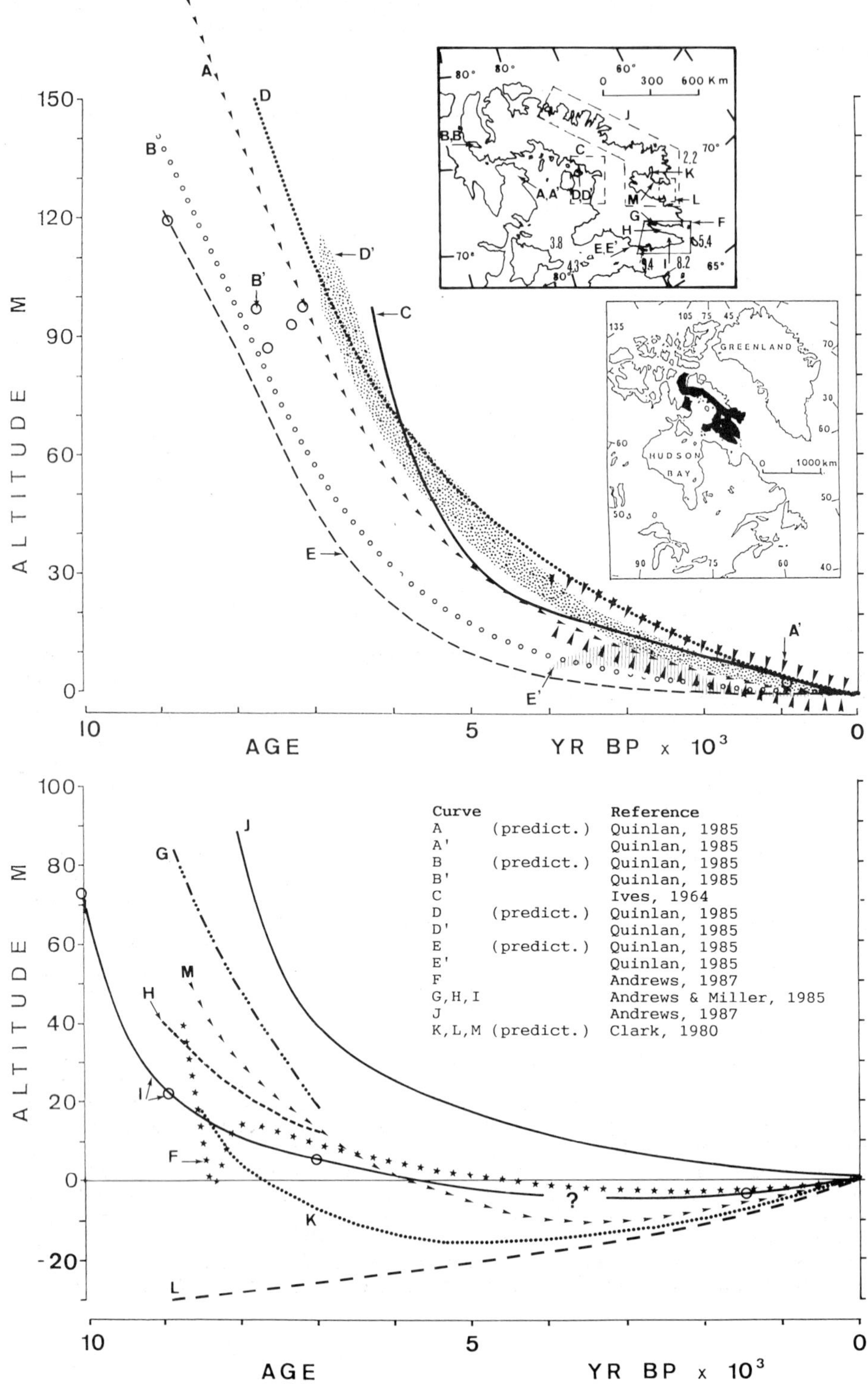

attained an elevation well below the marine limit but for a longer duration. Such fluctuations are indeed difficult to explain as eustatic. They are interpreted by Andrews and Miller (1985) as due to glacial loading and unloading which may have been accommodated along pre-existing fault systems.

In the Frobisher Bay area the outer sections of the bay were deglaciated between 10,700 and 8700 yr BP, whereas the inner bay became ice free by around 7000 yr BP. Here Andrews and Miller (1985) proposed three emergence curves for localities at 50 km (curve **G**), 100 km (curve **H**) and 150 km (curve **I**) distant from the Foul Inlet (at the head of the Bay).

According to Andrews (1987), curve **J** summarizes relative sea-level changes in "E. Baffin Is.". This is an oversimplification, however, since in this area local sea-level histories show significant changes from place to place. North of Frobisher Bay and in Cumberland Peninsula, for example, according to Andrews and Miller (1985), sea level changed little between 10,000 and 8000 yr BP; an early Holocene raised marine shoreline, dated between 8600 and 8000 yr BP, slopes seaward from a maximum present elevation of about +70 m in the inner fiords to very near or below present sea level near the outer capes, where no Holocene raised beaches have been recognized; lastly a series of peats and buried soils indicate a 0.5 to 1 m rise in sea level during the last 1000 to 3000 years.

Clark (1980) published several curves, each one apparently based on only one or two points, quoting Dyke (1977), which are called "observed relative sea level curves for Cumberland Peninsula". Three of these curves (**K** for "Kingnait F.", **L** for "outer Cumberland Sound" and **M** for "Pangnirtung F.") have been reproduced in Plate 74. As the data points given by Clark seem insufficient to constrain the trends indicated by the curves, the latter are considered "predicted" rather than "observed". This is in agreement with Dyke (1979b), who recognized that "at no site was a suite of samples found which would allow the construction of a well-controlled relative sea-level curve" and that the curves he proposed were therefore "speculative".

In the northern coasts of Baffin Island a comparison between sea-level obvervations (band **A**, **Plate 74**) and predictions (curve **B**) at Tay Sound, was given by Quinlan (1985). At Inugsuin Fiord, predictions by Quinlan (1985) in the inner fiord area (curve **C**) expect former sea-level positions much higher than in the outer fiord (curve **D**). Field observations at the south end of Inugsuin Fiord by LØken (1965) (curve **E**) are in agreement with predictions **C**.

In the Henry Kater Peninsula, according to King (1969), deglaciation took place at about 10,000 yr BP at the outer coast (emergence curve **G**), at about 8500 yr BP in Isabella Bay on the north side (curve **F**), and 8000 yr BP in Itirbilung Bay on the south side (curve **H**). In the inner part of the peninsula the marine limit is however higher than the series of fossiliferous deltas, built out into the late glacial sea, on which curves **F** and **H** are based; here sea level must have fallen quickly from the marine limit at +50 m (Itirbilung Bay) and at +42 m (Isabella Bay), before delta deposition started and mollusc life became abundant when sea level had reached about +30 m around 8000 yr BP.

Plate 74: BAFFIN ISLAND II

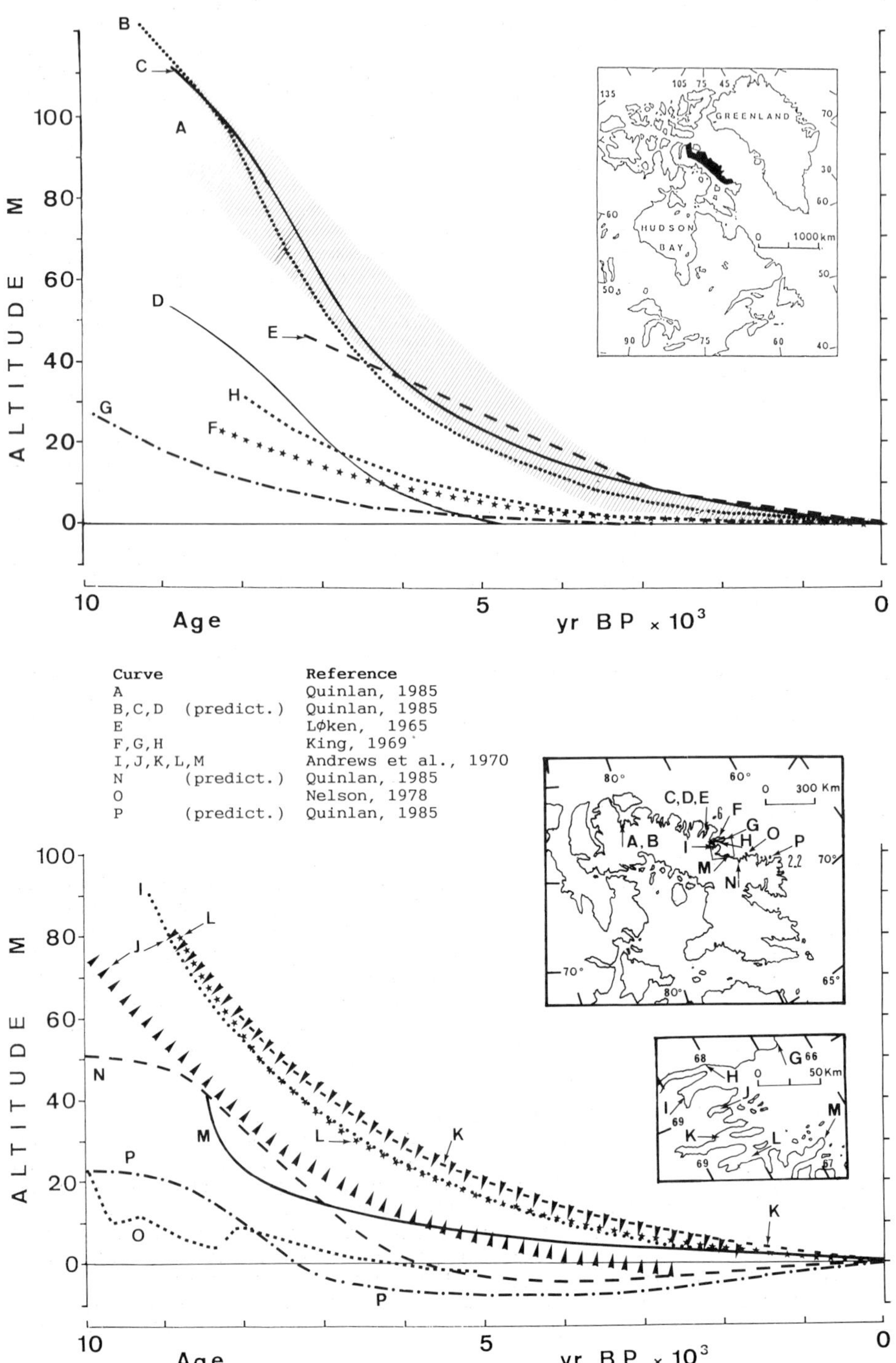

In Home Bay several emergence curves, based on 32 dates of marine shells, were constructed by Andrews et al. (1970). The construction of emergence curves is based on the assumptions that: (1) local uplift curves all had a similar proportional response, (2) the elevation and age of a marine limit is known, and (3) Shepard (1963) sea-level curve is appropriate to estimate the eustatic changes which have to be deducted from local uplift curves in order to obtain emergence curves. The latter assumption is certainly unacceptable from a methodological point of view (see the Introduction of this Atlas) and probably underestimates Holocene sea-level positions.

In the case of the Home Bay area, marine limit elevations were found to be a function of distance from the continental shelf (which can be related to former ice thickness) and date of deglaciation. They first incline inland, from about +50 m near the outer capes to a maximum of +91 m at the outer limit of the Ekalugad Moraine, dated 8000 yr BP, and then decline further inland, reflecting the slow deglaciation and the importance of restrained rebound. Eleven radiocarbon dates related to marine limits enabled Andrews et al. (1970) to determine with reasonable accuracy (±5 m to ±10 m?) five emergence curves for Tingin Fiord (**I**), Pitchforth Fiord (band **J**), Saaturaatuq (**K**), outer Kangok Fiord (**L**) and Cape Hooper (**M**) respectively. Of the five curves, **M** is based on a radiocarbon age of the local marine limit, **J** and **L** are relatively well controlled in their lower positions, but the age of the respective marine limits is based on the use of one of the lower dated shorelines to predict the age of the marine limits. The age of the marine limit of **K** was estimated from field correlations.

On the northern coasts of the Cumberland Peninsula, relative sea-level curves were predicted by Quinlan (1985) model at Okoa Fiord (**N**) and Broughton Island (**P**). Finally Nelson (1978) proposed an oscillating relative sea-level pattern at Qivitu (curve **O**).

BEAUFORT SEA - MELVILLE I. - SOMERSET I. (Plate 75)

A dozen radiocarbon-dated peat and peaty clay samples from geotechnical boreholes in the Canadian Beaufort continental shelf have been used by Hill et al. (1985) to reconstruct a relative sea-level curve covering the last 30,000 years, whose Holocene part is reproduced in **Plate 75** (curve **A**). A fairly constant rate of sea-level rise (6 mm/yr) is suggested by curve **A** during the Holocene. Hill et al. (1985) ascribe about 1.2 mm/yr to the effects of basin subsidence, sediment loading and consolidation subsidence, and the remaining sea-level rise to glacio-eustatic effects at the margin of the former ice sheet.

The coasts of Melville Island, except the west coast, were studied by Henoch (1964). The highest raised shore features were found between 50 and 70 m in the northeast part of the island, from where they decrease in altitude westwards. Emergence curve **B** by Henoch (1964) is based on five radiocarbon dates and suggests that the uplift rate was almost negligible over the past 2000 years, implying that Melville Island is close to isostatic equilibrium. McLaren and Barnett (1978), after having assembled 25 dated samples (marine shells, driftwood, peat and organic detritus, from raised marine and delta deposits and from archaeological sites),

Plate 75: BEAUFORT SEA - MELVILLE I. SOMERSET I.

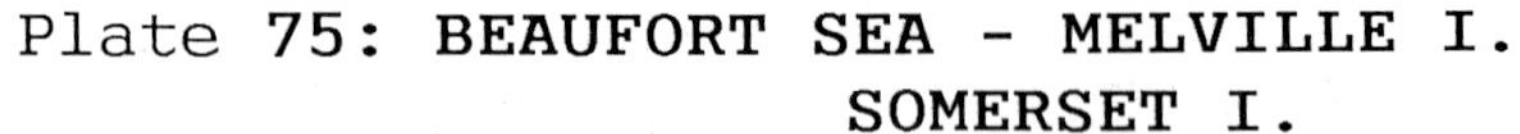

Curve Reference
A Hill et al., 1985
B Henoch, 1964
C McLaren & Barnett, 1978
E Dyke, 1979a
F Quinlan, 1985
F' (predict.) Quinlan, 1985
G-J Dyke, 1979a
K Quinlan, 1985
K' (predict.) Quinlan, 1985
Altitude m
Age yr BP × 10³
100
80
60
40
20
0
-20
-40
-60
10
5
0
A
B
C
D
F
F'

Altitude m
Age yr BP × 10³
150
120
90
60
30
0
10
5
0
J
K'
K
I
E
G
H
0 400 Km
140°
120°
100°
130°
c.a.
0.6
1.1
1.7
1.8
70°
0 1000 Km
150°
135°
80°
105°
60°
70°
120°
90°
75°

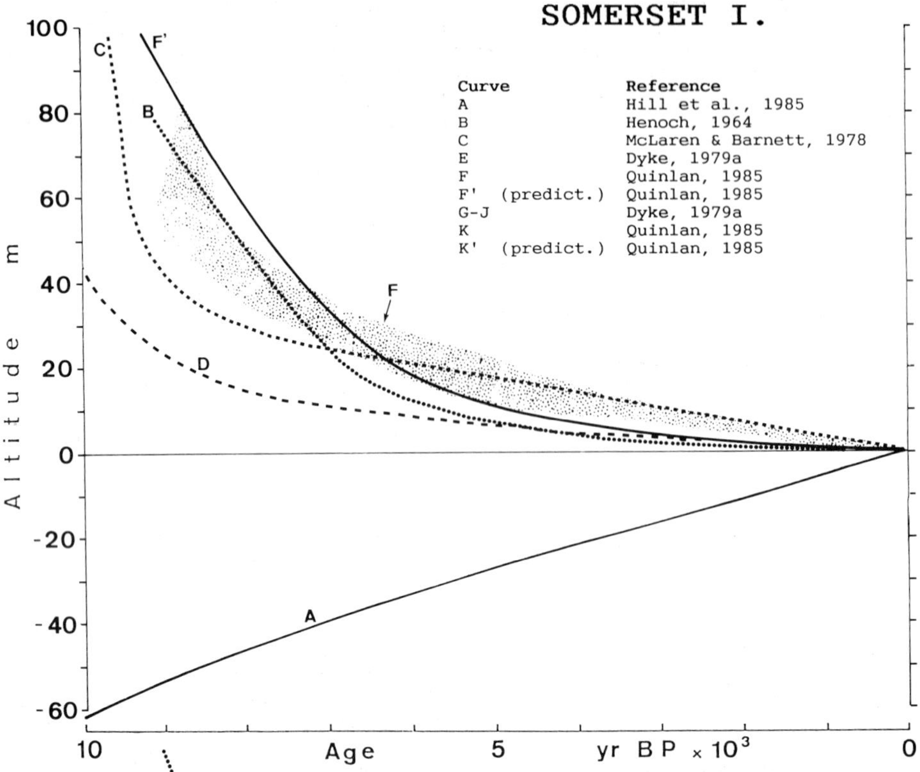

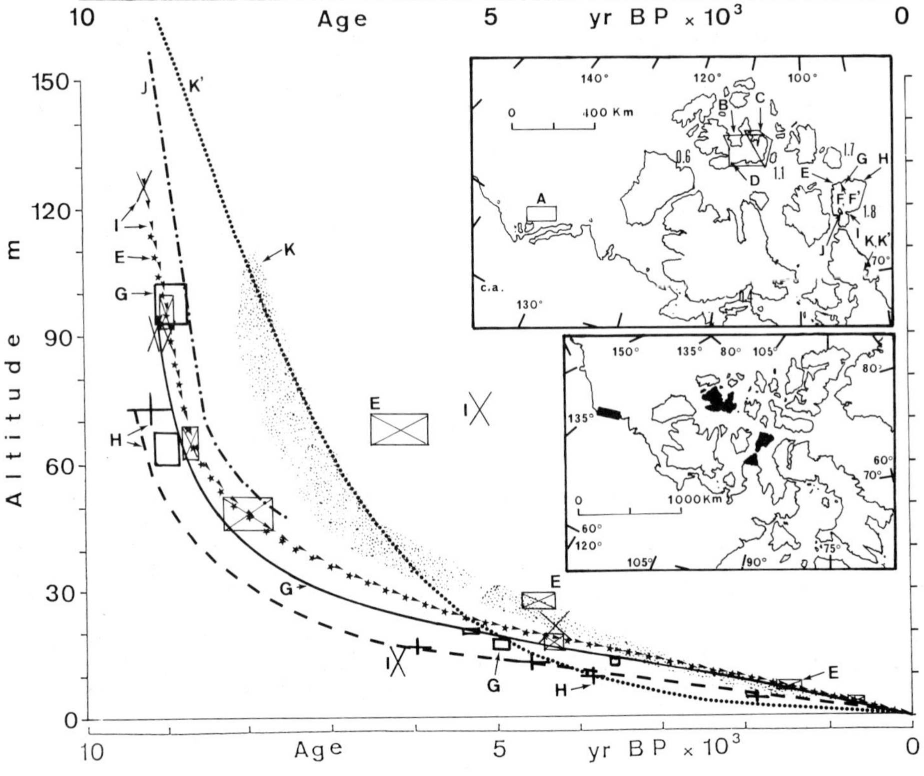

proposed two emergence curves: **C** for the east coast and **D** for the south coast. According to them, emergence is essentially complete on the south coast near the Winter Harbour moraine, which is thought to mark the maximum northward advance of the Laurentide ice sheet. The northeast coast of Melville Island, on the other hand, appears to be still recovering at a rate of approximately 3.5 mm/yr. This may imply the existence of ice cover in the north part of the island which has been thicker or has lasted longer than in the peripheral areas of the Laurentide ice sheet masses.

Sea-level changes in Somerset Island were studied mostly by Dyke (1979a), who used 36 radiocarbon dates on marine shells, whale and walrus bones, and driftwood to reconstruct four local emergence curves. At Cape Anne seven samples (rectangles with diagonals **E**, the height of which is the estimated error on elevation measurements, and the width is 2σ error on the radiocarbon age determination) define smooth curve **E**, the lower part of which is controlled by 4 dates on whale bone embedded in raised gravel beach ridges. The basic assumption here is that the bones arrived at the shoreline as a result of stranding of whales in the intertidal zone or slightly below it. At Cunningham Inlet six samples (straight crosses **G**), mostly of driftwood, define emergence curve **G**. At Rodd Bay four samples of whale bone and one of driftwood (straight crosses **H**) were used to construct emergence curve **H**, the shape of which in the 6000 to 9000 yr BP range was drawn to mimic curves **E** and **G**. In the Creswell River lowland several indications suggest that local emergence curve **I** is almost the same as the Cape Anne curve **E**, which lies on the same isobase during the early and middle Holocene. Finally emergence curve **J** was proposed by Dyke (1979a) for the Stanwell-Fletcher Lake area during the early Holocene. Nine of the dates provided by Dyke, all around 9200 ± 100 yr BP, pertain to the marine limit, which lies at +76 m in the northeast and +157 m in the southwest part of the island. Early emergence rates were 80 to 110 mm/yr and 56% of total emergence was accomplished in the first 1000 years. During the last 5000 to 6000 years emergence has proceeded at a constant rate at each site, but has varied spatially from 4.6 mm/yr in the west to 2.8 mm/yr in the east.

In northern Somerset Island, Quinlan (1985) found a relatively good agreement between field observations (summarized with band **F**) and emergence curve **F'**predicted by his model. A similar result was obtained by the same author in Boothia Peninsula, between field data (band **K**) and predictions (curve **K'**).

ELLESMERE AND DEVON ISLANDS (Plate 76)

On northeast Ellesmere Island, which is separated from northwest Greenland by a 20-40 km wide sea corridor, Holocene emergence reaches up to about 120 m. England (1983) obtained from here over 30 new radiocarbon dates on marine pelecypods and driftwood, which were assembled with earlier published data to obtain several local relative sea-level curves (**A-G**). The curves proposed by England have a form unlike any other published emergence curve from Arctic Canada or from Fennoscandia, showing three fragments: (1) an interval of almost stable relative sea level near the marine limit between at least 11,000 and 8000 yr BP; (2) an interval of slow emergence from 8000 to 6200 yr BP during which northeast Ellesmere Island ice seems to have retreated slowly; and (3) an interval of

Plate 76: ELLESMERE & DEVON ISLANDS

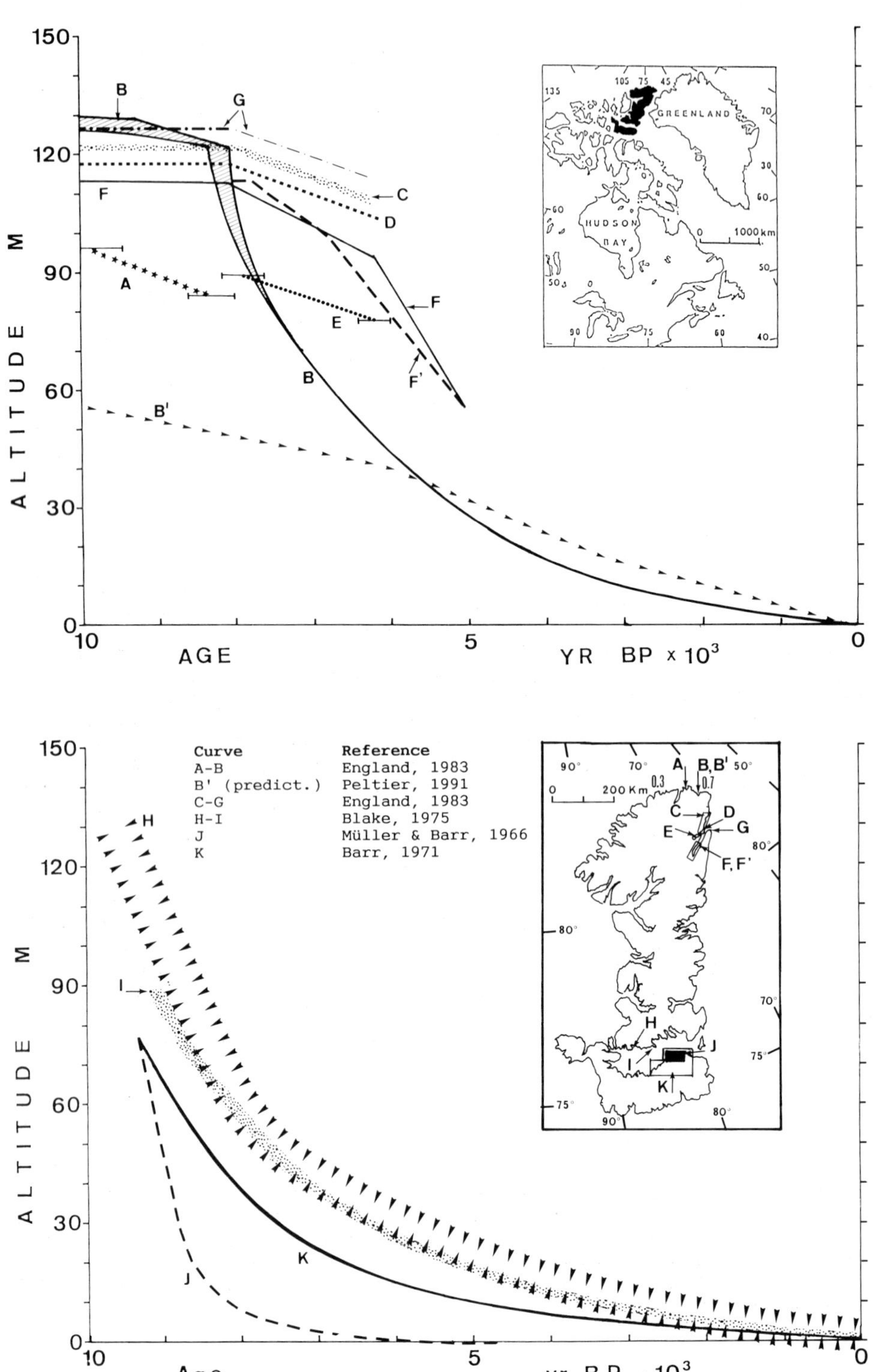

more rapid emergence after 6200 yr BP, apparently caused by glacial unloading.

At James Ross Bay, which now appears to be undergoing submergence, only the early Holocene emergence trend could be outlined (curve **A**). Near Alert relative sea-level envelop **B** proposed by England (1983) parallels the slow rate of initial emergence in James Ross Bay, before a sudden acceleration postulated around 8200 yr BP. The physical meaning of such a sudden acceleration remains to be explained. Relative sea-level predictions made by Peltier (1991) at Alert, using the ICE-3G model (curve **B'**), seem to underestimate the local uplift rate during the early Holocene.

The coastline between the Beaufort Lakes and Sun Cape Peninsula, where the marine limit is at about +115 m, is represented by relative sea-level band **C** (England, 1983); Muskox Bay's curve **D** is assumed to be parallel to **C**, though at a slightly lower level. In the Chandler Fiord area, which seems to have been ice-occupied until about 8000 yr BP, a slow rate of initial emergence is suggested by curve **E**. In the Archer Fiord, which parallels the postglacial isobases, data from various sites were plotted on one curve (**F**); an alternate curve for Archer Fiord suggested by the same author (England, 1983) is indicated by **F'**. Finally at Cape Baird the marine limit rises to +120 m (curve **G**).

In southern Ellesmere Island over 50 radiocarbon dates of whale bones, marine shells, driftwood and plant detritus and various pumice occurrences enabled Blake (1975) to construct a very well documented emergence band (**H**) at Cape Storm. During the first 2000 years after deglaciation, i.e. between 9500 and 7500 yr BP, sea level fell approximately 80 m in relation to the land, at an average rate of 40 mm/yr, despite the fact that the level of the sea was actually rising rapidly, probably at a rate of the order of 70 mm/yr between 9500 and 8500 yr BP. By contrast, during the 2000 years between 6500 and 4500 yr BP, only 16 m of emergence occurred, at an average rate of 8 mm/yr. Over the past 2400 years, emergence has taken place at an average rate of <3 mm/yr. On the west side of South Cape Fiord, emergence band **I** by Blake (1975), based on nine dated samples, shows a pattern similar to that of band **H**, though at a slightly lower rate of emergence.

In northern Devon Island, near Cape Sparbo, where the marine limit varies considerably in altitude (between +65 m and +82 m over 30 km of coast surveyed), Müller and Barr (1966) proposed emergence curve **J**, obtained by plotting six marine shell dates. Subsequently, however, the shape of the emergence curve for this area was considerably modified by Barr (1971) on the basis of more detailed field work and new radiocarbon dates of more reliable samples (curve **K**).

GREENLAND (Plate 77)

In Greenland, sea-level investigations have been carried out mostly in two regions: the East Greenland fjord zone and, in West Greenland, the coastal area between 66° and 69° N. In central East Greenland, Petersen (1986) used radiocarbon dates of scattered occurrences of *Mytilus edulis* reported by Hjort and Funder (1974) to construct emergence band **A**.

Plate 77: GREENLAND

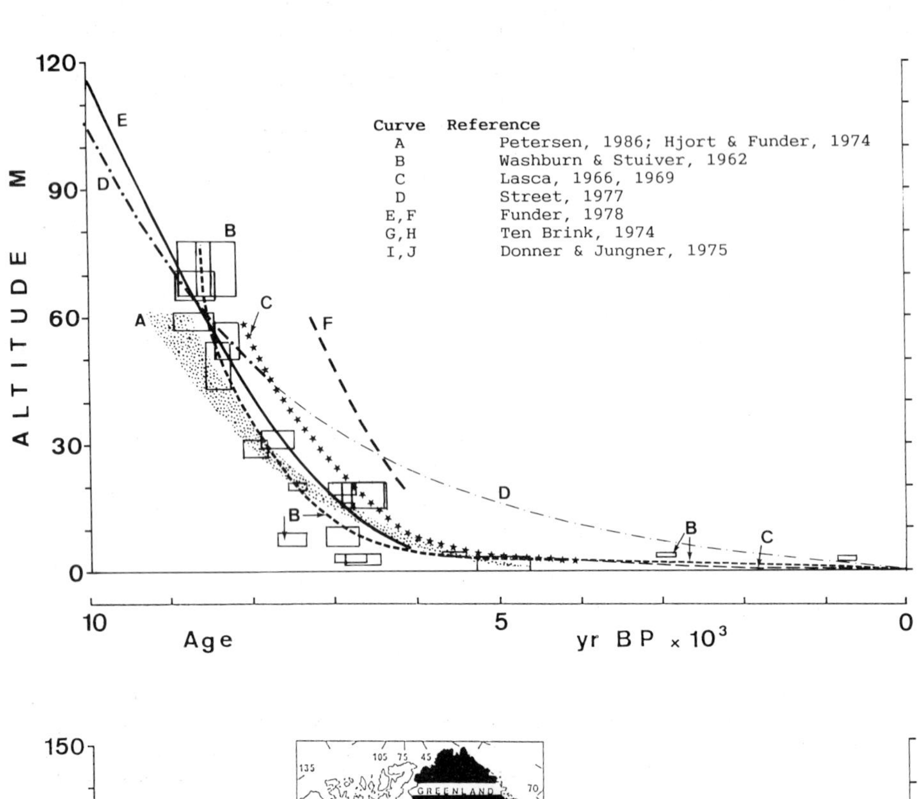

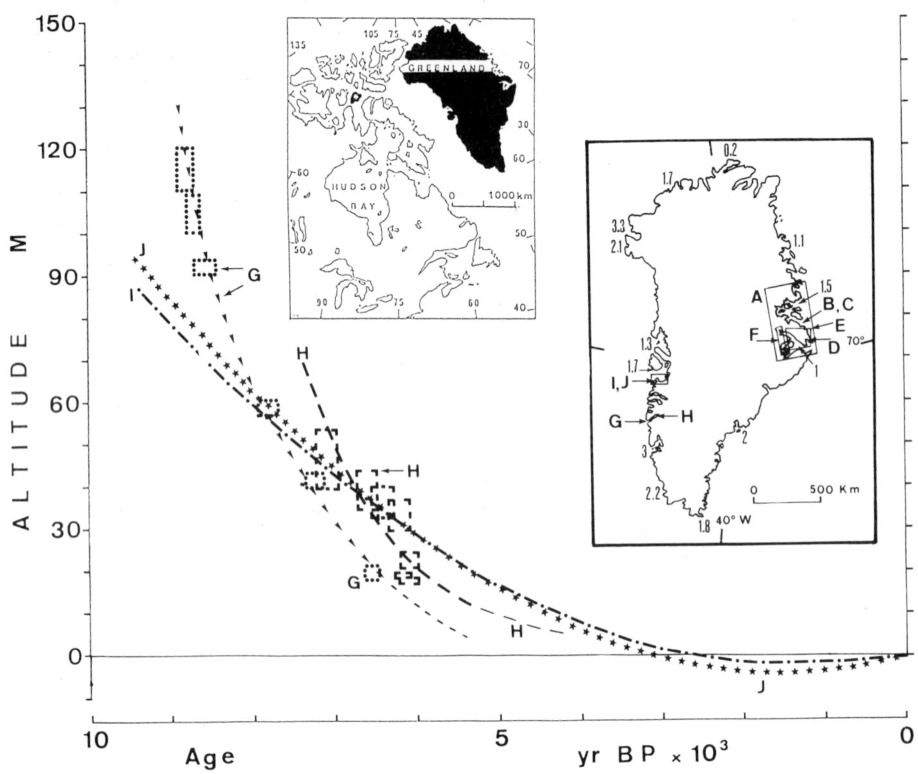

At Mesters Vig, on the south side of Kong Oscars Fjord, Washburn and Stuiver (1962) studied till-like deposits containing well-preserved shells, laid down by an advancing glacier prior to deglaciation. About twenty individual sea-level data, for which accuracy in altitude and time are represented in **Plate 77** by boxes **B**, include marine shells and a few driftwood specimens. The overall relative sea-level trend is summarized by curve **B**. The Mesters Vig district was open to the sea and, therefore, deglaciated at least in part by 9000-8500 yr BP and has remained largely free of glaciers since that time. The rate of emergence was high initially, of the order of 90 mm/yr, decreased exponentially to about 6 mm/yr for the interval 9000 to 6000 yr BP, and has remained perhaps as low as 0.7 mm/yr since 6000 yr BP. As pointed out by Ten Brink (1974) and Funder (1978), however, there seem to be reasons for believing that curve **B** is subject to "false steepening", due to the fact that most shells provide only a minimum value for the sea level and that old shells appear at too low a level.

At Skeldal, in another part of the Mesters Vig district, Lasca (1966, 1969) obtained from 13 radiocarbon dates of shells and peat, most of which were collected from emerged delta deposits, relative sea-level curve **C**. Lasca estimated the emergence rate at approximately 30 mm/yr between 8000 and 7000 yr BP. However, according to Funder (1978), some of the data by Lasca would include an erroneous correction which makes their ages 400 years too young. At Schuchert Dal, Street (1977) plotted four dates, one of which was new, to obtain relative sea-level curve **D**, which suggests an initial rate of rebound of 48 mm/yr.

In the Scoresby Sund region a compilation of seventeen radiocarbon dates of Holocene marine bivalve shells, some of which were new, enabled Funder (1978) to construct two emergence curves, for the central area where marine limits reach altitudes ≥90 m (curve **E**), and for the interior fjord area where marine limits are ≤60 m (curve **F**).

On the west coast of Greenland, Ten Brink provided two emergence curves: **G** for the outer SØndre StrØmfjord, representing the zone of maximum uplift, 125 km west of the present ice sheet margin, where emergence since ≈8000 yr BP is 145 ± 5 m, with an initial rate of about 105 mm/yr, the highest reported from Greenland; and curve **H** for the inner SØndre StrØmfjord, 50 km west of the present ice sheet, where 75 ± 5 m of postglacial uplift began ≈7300 yr BP, with an initial rate of about 60 mm/yr. Radiocarbon-dated samples used to construct curves **G** and **H** are represented by rectangles, the width of which being equal, as usually, to the standard error of the date, and height being equal to both the estimated uncertainty of field altitude measurement and the relation to relative sea level.

In the Disko Bugt area Donner and Jungner (1975) traced several least square curves, two of which (**I** and **J**) fit best to dated shell samples and are therefore considered to describe emergence trends in the area.

PART 3: CONCLUSIONS

CONCLUSIONS

The study of sea-level changes has fascinated generations of geologists, geophysicists and oceanographers. The changes which occurred during the Holocene (last 10,000 years) also have important implications for prehistorians, archaeologists, historians, palaeoecologists and geographers.

This Atlas has been produced to provide a visual assessment of geological trends in sea level during the Holocene. By comparing them with trends deduced from tide gauges and near-future trends predicted by climate models, this should help the assessment of near-future local sea-level changes at a local scale. Other aims of this Atlas are to give: (1) a global review of the state of the art in the field, (2) a reference compilation of Holocene sea-level curves, most useful when estimating vertical earth movements, establishing palaeogeographic maps and testing geophysical models, (3) an evaluation of the adequacy of current knowledge and of the appropriateness of certain approaches, methods and assumptions used in sea-level data production and interpretation as a source of the information required to assess future coastal changes.

The consequences of a sea-level rise during the next decades would be drastic for mankind, since two-thirds of the world population live in the coastal zone, as defined by United Nations experts. To assess the impacts of a possible near-future sea-level rise in each place, it is essential to be aware of the behaviour of the geological trend there.

The research work necessary to obtain the results summarized in this Atlas has been an intricate and laborious job, involving thousands of people, many working at the boundaries between various disciplines, most often in the framework of small to very small research projects. This research was undertaken in most cases merely to satisfy scientific curiosity and attempt to understand how and when the relative sea level varied at each place. Most of these small-scale studies were only of local interest and unconnected to other studies elsewhere. Nevertheless, as each sea-level curve is usually based on several radiocarbon dates and represents several months of field and laboratory work, some 800 sea-level curves correspond, once assembled together, to a significant financial effort, equivalent to that of a major research programme, which can be estimated for the whole of the countries involved at several million dollars over the last thirty years. This vast effort has been successful not only in achieving a better understanding of local sea-level changes in many areas, but also by enabling us to gain more knowledge of the postglacial history of the earth, and even of the viscosity and the rheology of terrestrial materials.

The sea-level curves reproduced in Part 2 vary considerably in precision, of a decimetric order in the best cases, but only of metric or even decametric order in other cases (e.g. for certain material dredged from the continental shelf, for correlations in rapidly uplifting areas, etc.). If a metric or decametric precision may be useful to outline glacio-isostatic trends, it is certainly inadequate to use sea-level changes as climate proxydata

(Shennan, 1982b; Pirazzoli and Pluet, 1991) or to infer detailed conclusions in other related fields. In spite of this difference in precision between plates, and often between sea-level curves in the same plate, the Atlas gives an overall impression of relative consistency, though more work is indeed required on addressing the accuracy of local relative sea-level histories.

As many of the sea-level curves assembled lack uncertainty margins, it has not been possible to specify the degree of reliability for each one of them. A selection based on strict criteria, with a rejection of unreliable data at an early stage in the preparation of this Atlas would certainly have been useful. However one single person cannot possibly be expected to make such a selection for all areas. This would be a more feasible aim for an internationally coordinated research project, backed by authoritative sea-level databases (see Part 1). Nevertheless, a simple visual comparison between curves, each of them considered as the middle part of an uncertainty band of unknown width, is often sufficient to disclose which data may be less reliable or set a problem. From this point of view, this Atlas can be useful and encourage a more critical evaluation of sea-level data.

As sea-level history has varied greatly from place to place, drawing a single relative sea-level time/depth curve for wide areas is a misleading generalization. This remark may apply to several curves reproduced in this Atlas, particularly to composite curves deduced from sparse data in broad regions (e.g., for the continental shelf, Plates 35, 59). Concerning shelf data in particular, doubts about the reliability of dated shells and many conflicting results help to show the difficulties involved when assessing submerged materials in sea-level studies. Shells can undergo landward transport by sediment action on the continental shelf, so that dated shells do not always represent in situ material (MacIntyre et al., 1978).

The geographical distribution of the areas studied is very uneven (Table 3.1). Though some regions (Europe, North America, Japan, Australia, New Zealand) and especially the coasts of the North Atlantic (including marginal seas) are relatively well covered, many wide gaps exist. Among the less studied regions mention can be made of Central and South America (3% of the data), Africa (4%), the coasts of the Indian Ocean (4%) and the Southern Hemisphere in general. These gaps are almost the same as those found in a previous study concerning sea-level data of the last 2000 years (Pirazzoli, 1976b, 1977).

Important gaps exist also in some coastal areas of relatively well studied continents. In Europe this in the case for Iceland, Spain, Portugal and the coasts of the Adriatic Sea, where no relative sea-level curve is available. In the East Baltic (Plate 13) and the Black Sea (Plates 26-27), though several curves have been proposed, their oscillations differ so much from one curve to the other that their reliability may be questionable. In South Italy and Greece, though some data exist, they are certainly insufficient to assess local tectonic trends, which may change from place to place. In the United States, the number of sea-level curves for the Atlantic coast is almost 20 times greater than that for the west coast. There are almost no data at all for the west coasts of Australia and New Zealand.

In Africa most of the data available are concentrated in a few areas on the west coast, whereas only one curve has been plotted for the Mediterranean coasts of Africa and one for the east coasts of the continent. In particular, no sea-level curve has been found for Namibia, South Africa, Madagascar, Tanzania, Kenya, Somalia, Ethiopia and the Red Sea Coasts.

In South Asia, the data available are concentrated on the western side of the Persian Gulf, on the west coast of India (1 curve) on southern Sri Lanka (1 curve), and Bangladesh (1 curve) with no sea-level curves from Iran, Pakistan and Burma.

In Southeast Asia no data are available from Thailand and Cambodia and only one composite sea-level curve is reported for all the coasts of Vietnam; Malaysia is relatively well studied and a few data have also been reported from Indonesia, though they are certainly insufficient to define the prevailing relative sea-level trends in a tectonically active country consisting of over 10,000 islands.

On the coasts of China, several sea-level curves have been proposed, but they diverge so much for the same areas that their accuracy is questionable. The involvement of Chinese scholars in international sea-level projets is quite recent, however, and rapid progress in methodology and interpretation can be expected here in the near future. All the Pacific coast of the USSR between Lat. 50° and 65°N seems still devoid of data. Finally in Micronesia, Melanesia and Polynesia, though information is still scanty in many wide areas, good progress has been made, especially in the 1980s.

The great variability of sea-level situations shown by the curves available can be summarized for three different lengths of time. The relative sea-level changes since 10,000 yr BP are indicated by 262 curves or bands based on field data (thus excluding predictions; for wide uncertainty bands, only their median part has been considered) which go back as far as this date. Emergence since 10,000 yr BP is shown by 44% of the curves (mainly in formerly glaciated areas, but in a few cases also in tectonically uplifting sites) and submergence by 56%. Emergence since 10,000 yr BP was greater than 100 m for 11% of the curves, between 100 and 50 m for 12%, 50-30 m for 7%, 30-10 m for 9%, less than 10 m for 5%. Submergence since 10,000 yr BP was less than 10 m for 5% of the curves, 10-20 m for 7%, 20-30 m for 13%, 30-40 m for 15%, 40-50 m for 8%, and over 50 m for 8%. The average of all curves since 10,000 yr BP is a relative sea-level drop of about 8 m. This value, which reflects the predominance of glacio-isostatically uplifted areas in the sample of curves available, is obviously no more representative of the global eustatic situation than would be, for the last century, the average of all tide-gauge records (Pirazzoli, 1986b). The high variability from place to place confirms indubitably that it is impossible to estimate directly, from field data alone, the amount of the eustatic change since 10,000 yr BP (or since any other time).

Since about 5000 yr BP, when the post-glacial eustatic rise is usually assumed to be complete, 594 relative sea-level curves based on field data show that emergence prevailed in 56% of the cases studied, submergence in 44%, and that for 55% of the cases the relative sea-level change was greater than 5 m. More precisely, emergence since 5000 yr BP was greater than 50 m in 5%

of the cases, between 50 and 30 m in 4%, 30-10 m in 17%, 10-5 m in 10%, and less than 5 m in 20%. Submergence since 5000 yr BP was less than 5 m in 25% of the cases, 5-10 m in 13%, 10-20 m in 5%, and 20-30 m in 1%. The average of all curves is a relative sea-level drop of about 7 m since 5000 yr BP, which is no more representative of the world situation than the average sea-level drop of 8 m since 10,000 yr BP.

The changes in the relative sea-level occurring since 2000 yr BP can be estimated from 501 curves or bands deduced from field data.

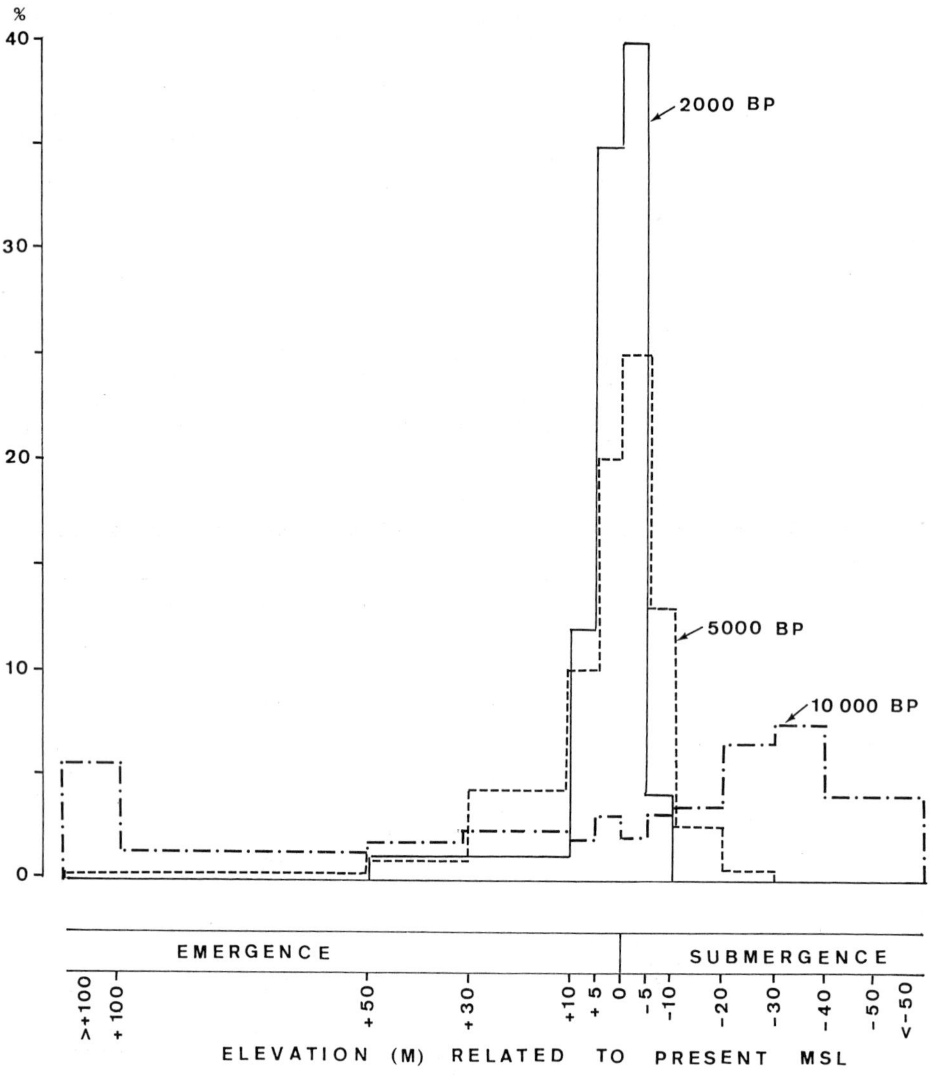

Fig. 3-1 Isometric curves summarizing the distribution of relative sea-level changes since 2000 yr BP, 5000 yr BP and 10,000 yr BP, as shown by all the curves compiled in this Atlas.

Emergence prevailed in 56% of the cases and submergence in 44%, generally showing the same trends as the period since 5000 yr BP, though at a gradually decreasing rate in formerly glaciated areas. The frequency of submergence since 2000 yr BP (44%) is very similar to the value of 43% obtained in a previous study based on 708 individual sea-level data (Pirazzoli, 1976b, 1977). In 25% of the cases, the relative sea-level change was still greater than 5 m. More precisely, emergence has been greater than 10 m since 2000 yr BP in 9% of the cases, between 10 and 5 m in 12% and less than 5 m in 35%; submergence has been less than 5 m in 40% and between 5 and 10 m in 4%. The average of all the curves is a relative sea-level drop of about 1.5 m since 2000 yr BP. The distribution of sea-level changes since 10,000, 5000 and 2000 yr BP has been summarized by three isometric curves in Fig. 3-1.

The above results are clearly not representative of the world situation, due to the very uneven distribution of the data available (Table 3.1): formerly glaciated regions are over-represented in relation to tropical regions, uplifting areas in relation to subsiding ones (except on both sides of the North Atlantic) and industrially developed countries in relation to developing countries.

Over one hundred relative sea-level curves or bands predicted by geophysical models have been compared to field data. Models are partly responsible for completely upsetting formerly accepted ideas in sea-level research, demonstrating that the disappearance of continental ice sheets and the redistribution of water in the world ocean could explain most of the regional variations observed in post-glacial sea-level changes. The main value of geophysical

TABLE 3.1

Geographical distribution of the sea-level curves reproduced in Plates 1-77

		Number of curves	%
Continents			
	Europe	292	32
	Africa and Indian Ocean islands	37	4
	Asia	135	15
	Oceania	105	12
	South and Central America	31	3
	North America	286	32
	Antarctica	9	1
	Greenland	10	1
	total:	905	100
Oceans			
	Atlantic Ocean	512	57
	Pacific Ocean	247	27
	Indian Ocean	33	4
	Arctic Ocean	113	12
	total:	905	100
Hemispheres			
	Northern Hemisphere	776	86
	Southern Hemisphere	129	14

models is that they make it possible to arrive at a first approximation in areas where no field data are available, and that they can be used to estimate which component of available data may be due to glacio-isostasy and which to hydro-isostasy. However, when predictions are compared to field data, discrepancies usually exist and it is difficult to ascertain whether they are due to non isostatic causes (tectonics, ocean dynamics changes, etc.) or to insufficiently accurate assumptions used by the models.

In **Plate 49** (lower graph) predictions by three different models produce very similar results in the Moruya area during most of the Holocene period. This is certainly encouraging, but also astonishing, if one considers the assumptions on which these models are based: different melting chronologies, different lithospheric thicknesses and different mantle viscosities. This means that if the assumptions made by one of the three models are correct, they will necessarily be inaccurate for the two other models. Nevertheless, deviations between predictions and field data can still be found in several areas for all models available and this confirms that the earth's system is too complex and the models and the assumptions made still too simple, in spite of their growing mathematical complexity, to represent the physical reality with sufficient accuracy. The fact that three different models based on different assumptions can give similar predictions in the same area, is therefore disturbing, and one may wonder about the reliability of the same models when their predictions diverge in other places. Indeed much progress has to be made and greater precision achieved, before a consensus can be reached on matters such as the detailed global ice-melting history, the lithosphere thickness in each area and the upper and lower mantle viscosities. To obtain this consensus, models have to be improved constantly. This implies that their assumptions will necessarily be subject to change and that their predictions have to be considered as essentially perishable results, valid only until next publication, which in addition have to be tested constantly against reliable field data.

Even though Peltier (1991) considers that using the new ICE-3G model "the fits to all of the data shown are of very high quality in general", misfits are necessary to stimulate further refining and progress. This can be the case, for example, with the sea-level predictions made at Boston (**Plate 63**): curve **K'** obtained using "the new high resolution" ICE-3G model does not seem to fit the field data much better than band **K**, which had been obtained using the previous "low resolution" ICE-2 model. The most difficult point is usually to find the cause of a misfit. In Boston, during the early Holocene, **K** and **K'** diverge from field data by about the same amount, but with opposite signs; the cause of this misfit can either be in some still unknown details in the ice history, or in local unhomogeneities of the earth's materials, or in the inappropriateness of one of the assumptions made by the model.

In conclusion, models can be a very useful tool, but their predictions should not be overestimated, since they do not represent the reality, as field data do, but only more or less careful approximations of this reality based on simplified assumptions. From this point of view, the IGBP (1988, p. 93; 1989, p. 17) statement that "we need to collect sea level data from tectonically stable coasts, take isostatic adjustments into consideration, and then obtain water volume changes" certainly oversimplifies the problem. Such an approach, if applied, would provide values of global water volume change varying with the assumption used (which are usually modified in new isostatic models) and, for the same model and assumptions, with each "stable" area considered. It should be noted that the location of the "stable coasts" to be considered is not specified in the IGBP reports and that, as confirmed by this Atlas, "stable" areas probably do not exist.

A recent correction which has affected most of the previous results obtained by post-glacial sea-level models concerns the calibration of the radiocarbon time scale. A new calibration based on mass spectrometry U-Th dating of Barbados corals (Bard et al., 1990) has shown that U-Th ages agree with the dendrochronological calibration and that before 9000 yr BP the radiocarbon ages are systematically younger than the U-Th ages, with a maximum difference of about 3500 yr at 20,000 yr BP. In other words, the maximum of last glaciation, dated c. 18,000 yr BP, occurred in fact c. 21,500 sideral years ago. As the models by Peltier were based, until 1990, on the assumption that the deviation between radiocarbon and sideral dates was reduced to zero by 15,000 yr BP (Peltier, 1991, Fig. 2), most of the predictions obtained by these models need to be recalibrated. The Holocene part of the predictions should not be affected by this recalibration, however, or almost not affected, since the deviation could already be calibrated satisfactorily for marine samples until 9000 yr BP and for terrestrial samples for the entire Holocene (Stuiver and Kra, 1986).

The areas where further studies are a priority can now be defined unambigously: Africa, Central and South America, South and Southeast Asia and, in general, the Indo-Pacific region and the Southern Hemisphere. Regarding the areas which would be most useful for testing geophysical models, K. Lambeck has suggested "the mainly north-south margins of Africa, or Atlantic South America in order to examine the trends of the late Holocene highstand amplitudes with distance from the ice loads. Another prioritary area is the Antarctic margin and far south latitude islands... to determine the role that changes in this ice sheet may have played in sea-level change" (Sabadini et al., 1991, p. 706).

As shown above, almost one half of the sites studied reveal the occurrence of a local, relative sea-level rise since 5000 yr BP, as well as since 2000 yr BP. In almost the same regions, tide-gauge records usually show that the relative sea-level has been rising during the last century more rapidly than average (Pirazzoli, 1986b). This helps to define the areas which would be more at risk in the case of a near-future sea-level rise. In fact, to predict future changes of sea level at any location, the local (natural or anthropogenic) sea-level trend should be added to the figures predicted due to global warming. In the near future, Holocene relative sea-level data will probably find useful applications and help in predicting the impacts of a "greenhouse" sea-level rise, as well as guiding decisions concerning coastal land use and management.

The regions more at risk are therefore primarily all those which are already subsiding and include:

1) Most major delta areas of the world, especially those densely populated where natural subsidence may be increased

drastically by human activities (Bangladesh, Nile, etc.) (Milliman et al., 1989). Unfortunately, reliable Holocene sea-level data from these regions are extremely rare. These should be highly prioritary regions for further Holocene studies.

2) The collapsing forebulge zone around former ice sheets which

Fig. 3-2. Effects of the 1983 "El Niño" cyclones in Anaa Atoll (Tuamotus). **a)** A small building lies upside down in the middle of recent coral debris, whereas small scattered mounds of rootlets indicate former sites of coconut trees uprooted by wind and waves. **b)** Among many reef blocks projected on the reef flat by cyclone waves, this one near the Village is 40-45 m^3 in volume.

Fig. 3-3. Increased storminess may cause significant coastal erosion and loss of soil, exposing bare vegetation roots. **a**) Orbetello (Italy). **b**) Tupai Atoll, Society Islands.

has been defined by geophysical models [e.g. Zone II of Clark et al. (1978) and Clark and Lingle (1979)], comprehending most of western Europe, of the East coast of the United States and of Alaska; this zone is often well represented in this Atlas. It should be stressed however that the relative sea-level trends determined in this zone by basal peats for the Holocene, and by tide gauges for the last century, will necessarily be different, since they are referred to independent datum levels: the pre-

Holocene bedrock for basal peats, and the surface of Holocene sediments for tide gauges. The former will follow the general tectonic and isostatic behaviour of the region. The latter will include, in addition, the compaction of Holocene sediments, which varies greatly from one place to another and may be increased sharply by human activities. It would not be surprising therefore if the subsidence rates indicated by tide-gauges were systematically higher than those suggested by basal peats and it would be incorrect to ascribe the apparent recent acceleration merely to a global "greenhouse"-induced sea-level rise.

3) The zone of oceanic submergence defined by geophysical models [e.g. Zone IV of Clark et al. (1978) and Clark and Lingle (1979)], especially in low coral islands, where there would be almost no defence against a relative sea-level rise or against storm surges caused by more frequent tropical cyclones which are expected to accompany a climatic warming. (Fig. 3-2, 3-3).

4) Any area affected by tectonic subsidence, gradual, or occurring in steps at the time of major earthquakes. In these areas studies of relative sea-level changes during the Holocene are essential and still too rare. Interseismic vertical trends revealed by tide gauges and levellings may be unrelated or even contrary to much greater seismic displacements which determine the local long-term trend [e.g. curves E and F, Plate 38; C and D, Plate 43; G and K, Plate 45; F, Plate 46; see also Darienzo and Peterson (1990); Savage and Plafker (1991)]. In active tectonic areas, Holocene relative sea-level data can be very useful to estimate rates of vertical displacement and even the recurrence time of major earthquakes.

It is often estimated that the total amount of land ice released from the late Pleistocene ice sheets raised the sea level by about 120 ± 20 m (e.g. Chappell and Shackkleton, 1986; Fairbanks, 1989). This is greater than the ice volume contained in most models of former ice sheets. Plausible regions for the missing meltwater include, according to Lambeck (1990), Antarctica, the Barents-Kara Seas, possibly western Laurentia and the Siberian platform. Peltier (1991), in the ICE-3G model, suggests the possibility of a thick ice cover over the Barents and Kara Seas, the Queen Elizabeth Islands and the East Siberian Sea (Fig. 1-27). As interpretations concerning these glacial histories still diverge however, further sea-level observations are required in these areas too.

PART 4 : REFERENCES

BIBLIOGRAPHY

Aaby, B., 1974. Cyckliske klimavariationer de sidste 7500 år påvist ved undersøgelser af højmoser og marine transgressionsfaser. Danm. Geol. Unders., Årb.: 91-104.
Adey, W.H., 1978. Coral reef morphogenesis: a multidimensional model. Science, 202, 4370: 831-837.
Adey, W.H., 1986. Coralline algae as indicators of sea-level. In: O. van de Plassche (Editor), Sea-Level Research. Geo Books, Norwich, pp. 229-280.
Alhonen, P., 1979. The Quaternary history of the Baltic, Finland. In: V. Gudelis and L.K. Königsson (Editors), The Quaternary History of the Baltic. Univ. Uppsala, pp. 101-113.
Allard, M. and Tremblay, G., 1983. La dynamique littorale des îles Manitounuk durant l'Holocène. Z. Geomorph., Suppl., 47: 61-95.
Aloïsi J.C., Monaco A., Planchais N., Thommeret J. and Thommeret Y., 1978. The Holocene transgression in the Golfe du Lion, southwestern France: paleogeographic and paleobotanical evolution. Géogr. phys. Quat., 32 (2): 145-162.
Andel, T.H. van and Lianos, N., 1983. Prehistoric and historic shorelines of the southern Argolid Peninsula: a subbottom profiler study. Int. J. Naut. Archaeol. Underw. Explor., 12 (4): 303-324.
Andel, T.H. van and Lianos, N., 1984. High-resolution seismic reflection profiles for the reconstruction of postglacial transgressive shorelines: an example from Greece. Quat. Res., 22: 31-45.
Andrews, J.T., 1970. A Geomorphological Study of Post-Glacial Uplift with Particular Reference to Arctic Canada. Alden and Mowbray, Oxford, 156 pp. (quoted by Møller, 1987).
Andrews, J.T., 1986. Elevation and age relationships: raised marine deposits and landforms in glaciated areas, examples based on Canadian Arctic data. In: O. van de Plassche (Editor), Sea-Level Research. Geo Books, Norwich, pp. 67-95.
Andrews, J.T., 1987. Glaciation and sea level: a case study. In: R.J.N. Devoy (Editor), Sea Surface Studies: A Global View. Croom Helm, London, pp. 95-126.
Andrews, J.T., 1989. Postglacial emergence and submergence. In: R.J. Fulton (Editor), Quaternary Geology of Canada and Greenland. Geol. Surv. Can., n° 1, pp. 546-562.
Andrews, J.T. and Falconer, G., 1969. Late glacial and post-glacial history and emergence of the Ottawa Islands, Hudson Bay, Northwest Territories: evidence on the deglaciation of Hudson Bay. Can. J. Earth Sci. 6 (2): 1263-1276.
Andrews, J.T. and Miller, G.H., 1985. Holocene sea level variations within Frobisher Bay. In: J.T. Andrews (Editor), Quaternary Environments: Eastern Canadian Arctic, Baffin Bay, and Western Greenland. Allen & Unwin, London, pp. 585-607.
Andrews, J.T. and Peltier, W.R., 1989. Quaternary geodynamics in Canada. In: R.J. Fulton (Editor), Quaternary Geology of Canada and Greenland. Geol Surv. Can., n° 1, pp. 543-572.
Andrews, J.T. and Retherford, R.M., 1978. A reconnaissance survey of late Quaternary sea levels, Bella-Bella Bella-Coola region, central British Columbia coast. Can. J. Earth Sci., 15: 341-350.
Andrews, J.T., Buckley, J.T. and England, J.H., 1970. Late-glacial chronology and glacio-isostatic recovery, Home Bay, East Baffin Island, Canada. Geol. Soc. Am. Bull. 81: 1123-1148.

Andrews, J.T., King, C.A.M. and Stuiver, M., 1973. Holocene sea level changes, Cumberland coast, northwest England: eustatic and glacio-isostatic movements. Geol. Mijnb., 52: 1-12.
Antonioli, F. and Frezzotti, M., 1991. I sedimenti tardo-pleistocenici e olocenici compresi nella fascia costiera tra Sabaudia e Sperlonga. Mem. Soc. Geol. Ital. (in press).
Anundsen, K., 1985. Changes in shore-level and ice-front position in Late Weischel and Holocene, southern Norway. Nor. Geogr. Tidsskr., 39: 205-225.
Ariga, T., 1984. Geomorphological development of Shonai coastal plain, northeastern Japan: the behavior of alluvial fan's front since latest Pleistocene (in Japanese, with English abstract). Ann. Tohoku Geogr. Assoc., 36, 1: 13-24.
Armstrong, J., 1976. Post-Vashon Wisconsin glaciation, Fraser Lowland, British Columbia, Canada. In: D.J. Easterbrook and V. Sibrava (Editors), Quaternary Glaciations in the Northern Hemisphere. Belliingham, Prague, pp. 13-27.
Åse, L.E., 1980. Shore displacement at Stockholm during the last 1000 years. Geogr. Ann., 62 A(1-2): 83-91.
Åse, L.E. and Bergström, E., 1982. The ancient shorelines of the Uppsala esker around Uppsala and the shore displacement. Geogr. Ann., 64 A(3-4): 229-244.
Ash, J., 1987. Holocene sea levels in northern Viti Levu, Fiji. N. Z. J. Geol. Geophys., 30: 431-435.
Atwater, B.F., Hedel, C.W. and Helley, E.J., 1977. Late Quaternary depositional history, Holocene sea-level changes, and vertical crustal movement, southern San Francisco Bay, California. U.S. Geol. Surv. Prof. Pap., 1014, 15 pp.
Auer, V., 1951. Preliminary results of studies on the Quaternary geology of Argentine. Ann. Acad. Sci. Fenn., A, III, 25: 3-11.
Badyukov, D.D., 1982. Influence of geoid form changes and solid Earth deformations on sea level fluctuations (in Russian). In: P.A. Kaplin, R.K. Klige and A.L. Chepalyga (Editors), Sea and Oceanic Level Fluctuations for 15 000 Years. Nauka, Moskow, pp. 51-77.
Badyukov, D.D. and Kaplin, P.A., 1979. Sea level changes on the coasts of the USSR during the last 15,000 years. In: K. Suguio T.R. Fairchild, L. Martin and J.M. Flexor (Editors), Proc. 1978 Int. Symp. Coastal Evolution in the Quaternary. Univ. Sâo Paulo, pp. 135-169.
Balabanov J.P., Kvirkvelia B.D. and Ostrovsky A.B., 1981. The recent evolution in engineering geology and long-term shoreline changes in Pitzunda Peninsula (in Russian). Ministry of Geology of the USSR, 2nd Hydrogeological Division, Tbilisi, 202 pp.
Baltzer, F., 1970. Datation absolue de la transgression holocène sur la côte ouest de Nouvelle-Calédonie sur des échantillons de tourbes à palétuviers, interprétation néotectonique. C.R. Acad. Sci. Paris, D, 271: 2251-2254.
Banerjee, M. and Sen, P.K., 1987. Palaeobiology in understanding the change of sea level and coast line in Bengal basin during Holocene period. Indian J. Earth Sci., 14, 3-4: 307-320.
Bard, E., Hamelin, B., Fairbanks, R.G. and Zindler, A., 1990. Calibration of the ^{14}C timescale over the past 30 000 years using mass spectrometric U-Th ages from Barbados corals. Nature, 345: 405-410.
Barnett, T.P., 1983. Recent changes in sea level and their possible causes. Clim. Change, 5: 15-38.
Baroni, C. and Orombelli, G., 1991. Holocene raised beaches at Terra Nova Bay (Victoria Land, Antarctica). Quat. Res. (in press).

Barr, W., 1971. Postglacial isostatic movement in Northeastern Devon Island: a reappraisal. Arctic, 24 (4): 249-268.

Bartlett, A.S. and Barghoorn, E.S., 1973. Phytogeographic history of the Isthmus of Panama during the past 12,000 years (A history of vegetation, climate, and sea-level changes). In: A Graham (Editor), Vegetation and Vegetational History of Northern Latin America. Elsevier, Amsterdam, pp. 204-299.

Barusseau, J.P., Descamps, C., Giresse, P., Monteillet, J. and Pazdur, M., 1989. Nouvelle définition des niveaux marins le long de la côte nord-mauritanienne (Sud du Banc d'Arguin) pendant les cinq derniers millénaires. C.R. Acad. Sci. Paris, II, 309: 1019-1024.

Behre, K.E., 1986. Analysis of botanical macro-remains. In: O. van de Plassche (Editor), Sea-Level Research. Geo Books, Norwich, pp. 413-433.

Behre, K.E., Menke, B. and Streif, H., 1979. The Quaternary geological development of the German part of the North Sea. In: E. Oele, R.T.E. Schüttenhelm and A.J. Wiggers (Editors), The Quaternary History of the North Sea. Univ. Uppsala, pp. 85-113.

Belknap, D.F. and Kraft, J.C., 1977. Holocene relative sea-level changes and coastal stratigraphic units on the northwest flank of the Baltimore Canyon trough geosyncline. J. Sediment. Petrol., 47 (2): 610-629.

Belknap, D.F., Andersen, B.G., Anderson, R.S., Anderson, W.A., Borns, H.W., Jr., Jacobson, G.L., Kelley, J.T., Shipp, R.C., Smith, D.C., Stuckenrath, R., Jr., Thompson, W.B. and Tyler, D.A., 1987. Late Quaternary sea-level changes in Maine. In: D. Nummedal, O.H. Pilkey and J.D. Howard (Editors), Sea-Level Fluctuation and Coastal Evolution. Soc. Econ. Paleontol. Mineral., Spec. Publ. 41, Tulsa, pp. 71-85.

Belperio, A.P., 1979. Negative evidence for a mid-Holocene high sea level along the coastal plain of the Great Barrier Reef Province. Mar. Geol., 32: M1-M9.

Belperio, T., 1989. The greenhouse debate and rising sea-levels. Search, 20, 2: 48-50.

Belperio, A.P., Hails, J.R. and Gostin, V.A., 1983. A review of Holocene sea levels in south Australia. In: (D. Hopley, Editor), Australian Sea Levels in the Last 15 000 Years: A Review. Dep. Geogr., James Cook Univ. North Qld., Monogr. Ser., Occasion. Pap. No. 3. Townsville, pp. 37-47.

Belperio, A.P., Hails, J.R., Gostin, V.A. and Polach, H.A., 1984. The stratigraphy of coastal carbonate banks and Holocene sea levels of northern Spencer Gulf, South Australia. Mar. Geol., 61: 297-313.

Bennema, J., 1954. Holocene movements of land and sea level in the coastal area of the Netherlands. Geol. Mijnb., 16: 254-264.

Berelson, W.M., 1979. Barrier Island Evolution and its Effects on Lagoonal Sedimentation: Shackleford Banks, Back Sound, and Harkers Island: Cape Lookout National Seashore. Thesis, Duke University, Durham, N.C., 132 pp. (unpublished, quoted by Moslow and Heron, 1981).

Berger, R., 1983. Sea levels and tree-ring calibrated radiocarbon dates. In: P.M. Masters and N.C. Flemming (Editors), Quaternary Coastlines and Marine Archaeology. Academic Press, London, pp.51-61.

Berglund, B.E., 1964. The post-glacial shore displacement in eastern Blekinge, southeastern Sweden. Sver. Geol. Unders., C, 599: 47 pp.

Berglund, B.E., 1971. Littorina transgressions in Blekinge, South Sweden, A preliminary survey. Geol. Förening. i Stockholm Förhandling., 93: 625-652.

Berglund, B.E. and Welinder, S., 1972. Stratigrafin vid Siretorp. Fornvänen., 67: 73-93.

Bergsten, F., 1954. The land uplift in Sweden from the evidence of old water marks. Geogr. Ann., 36: 81-111 (quoted by Renberg and Segerström, 1981).

Berryman, K., 1987a. Tectonic processes and their impact on the recording of relative sea-level changes. In: (R.J.N. Devoy, Editor), Sea Surface Studies: A Global View. Croom Helm, London, pp. 127-161.

Berryman, K., 1987b. Report on sea-level research in New Zealand in relation to IGCP Project 200. In: (P.A. Pirazzoli, Editor), IGCP Project 200 Summary Final Report. CNRS-Intergéo, Paris, p. 68-71.

Berzin, L.E., 1967. On the age and sea level oscillations of early and middle Holocene Baltic Sea transgressions in the upper part of the Gulf of Riga (in Russian with English and German abstracts). Baltica, 3: 87-104.

Bigras, P. and Dubois, J.M.M., 1987. Répertoire commenté des datations ^{14}C du nord de l'estuaire et du golfe du Saint-Laurent, Québec et Labrador. Bull. Rech. Dép. Géogr. Univ. Sherbrooke, 94-95-96, 166 pp.

Bird, E.C.F. and Klemsdal, T., 1986. Shore displacement and the origin of the lagoon at Brusand, southwestern Norway. Nor. Geogr. Tidsskr., 40: 27-35.

Biryukov, V.Y., Faustova, M.A., Kaplin, P.A., Pavlidis, Y.A., Romanova, E.A. and Velichko, A.A., 1988. The paleogeography of the Arctic shelf and coastal zone of Eurasia at the time of the last glaciation (18,000 yr B.P.). Palaeogeogr., Palaeoclim., Palaeoecol., 68: 117-125.

Björck, S., 1979. Late Weichselian Stratigraphy of Blekinge, SE Sweden, and Water Level Changes in the Baltic Ice Lake. Dep. Quat. Geol., Thesis 7, Lund, 248 pp.

Black, R.F., 1974. Late Quaternary sea level changes, Umnak Island, Aleutians, their effects on ancient Aleuts and their causes. Quat. Res., 4, 3: 264-281.

Black, R.F., 1982. Holocene sea-level changes in the Aleutian Islands: new data from Atka Island. In: Holocene Sea Level Fluctuations, Magnitude and Causes. Dep. Geol. Univ. S.C., Columbia, pp. 1-12.

Blackwelder, B.W., 1980. Late Wisconsin and Holocene tectonic stability of the United States mid-Atlantic coastal region. Geol., 8: 534-537.

Blake, W. Jr, 1975. Radiocarbon age determinations and postglacial emergence at Cape Storm, southern Ellesmere Island, Arctic Canada. Geogr. Ann., A, 57 (1-2): 1-71.

Blanc A.C., 1936. La stratigraphie de la plaine côtière de la Basse Versilia (Italie) et la transgression flandrienne en Italie. Rev. Géogr. Phys. Géol. Dyn., 9: 129-160.

Bloom, A.L., 1963. Late-Pleistocene fluctuations of sealevel and postglacial crustal rebound in coastal Maine. Am. J. Sci., 261: 862-879.

Bloom, A.L., 1967. Pleistocene shorelines: a new test of isostasy. Geol. Soc. Am. Bull., 78: 1477-1494.

Bloom, A.L., 1970a. Holocene submergence in Micronesia as the standard for eustatic sea-level changes. Quaternaria, 12: 145-154.

Bloom, A.L., 1970b. Paludal stratigraphy of Truk, Ponape, and Kusaie, eastern Caroline Islands. Bull. Geol. Soc. Am., 81: 1895-1904.

Bloom, A.L., 1971. Glacial-eustatic and isostatic controls of sea level since last glaciation. In: K.K. Turekian (Editor), The Late Cenozoic Glacial Ages, Yale Univ. Press, pp. 355-379.

Bloom, A.L., 1977. Atlas of Sea-Level Curves. Cornell Univ., Ithaca.

Bloom, A.L., 1980. Late Quaternary sea level changes on south Pacific coasts: a study in tectonic diversity. In (N.A. Mörner, Editor), Earth Rheology, Isostasy, Eustasy. Wiley, New York, pp. 505-516.

Bloom, A.L. and Park, Y.A., 1985. Holocene sea-level history and tectonic movements, Republic of Korea. Quat. Res. Japan, 24: 77-83 (quoted by Ota, 1987).

Bloom, A.L. and Stuiver, M., 1963. Submergence of the Connecticut coast. Science, 139: 332-334.

Bloom, A.L. and Yonekura, N., 1985. Coastal terraces generated by sea-level change and tectonic uplift. In: M.J. Woldenberg (Editor), Models in Geomorphology. Binghamton Symp. Geomorphol. Int. Ser., 14, Allen & Unwin, Boston, pp. 139-154.

Bloom, A.L., Broecker, W.S., Chappell, J.M.A., Matthews, R.K. and Mesolella, K.J., 1974. Quaternary sea level fluctuations on a tectonic coast: new $^{230}Th/^{234}U$ dates from the Huon Peninsula, New Guinea. Quat. Res., 4: 185-205.

Bortolami G.C., Fontes J.C., Markgraf V. and Saliège J.F., 1977. Land, sea and climate in the northern Adriatic region during the late Pleistocene and Holocene. Palaeogeogr., Palaeoclim., Palaeoecol., 21: 139-156.

Bosch, J.H.A., 1988. The Quaternary Deposits in the Coastal Plains of Peninsular Malaysia. Geological Survey, Malaysia, Ipoh, Rep. QG/1.

Boulton, G.S., 1979. Glacial history of the Spitsbergen archipelago and the problem of a Barents Shelf ice sheet. Boreas, 8: 31-57.

Bourrouilh-Le Jan, F.G., Talandier, J. and Salvat, B., 1985. La diagénèse précoce depuis 6000 ans et la géomorphologie des couronnes atolliennes aux Tuamotu. Proc. 5th Int. Coral Reef Congr., Tahiti, vol. 3, pp. 235-240.

Brandt, K., 1980. Die Höhenlage ur- und frühgeschichtlicher Wohnniveaus in nordwestdeutschen Marschengebieten als Hölenmarken ehemaliger Wasserstände. Eiszeitalt. Ggw., 30: 161-170.

Broadbent, N., 1979. Coastal Resources and Settlement Stability, A Critical Study of a Mesolithic Site Complex in Northen Sweden. Archaeol. Stud. Aun 3. Uppsala Univ., Inst. North Eur. Archaeol., Uppsala, 268 pp.

Brookes, I.A., Scott, D.B. and McAndrews, J.H., 1985. Postglacial relative sea-level change, Port au Port area, west Newfoundland. Can. J. Earth Sci. 22: 1039-1047.

Brooks, M.J., Colquhoun, D.J., Pardi, R.R., Newman, W. and Abbott, W.H., 1979. Preliminary archeological and geological evidence for Holocene sea level fluctuations in the Lower Cooper River Valley, S.C. Fla. Anthr., 32, 3: 85-103.

Buddemeier, R.W., Smith, S.V. and Kinzie, R.A., 1975. Holocene windward reef-flat history, Enewetak Atoll. Geol. Soc. Am. Bull., 86: 1581-1584.

Bujalesky G.G. and González Bonorino, G., 1990. Evidence for stable sea level in the late Holocene in San Sebastián Bay, Tierra del Fuego, southernmost Argentina. In: Int. Symp. Quaternary Shorelines: Evolution, Processes and Future Changes (IGCP Project 274 and INQUA), Abstracts. La Plata, 19-20 Nov. 1990, p. 9.

Burne, R.V., 1982. Relative fall of Holocene sea level and coastal progradation, northern Spencer Gulf, south Australia. BMR J. Aust. Geol. Geophys., 7: 35-45.

Cabioch, G., Thomassin, B.A. and Lecolle, J.F., 1989. Age d'émersion des récifs frangeants holocènes autour de la «Grande Terre» de Nouvelle-Calédonie (SO Pacifique); nouvelle interprétation de la courbe des niveaux marins depuis 8000 ans. C.R. Acad. Sci. Paris, II, 308: 419-425.

Carrera, G. and Vanicek, P., 1988. A comparison of present sea level linear trends from tide gauge data and radiocarbon curves in eastern Canada. Palaeogeogr. Palaeoclimatol. Palaeoecol. 68 (2-4): 127-134.

Carter, R.W.G., 1982. Sea-level changes in Northern Ireland. Proc. Geol. Ass., 93: 7-23.

Carter, R.W.G., Johnston, T.W., McKenna, J. and Orford, J.D., 1987. Sea-level, sediment supply and coastal changes: examples from the coast of Ireland. Progr. Oceanogr., 18: 79-101.

Carter, R.W.G., Devoy, R.J.N. and Shaw, J., 1989. Late Holocene sea levels in Ireland. J. Quat. Sci., 4: 7-24.

Cathles, L.M., 1975. The Viscosity of the Earth's Mantle. Princeton Univ. Press, Princeton, N.J., 386 pp.

Chappell, J., 1974. Geology of coral terraces, Huon Peninsula, New Guinea: a study of Quaternary tectonic movements and sea-level changes. Geol. Soc. Am. Bull., 85: 553-570.

Chappell, J., 1983a. A revised sea-level record for the last 300,000 years from Papua New Guinea. Search, 14, 99-101.

Chappell, J., 1983b. Evidence for smoothly falling sea level relative to north Queensland, Australia, during the past 6000 yr. Nature, 302: 406-408.

Chappell, J., 1987. Late Quaternary sea-level changes in the Australian region. In: (M.J. Tooley and I. Shennan, Editors), Sea-Level Changes. Basil Blackwell, Oxford, pp. 296-331.

Chappell, J. and Polach, H.A., 1976. Holocene sea-level change and coral-reef growth at Huon Peninsula, Papua New Guinea. Geol. Soc. Am. Bull., 87: 235-240.

Chappell, J. and Shackleton, N.J., 1986. Oxygen isotopes and sea level. Nature, 324: 137-140.

Chappell, J., Rhodes, E.G., Thom B.G. and Wallensky, E, 1982. Hydro-isostasy and the sea-level isobase of 5500 B.P. in north Queensland, Australia. Mar. Geol., 49: 81-90.

Chappell, J., Chivas, A., Wallensky, E., Polach, H.A. and Aharon, P., 1983. Holocene palaeo-environmental changes, central to north Great Barrier Reef inner zone. BMR J. Aust. Geol. Geophys., 8: 223-235.

Chepalyga A.L., 1984. Inland sea basins. In: A.A. Velichleo, H.E. Wright and C.W. Barnosky (Editors), Late Quaternary Environments of the Soviet Union. London, Longman, pp. 229-247.

Chida, N., Matsumoto, H. and Obara, S., 1984. Recent alluvial deposit and Holocene sea-level change on Rikuzentakata coastal plain, northeast Japan (in Japanese, with English abstract). Ann. Tohoku Geogr. Assoc., 36, 4: 232-239.

Clague, J.J., 1983. Glacio-isostatic effects of the Cordilleran ice sheet, British Columbia, Canada. In: D.E. Smith and A.G. Dawson (Editors), Shorelines and Isostasy. Inst. Br. Geogr. Sp. Publ. 16, Acad. Press, London, pp. 321-343.

Clague, J.J., 1989. Sea levels on Canada's Pacific coast: past and future trends. Episodes, 12, 1: 29-33.

Clague, J.J., Harper, J.R., Hebda, R.J. and Howes, D.E., 1982. Late Quaternary sea levels and crustal movements, coastal British Columbia. Can. J. Earth Sci., 19, 3: 597-618.

Clark, J.A., 1976. Greenland's rapid postglacial emergence: a result of ice-water gravitational attraction. Geology, 4 (5): 310-312.

Clark, J.A., 1977a. Global Sea Level Changes Since the Last Glacial Maximum and Sea Level Constraints on the Ice Sheet Disintegration History. Unpubl. Ph.D. Thesis, Univ. Colorado, 105 p. (quoted by Ridley and Seeley, 1979).

Clark, J.A., 1977b. A numerical model of worlwide sea level changes on a viscoelastic Earth since 18,000 B.P. In: Earth Rheology and Late Cenozoic Isostatic Movements, Abstracts. Stockholm Univ., Geol. Inst., pp. 32-42.

Clark, J.A., 1980. A numerical model of worldwide sea level changes on a viscoelastic earth. In: N.A. Mörner (Editor), Earth Rheology, Isostasy and Eustasy. Wiley, New York, N.Y., pp. 525-534.

Clark, J.A., 1980a. The reconstruction of the Laurentide Ice Sheet of North America from sea-level data: methods and preliminary results. J. Geophys. Res., 85: 4307-4323.

Clark, J.A. and Bloom, A., 1979. The effect of the Patagonian ice sheet on relative sea levels along the Atlantic coast of South America: a numerical calculation. In: Proc. Int. Symp. Coastal Evol. in the Quat., Sâo Paulo, pp. 61- 76.

Clark, J.A. and Lingle, C.S., 1979. Predicted relative sea-level changes (18,000 years B.P. to Present) caused by late-glacial retreat of the Antarctic ice sheet. Quat. Res., 11: 279-298.

Clark, J.A., Farrell, W.E. and Peltier, W.R., 1978. Global changes in postglacial sea level: a numerical calculation. Quat. Res., 9: 265-278.

Clarke, R.D., 1970. Quaternary sediments off south-east Devon. Q. J. Geol. Soc. Lond., 125: 277-318.

Clet-Pellerin, M., Lautridou, J.P. and Delibrias, G., 1981. Les formations holocènes et pléistocènes de la partie orientale de la baie du Mont-Saint-Michel. Bull. Soc. Linn. Normandie, 109: 3-20.

Coleman, J.M. and Smith, W.G., 1964. Late recent rise of sea level. Geol. Soc. Am. Bull., 75: 833-840.

Colquhoun, D.J. and Brooks, M.J., 1987. New evidence for eustatic components in late Holocene sea levels. In: M.R. Rampino, J.E. Sanders, W.S. Newman and L.K. Königsson (Editors), Climate: History, Periodicity, and Predictability. Van Nostrand Reinhold, New York, pp. 143-156.

Cornen, G., Giresse, P., Kouyoumontzakis, G. and Moguedet, G, 1977. La fin de la transgression holocène sur les littoraux atlantiques d'Afrique équatoriale et australe, (Gabon, Congo, Angola, Sao Thomé, Annobon), rôles eustatiques et néotectoniques. Assoc. Sénégal. Et. Quat. Afr., Bull. Liaison, 50: 59-83.

Corner, G.D., 1980. Preboreal deglaciation chronology and marine limits of the Lyngen-Storfjord area, Troms, North Norway. Boreas, 9: 239-249.

Coudray, J. and Delibrias, G., 1972. Variations du niveau marin au-dessus de l'actuel en Nouvelle-Calédonie depuis 6000 ans. C.R. Acad. Sci. Paris, D, 275: 2623-2626.

Coudray, J. and Montaggioni, L., 1986. The diagenetic products of marine carbonates as sea-level indicators. In: O. van de Plassche (Editor), Sea-Level Research. Geo Books, Norwich, pp. 311-360.

Craig, B.G., 1969. Late-glacial and postglacial history of the Hudson Bay region. Geol. Surv. Can. Pap., 68-53: 63-77.

Cullen, D.J., 1967. Submarine evidence from New Zealand of a rapid rise in sea level about 11,000 years B.P. Palaeogeogr., Palaeoclim., Palaeoecol., 3: 289-298.
Cullingford, R.A., Caseldine, C.J. and Gotts, P.E., 1980. Early Flandrian land and sea-level changes in Lower Strathearn. Nature, 284: 159-161.
Curray, J.R., 1960. Sediments and history of Holocene transgression, continental shelf, southwest Gulf of Mexico. In: F.P. Shepard (Editor), Recent Sediments, Northwest Gulf of Mexico. Am. Assoc. Pet. Geol., pp. 221-266 (quoted by Curray, 1965).
Curray, J.R., 1961. Late Quaternary sea level: a discussion. Geol. Soc. Am. Bull., 72: 1707-1712.
Curray, J.R., 1965. Late Quaternary history, continental shelves of the United States. In: H.E. Whuglit and D.C. Frey (Editors), The Quaternary of the United States, VII° INQUA Congr., Princeton Univ. Press, pp. 723-735.
Dai Pra, G. and Hearty, P.J., 1991. Variazioni del livello del mare sulla costa ionica salentina durante l'Olocene - Epimerizzazione dell'isoleucina in *Helix sp.* Mem. Soc. Geol. Ital. (in press).
Dalongeville R., 1983. Variations de la ligne de rivage en Méditerranée orientale au cours de l'Holocène: témoins morphologiques et archéologiques. In: Les Cyclades. CNRS, Paris, pp. 89-98.
Dalongeville, R. and Sanlaville, P., 1987. Confrontation des datations isotopiques avec les données géomorphologiques et archéologiques à propos des variations relatives du niveau marin sur la rive arabe du Golfe Persique. In: O. Aurenche, J. Evin and F. Hours (Editors), Chronologies in the Near East. B.A.R. Int. Ser., Oxford, vol. 379, pp. 567-583.
Daly, R.A., 1934. The Changing World of the Ice Age. Hafner, New York, 271.
Danielsen, A., 1970. Pollen-analytical late Quaternary studies in the Ra district of Østfold, southeast Norway. Årb. Univ. Bergen., Math. Nat. Vitensko, 14, 143 pp.
Darienzo, M. and Peterson, C., 1990. Episodic tectonic subsidence of late Holocene salt marshes, northern Oregon and central Cascadia margin. Tectonics, 9: 1-22.
Davies, P.J. and Montaggioni, L., 1985. Reef growth and sea-level change: the environmental signature. Proc. 6th Int. Coral Reef Symp., Australia, vol. 3, pp. 477-511.
Degtyarenko, Y.P., Puminov, A.P. and Blagoveshensky, M.G., 1982. Late Pleistocene and Holocene shorelines of the eastern Arctic seas (in Russian). In: P.A. Kaplin, R.K. Klige and A.L. Chepalyga (Editors), Sea and Oceanic Level Fluctuations for 15 000 Years. Nauka, Moscow, pp. 179-185.
Delibrias, G., 1974. Variation du niveau de la mer, sur la côte ouest africaine, depuis 26000 ans. In: Méthodes Quantitatives d'Etudes des Variations du Climat au Cours du Pléistocène. Colloq. Int. C.N.R.S., N° 219, Paris, pp. 127-134.
Delibrias, G. and Guillier, M.T., 1971. The sea level on the Atlantic coast and the Channel for the last 10,000 years by the ^{14}C method. Quaternaria, 14: 131-135.
Delibrias, G. and Laborel, J., 1971. Recent variations of the sea level along the Brazilian coast. Quaternaria, 14: 45-49.
Delibrias, G. and Pirazzoli, P.A., 1983. Late Holocene sea-level changes in Yoron island, the Ryukyus, Japan. Mar. Geol., 53: M7-M16.

Delibrias, G., Giot, P.R., Gouletquer, P.L. and Morzadec-Kerfourn, M.T., 1971. Evolution de la ligne de rivage le long du littoral armoricain depuis le Néolithique. Quaternaria, 14: 175-179.
Denton, G.H., Borns, H.W. Jr, Grosswald, M.G., Stuiver, M. and Nichols, R.L., 1975. Glacial history of the Ross Sea. Antarct. J. U.S., 10: 160-164 (quoted by Lingle and Clark, 1979).
Devoy, R.J.N., 1977. Flandrian sealevel changes in the Thames Estuary and implications for land subsidence in England and Wales. Nature, 270, 5639: 712-715.
Devoy, R.J.N., 1979. Flandrian sea level changes and vegetational history of the lower Thames estuary. Phil. Trans. R. Soc. Lond., B, 285, 1010: 355-410.
Devoy, R.J.N., 1983. Late Quaternary shorelines in Ireland: an assessment of their implications for isostatic land movement and relative sea-level changes. In: D.E. Smith and A.G. Dawson (Editors), Shorelines and Isostasy. Academic Press, London, pp. 227-254.
Devoy, R.J.N., 1987. Introduction: first principles and the scope of sea-surface studies. In: R.J.N. Devoy (Editor), Sea Surface Studies, Croom Helm, London, pp. 1-30.
Digerfeldt, G., 1972. A preliminary report of an investigation of Littorina transgressions in the Barsebäck area, western Skåne. Geol. Fören. Stockh. Förh., 94: 537-548 (quoted by Fredén, 1979).
Digerfeldt, G., 1975. A standart profile for Littorina transgressions in western Skåne, South Sweden. Boreas, 4: 125-142.
Digerfeldt, G. and Enell, M., 1984. Paleoecological studies of the past development of the Negril and Black River Morasses, Jamaica. Dep. Quat. Geol. Lund and Pet. Corp. Jam. Kingston, 145 p.
Digerfeldt, G. and Hendry, M.D., 1987. An 8000 year Holocene sea-level record from Jamaica: implications for interpretation of Caribbean reef and coastal history. Coral Reefs, 5: 165-169.
Dillon, W.P. and Oldale, R.N., 1978. Late Quaternary sea-level curve: reinterpretation based on glaciotectonic influence. Geology, 6: 56-60.
Dionne, J.C., 1972. Le Quaternaire de la région de Rivière-du-Loup/Trois-Pistoles, côte sud de l'estuaire du Saint-Laurent. Cent. Rech. Forest. Laur., Québec, Rapp. Inf. Q-F-X-27, 95 pp. (quoted by Locat, 1977).
Dionne, J.C., 1988. Holocene relative sea-level fluctuations in the St. Lawrence estuary, Québec, Canada. Quat. Res. 29: 233-244.
Dionne, J.C., 1988a. Evidence d'un bas niveau marin durant l'Holocène à Saint-Fabien-sur-Mer, estuaire maritime du Saint-Laurent. Norois, 35, 137: 19-34.
Diop, S., 1990. La côte Ouest-Africaine: Du Saloum (Sénégal) à la Mellacorée (Rep. de Guinée). Etudes et Thèses, ORSTOM, Paris, 310 pp., Annexes.
Diop, E.S. and Sall, M., 1986. Estuaires et mangroves en Afrique de l'Ouest: évolution et changements - du Quaternaire récent à l'actuel-. In: Changements Globaux en Afrique Durant le Quaternaire: Passé-Présent-Futur, Symp. Int. INQUA-ASEQUA, Dakar 21-28 avril 1986. ORSTOM, Paris, pp. 109-114.
Dolukhanov, P.M., 1979. The Quaternary history of the Baltic, Leningrad and Soviet Carelia. In: V. Gudelis and L.K. Königsson (Editors), The Quaternary History of the Baltic. Univ. Uppsala, pp. 115-125.
Donner, J., 1965. Shoreline diagrams in Finnish Quaternary research. Baltica, 2: 11-20 (quoted by Møller, 1987).

Donner, J.J., 1968. The late-glacial and postglacial shoreline displacement in southwestern Finland. In: Means of Correlation of Quaternary Successions. Proc. VII INQUA Congr., vol. 8, Univ. Utah Press, pp. 367-373.

Donner, J.J., 1970. Land/sea level changes in Scotland. In: D. Walker and R.G. West (Editors), Studies in the Vegetational History of the British Isles. Cambridge Univ. Press, pp. 23-39.

Donner, J.J., 1980. The determination and dating of synchronous Late Quaternary shorelines in Fennoscandia. In: N.A. Mörner (Editor), Earth Rheology, Isostasy and Eustasy. Wiley, New York, N.Y., pp. 285-293.

Donner, J.J., 1983. The identification of Eemian interglacial and Weichselian interstadial deposits in Finland. Ann. Acad. Sci. Fenn., A, 136: 1-38.

Donner, J. and Jungner, H., 1975. Radiocarbon dating of shells from marine Holocene deposits in the Disko Bugt area, West Greenland. Boreas, 4: 25-45.

Donner, J. and Jungner, H., 1981. Radiocarbon dating of marine shells from southeastern Australia as a means of dating relative sea-level changes. Ann. Acad. Sci. Fenn., A, III, 131: 5-44.

Donner, J.J., Eronen, M. and Junger, H., 1977. The dating of the Holocene relative sea level changes in Finnmark, North Norway. Nor. Geogr. Tidsskr., 31: 103-128.

Dubar, M., 1987. Données nouvelles sur la transgression holocène dans la région de Nice (France). Bull. Soc. géol. Fr., (8), 3, (1): 195-198.

Duphorn, K., 1979. The Quaternary history of the Baltic, The Federal Republic of Germany. In: V. Gudelis and L.K. Königsson (Editors), The Quaternary History of the Baltic. Univ. Uppsala, pp. 195-206.

Dyke, A.S., 1977. Quaternary Geomorphology, Glacial Chronology, and Climatic and Sea-Level History of Southwestern Cumberland Peninsula, Baffin Island, N.W.T., Canada. Unpublished Ph.D. Thesis, Univ. Colorado, Boulder, 184 pp. (quoted by Clark, 1980).

Dyke, A.S., 1979a. Radiocarbon-dated Holocene emergence of Somerset Island, central Canadian Arctic. Current Research, Part B., Geol. Surv. Can., Pap. 79 (1B): 307-318.

Dyke, A.S., 1979b. Glacial and sea-level history of southwestern Cumberland Peninsula, Baffin Island, N.W.T., Canada. Arct. Alp. Res., 11: 179-202.

Dyke, A.S., Vincent, J.S., Andrews, J.T., Dredge, L.A. and Cowan, W.R., 1989. The Laurentide Ice Sheet and an introduction to the Quaternary geology of the Canadian Shield. In: R.J. Fulton (Editor), Quaternary Geology of Canada and Greenland. Geol. Surv. Can., n° 1, pp. 178-189.

Easton, W.H. and Olson, E.A., 1976. Radiocarbon profile of Hanauma reef, Oahu, Hawaii. Geol. Soc. Am. Bull., 87: 711-719.

Einsele, G., Herm, D. and Schwarz, H.U., 1974. Sea level fluctuation during the past 6000 yr at the coast of Mauritania. Quat. Res., 4: 282-289.

Einsele, G., Herm, D. and Schwarz, H.U., 1977. Variations du niveau de la mer sur la plate-forme continentale et la côte mauritanienne vers la fin de la glaciation de Würm et à l'Holocène. Assoc. Sénégal. Et. Quat. Afr., Bull. Liaison, 51: 35-48.

Ekman, M., 1988. The world's longest continued series of sea level observations. Pure Appl. Geophys. (Pageoph.), 127, 1: 73-77.

Ellison, J.C., 1989. Pollen analysis of mangrove sediments as a sea level indicator: assessment from Tongatapu, Tonga. Palaeogeogr., Palaeoclim., Palaeoecol., 74: 327-341.

Ellison, J.C. and Stoddart, D.R., 1991. Mangrove ecosystem collapse during predicted sea-level rise: Holocene analogues and implications. J. Coast. Res., 7, 1: 151-165.

Elson, J.A., 1969. Late Quaternary marine submergence of Quebec. Rev. Géogr. Montréal, 23 (3): 247-258.

Emery, K.O. and Garrison, L.E., 1967. Sea levels 7,000 to 20,000 years ago. Science, 157: 684-687.

Emery, K.O. and Milliman, J.D., 1970 Quaternary sediments of the Atlantic continental shelf of the United States. Quaternaria, 12: 3-18.

Emery, K.O., Wigley, R.L., Bartlett, A.S., Rubin, M. and Barghoorn, E.S., 1967. Freshwater peat on the continental shelf. Science, 158: 1301-1307.

Endo, K., Sekimoto, K. and Takano, T., 1982. Holocene stratigraphy and paleoenvironments in the Kanto plain, in relation to the Jomon transgression. Proc. Inst. Nat. Sci., Coll. Humanit. Sci., Nihon Univ., Earth Sci. 17: 1-16.

England, J., 1983. Isostatic adjustments in a full glacial sea. Can. J. Earth Sci., 20 (6): 895-917.

Ernst, T., 1974. Die Hohwachter Bucht, Morphologische Entwicklung einer Küstenlandschaft Ostholsteins. Schrift. Naturwiss. Ver. Schl.-Holst., 44: 47-96 (quoted by Klug, 1980).

Erol O., 1981. Summary report from Turkey. Sea-Level (Inf. Bull. IGCP Proj. 61), 6: 20-21.

Eronen, M., 1974. The history of the Litorina Sea and associated Holocene events. Soc. Sci. Fenn., Comment. Phys.-Math., 44: 79-195 (quoted by Hyvärinen and Eronen, 1979).

Eronen, M., 1976. A radiocarbon-dated Ancylus transgression site in south-eastern Finland. Boreas, 5: 65-76.

Eronen, M., 1983. Late Weichselian and Holocene shore displacement in Finland. In: D.E. Smith and A.G. Dawson (editors), Shorelines and Isostasy. Academic Press, London, pp.183-207.

Eronen, M. and Haila, H., 1982. Shoreline displacement near Helsinki, southern Finland, during the Ancylus Lake stage. Ann. Acad. Sci. Fenn., A, III, 134: 111-129.

Eronen, M., Kankainen, T. and Tsukada, M., 1987. Late Holocene sea-level record in a core from the Puget Lowland, Washington. Quat. Res., 27: 147-159.

Faegri, K., 1944. Studies on the Pleistocene of western Norway, III. Bømlo. Bergen Mus. Årb. 1943, Nat. Rk., 8, 100 pp. (quoted by Kaland, 1984).

Fairbanks R.G., 1989. A 17,000-year glacio-eustatic sea level record: influence of glacial melting rates on the Younger Dryas event and deep-ocean circulation. Nature, 342: 637-642.

Fairbridge, R.W., 1948. The geology and geomorphology of Point Peron, Western Australia. J. R. Soc. West. Aust., 34: 35-72.

Fairbridge, R.W., 1961. Eustatic changes in sea level. Phys. Chem. Earth, 4: 99-185.

Farinati, E.A., 1984. Dataciones radiocarbonicas en depositos holocenos de los alrededores de Bahia Blanca, Provincia de Buenos Aires, Argentina. In: International Symposium on Late Quaternary Sea-Level Changes and Coastal Evolution. Centro de Geologia de Costas, Mar del Plata, pp. 27-31.

Farrell, W.E. and Clark, J.A., 1976. On postglacial sea level. Geophys. J. R. Astron. Soc., 46: 647-667.

Fasano, J.L., Isla, F.I. and Schnack, E.J., 1984. Un análisis comparativo sobre la evolución de ambientes litorales durante el Pleistoceno tardío-Holoceno: Laguna Mar Chiquita (Buenos Aires) - Caleta Valdés (Chubut). In: Oscilaciones del nivel del mar durante el último hemiciclo deglacial en la Argentina. Grupo de Trabajo Argentino Proyecto N° 61, Mar del Plata, pp. 27-47.

Faure, H., 1980. Late Cenozoic vertical movements in Africa. In: N.A. Mörner (Editor), Earth Rheology, Isostasy and Eustasy. Wiley, Chichester, pp. 465-469.

Faure, H. and Hebrard, L., 1977. Variations des lignes de rivages au Sénégal et en Mauritanie au cours de l'Holocène. Stud. Geol. Pol., 52: 143-156.

Faure, H., Vieillefon, J. and Diop, C.A., 1974. Evolution de la ligne de rivage holocène en Casamance (Sud du Sénégal). Assoc. Sénégal. Et. Quat. Afr., Bull. Liaison, 42-43: 91-99.

Faure, H., Fontes, J.C., Hebrard, L., Monteillet, J. and Pirazzoli, P.A., 1980. Geoidal change and shore-level tilt along Holocene estuaries: Sénégal River area, West Africa. Science, 210: 421-423.

Fedorov, P.V., 1971. Postglacial transgression of the Black Sea. Translated in: Int. Geol. Rev., 14, (2): 160-164.

Fedorov, P.V., 1977. Late Quaternary history of the Black Sea and evolution of the South Europe Seas (in Russian). Nauka, Moscow, pp. 25-32.

Fedorov, P.V., 1981. Guidebook for excursions A-8 & C-8, XI INQUA Congr., Moscow, 1982, 31 pp.

Feller, C., Fournier, M., Imbert, D., Caratini, C. and Martin, L., 1990. Datations ^{14}C et palynologie d'un sédiment tourbeux continu (0-7 m) dans la mangrove de Guadeloupe (F.W.I.) - Résultats préliminaires. In: Symp. Int. Evolution des Littoraux des Guyanes et de la Zone Caraïbe Méridionale Pendant le Quaternaire (9-14 Nov. 1990), Résumés. ORSTOM, Cayenne, pp. 69-72.

Feng, H. and Wang, Z., 1986. Holocene sea-level changes and coastline shifts in Zhejiang Province, China. Dep. Geogr. Hangzhou Univ., 23 pp.

Feyling-Hanssen, R.W. and Olsson, I., 1959. Five radiocarbon datings of postglacial shorelines in central Spitsbergen. Nor. Geogr. Tidsskr., 1-4: 122-131.

Finkelstein, K. and Ferland, M.A., 1987. Back-barrier response to sea-level rise, eastern shore of Virginia. In: D. Nummedal, O.H. Pilkey and J.D. Howard (Editors), Sea-Level Fluctuation and Coastal Evolution. Soc. Econ. Paleontol. Mineral., Spec. Publ. 41, Tulsa, pp. 145-155.

Firth, C.R. and Haggart, B.A., 1989. Loch Lomond stadial and Flandrian shorelines in the inner Moray Firth area, Scotland. J. Quat. Sci., 4: 37-50.

Fisk, H.N., 1956. 8th Texas Conf. Soil Mechanics and Found. Engrg., p. 1 (quoted by Shepard, 1964).

Fitzhugh, W., 1977. Population movement and culture change on the central Labrador coast. Ann. N.Y. Acad. Sci., 288: 481-497.

Flemming N.C., 1972. Eustatic and tectonic factors in the relative vertical displacement of the Aegean coast. In: D.J. Stanley (Editor), The Mediterranean. Dowden, Hutchison & Ross, Stroudsburg, Pa., pp. 189-201.

Flemming N.C., 1978. Holocene eustatic changes and coastal tectonics in the northeast Mediterranean: implications for models of crustal consumption. Phil. Trans. R. Soc. Lon., A, 289 (1362): 405-458 + appendix 1.

Flemming, N.C., 1979-80. Archaeological indicators of sea level. Oceanis, 5, Fasc. Hors-Sér.: 149-166.

Flemming, N.C. and Webb, C.O., 1986. Tectonic and eustatic coastal changes during the last 10,000 years derived from archaeological data. Z. Geomorphol., Suppl. 62: 1-29.

Flemming, N.C. and Woodworth, P.L., 1988. Monthly mean sea levels in Greece during 1969-1983 compared to relative vertical land movements measured over different timescales. Teconophysics, 148: 59-72.

Flemming N.C., Czartoryska N.M.G. and Hunter P.M., 1973. Archaeological evidence for eustatic and tectonic components of relative sea level change in the South Aegean. Colston Pap. 23: 1-66.

Flemming N.C., Raban A. and Goetschel C., 1978. Tectonic and eustatic changes on the Mediterranean coast of Israel in the last 9000 years. Prog. Underw. Sci., 3: 33-93.

Fletcher, C.H. III, 1987. Holocene sea level history and neotectonics of the United States mid-Atlantic region: applications and corrections. J. Geol., 96: 323-337.

Fletcher, C.H., Fairbridge, R.W.H., MØller, J.J. and Long, A.J., 1990. 15 000 years of warm to cold arctic climate transitions: a paleoclimate record from the raised beaches of northern Norway. In: Int. Symp. "Quaternary Shorelines: Evolution, Processes and Future Changes", La Plata, 19-20 Nov. 1990, Abstracts, pp. 27-30.

Flint, R.F., 1971. Glacial and Quaternary Geology. Wiley, New York, 892 pp.

Flood, P.G. and Frankel, E., 1989. Late Holocene higher sea level indicators from eastern Australia. Mar. Geol., 90: 193-195.

Fontaine, H. and Delibrias, G., 1974. Niveaux marins pendant le Quaternaire au Viet-Nam. Arch. Géol. Viet-Nam, 17: 35-44.

Forman, S.L., Mann, D. and Miller, G.H., 1987. Late Weichselian and Holocene relative sea-level history of Bröggerhalvöya, Spitsbergen. Quat. Res., 27: 41-50.

Forsström, L. and Eronen, M., 1985. Flandrian and Eemian shore levels in Finland and adjacent areas: a discussion. Eizeitalt. Ggw., 35: 135-145.

Fredén, C., 1979. The Quaternary history of the Baltic, The western part. In: V. Gudelis and L.K. Königsson (Editors), The Quaternary History of the Baltic. Univ. Uppsala, pp. 59-74.

Fredoux, A. and Tastet, J.P., 1976. Apport de la palynologie à la connaissance paléogéographique du littoral ivoirien entre 8.000 et 12.000 ans B.P. 7th Afr. Micropalaeontol. Colloq., Ile-Ife, Nigeria, 1976, 10 pp.

Frydl, P.M., 1982. Holocene ostracods in the southern Boso peninsula. Univ. Mus., Univ. Tokyo, Bull., 20: 61-140.

Fujii, S. and Fuji, N., 1982. Postglacial sea-level changes in the Hokuriku region, central Japan (in Japanese, with English abstract). Daiyonki Kenkyu (Quat. Res.), 21, 3: 183-193.

Fujii, S., Lin, C.C. and Tjia, H.D., 1971. Sea level changes in Asia during the past 11,000 years. Quaternaria, 14: 211-216.

Fujii, S. et al., 1986. Submerged forest off Nyuzen, Kurobegawa alluvial fan, Toyama Bay, Central Japan. Boreas, 15, 4: 265-277.

Fulton, R.J. (Co-ordinator), 1989. Quaternary geology of the Canadian Shield. In: R.J. Fulton (Editor), Quaternary Geology of Canada and Greenland. Geol. Surv. Can., n° 1, p. 175-318.

Fulton, R.J., Anderson, T.W., Gadd, N.R., Harington, C.R., Kettles, I.M., Richard, S.H., Rodrigues, C.G., Rust, B.R. and Shilts, W.W., 1987. Summary of the Quaternary of the Ottawa region. In: H.M. French and P. Richard (Editors), Quaternary of the Ottawa Region and Guides for Day Excursions, XII INQUA Congress. Natl. Res. Counc. Can., Ottawa, pp. 7-20.

Funder, S., 1978. Holocene stratigraphy and vegetation history in the Scoresby Sund area, East Greenland. GrØn. Geol. Unders., 129: 66 pp.

Furukawa, H., 1972. Alluvial deposits of the Nobi plain, central Japan (in Japanese, with English abstract). Mem. Geol. Soc. Japan, 7: 39-59.

Gadd, N.R., 1973. Quaternary geology of southwest New Brunswick with particular reference to Fredericton area. Geol. Surv. Can., Pap. 71-34: 31 pp. (quoted by Grant, 1980).

Gabet, C., 1971. La phase terminale de la transgression flandrienne sur le littoral charentais. Quaternaria, 14: 181-188.

Galili, E., Weinstein-Evron, M. and Ronen, A., 1988. Holocene sea-level changes based on submerged archaeological sites off the northern Carmel coast in Israel. Quat. Res. 29: 36-42.

Geng, X., Wang, Y. and Fu, M., 1987. Holocene sea-level oscillations around Shandong Peninsula. In: Y. Qin and S. Zhao (Editors), Late Quaternary Sea-Level Changes. China Ocean Press, Beijing, pp. 81-96.

Geyh, M.A., Kudrass, H.R. and Streif, H., 1979. Sea-level changes during the late Pleistocene and Holocene in the Strait of Malacca. Nature, 278, 5703: 441-443.

Gibb, J.G., 1983. Sea levels during the past 10 000 years B.P. from the New Zealand region - south Pacific Ocean. In: Int. Symp. Coast. Evol. Holocene, Abstracts. Komazawa Univ., Tokyo, pp. 28-31.

Gibb, J.G., 1986. A New Zealand regional Holocene eustatic sea-level curve and its application to determination of vertical tectonic movements. R. Soc. N.Z. Bull., 24: 377-395.

Gifford, J.A., 1980. Paleogeography of archaeological sites of the Larnaca lowlands, southeastern Cyprus. Nivmer Inf., 5: 6.7.

Giresse, P., 1975. Nouveaux aspects concernant le Quaternaire littoral et sous-marin du secteur Gabon-Congo-Cabinda-Zaire et accessoirement de l'Angola. Assoc. Sénégal. Et. Quat. Afr., Bull. Liaison, 46: 45-52.

Giresse, P., Kouyoumontzakis, G. and Delibrias, G., 1976. La transgression fini-holocène en Angola, aspects chronologique, eustatique, paléoclimatique et épirogénique. C.R. Acad. Sci. Paris, D, 283: 1157-1160.

Giresse, P., Hoang C.T. and Kouyoumontzakis, G., 1984. Analysis of vertical movements deduced from a geochronological study of marine Pleistocene deposits, southern coast of Angola. J. Afr. Earth Sci., 2, (2): 177-187.

Giresse, P., Malounguila-N'Ganga, D. and Barusseau, J.P., 1986. Submarine evidence of the successive shorefaces of the Holocene transgression off southern Gabon and Congo. J. Coast. Res., Spec. Is. 1: 61-71.

Glückert, G., 1976. Post-glacial shore-level displacement of the Baltic in SW Finland. Ann. Acad. Sci. Fenn., A, III: 1-92.

Glückert, G. and Ristaniemi, O., 1982. The Ancylus transgression west of Helsinki, south Finland, A preliminary report. Ann. Acad. Sci. Fenn., A, III, 134: 99-134.

Godwin, H., 1940. Studies in the post-glacial history of British vegetation. IV. Post-glacial changes of relative land- and sea-level in the English Fenland. Phil. Trans. R. Soc. Lond., B, 230: 239-303.

Godwin, H., 1978. Fenland: Its Ancient Past and Uncertain Future. Cambridge Univ. Press, 196 pp.

Godwin, H., Suggate, R.P. and Willis, E.H., 1958. Radiocarbon dating of the eustatic rise in ocean-level. Nature, 181: 1518-1519.

Gornitz, V., 1991. Global coastal hazards from future sea level rise. Global Planet. Change, 3, 4: 379-398.

Gornitz, V. and Seeber, L., 1990. Vertical crustal movements along the East Coast, North America, from historic and late Holocene sea-level data. Tectonophysics, 178: 127-150.

Granlund, E., 1932. De svenska högmossarnas geologi. Sveriges Geol. Unders., C, 373, 193 pp.
Granlund, E., 1943. Beskrivning till jordartskarta över Västerbottens Län nedanför odlings-gränsen. Sveriges Geol. Unders., Ca. 26, 165 pp.
Grant, D.R., 1970. Recent coastal submergence of the Maritime Provinces, Canada. Can. J. Earth Sci., 7: 676-689.
Grant, D.R., 1975. Recent coastal submergence of the Maritime Provinces. Proc. N.S. Inst. Sci., 27, Suppl. 3: 83-102.
Grant, D.R., 1977. Glacial style and ice limits, the Quaternary stratigraphic record, and changes of land in the Atlantic Provinces, Canada. Geogr. Phys. Quat., 31: 247-260 (quoted by Quinlan and Beaumont, 1981).
Grant, D.R., 1980. Quaternary sea-level change in Atlantic Canada as an indication of crustal delevelling. In: N.A. Mörner (Editor), Earth Rheology, Isostasy and Eustasy. John Wiley and Sons, London, pp. 201-214.
Grant, D.R., unpublished? Variation of postglacial relative sea-level change from North to South in Newfoundland. (graph circulated in 1989).
Graul, H., 1959. Der Verlauf des glazialeustatischen Meeres-spiegelanstiegs, berechnet an Hand von C14 Datierungen (quoted by Müller, 1962).
Gray, J.M., 1983. The measurement of shoreline altitudes in areas affected by glacio-isostasy, with particular reference to Scotland. In; D.E. Smith and A.G. Dawson (Editors), Shorelines and Isostasy. Academic Press, London, pp. 97-127.
Gray, J., Boutray B. de, Hillaire-Marcel, C. and Lauriol, B., 1980. Postglacial emergence of the west coast of Ungava Bay, Québec. Arct. Alp. Res., 12 (1): 19-30.
Greensmith, J.T. and Tucker, E.V., 1973. Holocene transgressions and regressions on the Essex coast outer Thames estuary. Geol. Mijnb. 52: 193-202.
Greensmith, J.T. and Tucker, E.V., 1986. Compaction and consolidation. In: O. van de Plassche (Editor), Sea-Level Research. Geo Books, Norwich, pp. 591-603.
Grønlie, A., 1981. The late and postglacial isostatic rebound, the eustatic rise of the sea level and the uncompensated depression in the area of the Blue Road Geotraverse. Earth. Evol. Sci., 1: 50-57.
Grønlie, O.T., 1940. On the traces of the ice ages in Nordland, Troms and south-western part of Finnmark in northern Norway. Nor. Geol. Tidsskr., 20: 1-70 (quoted by Møller, 1987).
Gross, M.G., Milliman, J.D., Tracey, J.I. Jr. and Ladd, H.S., 1969. Marine geology of Kure and Midway Atolls, Hawaii: a preliminary report. Pac. Sci., 23: 17-25.
Grosswald, M.G. et al., 1961 (quoted by Badyukov and Kaplin, 1979).
Gudelis, V., 1979. The Quaternary history of the Baltic, Lithuania. In: V. Gudelis and L.K. Königsson (Editors), The Quaternary History of the Baltic. Univ. Uppsala, pp. 159-173.
Guéry, R., Pirazzoli, P.A. and Trousset, P., 1981. Les variations du niveau de la mer depuis l'Antiquité à Marseille et à la Couronne. Hist. Archéol., Dossiers, 50: 8-27.
Hafsten, U., 1956. Pollen-analytic investigations on the late Quaternary development in the inner Oslofjord area. Univ. Bergen Arb., Naturvitensk. R. 8, 161 pp.
Hafsten, U., 1983. Shore-level changes in South Norway during the last 13,000 years, traced by biostratigraphical methods and radiometric datings. Nor. Geogr. Tidsskr., 37: 63-79.

Haggart, B.A., 1989. Variations in the pattern and rate of isostatic uplift indicated by a comparison of Holocene sea-level curves from Scotland. J. Quat. Sci., 4: 67-76.
Hails, J.R., Belperio, A.P. and Gostin, V.A., 1983. Holocene sea levels of Upper Spencer Gulf, South Australia. In: (D. Hopley, Editor), Australian Sea Levels in the Last 15 000 Years: A Review. Dep. Geogr., James Cook Univ. North Qld., Monogr. Ser., Occasion. Pap. No. 3. Townsville, pp. 48-53.
Hald, M. and Vorren, T.O., 1983. A shore displacement curve from the Tromsø district, North Norway. Nor. Geol. Tidsskr., 63: 103-110.
Hantoro, W.S. et al., 1988. Quoted in: News Letter, Working Group II (East and Southeast Asia) of INQUA-CQS, N°1, pp. 6-7.
Harrison, W. and Lyon, C.J., 1963. Sea-level and crustal movements along the New England-Acadian shore, 4,500-3,000 B.P. J. Geol., 71 (1): 96-108.
Harrison, W., Malloy, R.J., Rusnak, J.A. and Terasmae, J., 1965. Possible late Pleistocene uplift, Chesapeake Bay entrance. J. Geol., 73: 201-229 (quoted by Newman et al., 1971).
Harten, D. van, 1986. Ostracode options in sea-level studies. In: O. van de Plassche (Editor), Sea-Level Research. Geo Books, Norwich, pp. 489-501.
Hawkins, A.B., 1971. The late Weichselian and Flandrian transgression of south west Britain. Quaternaria, 14: 115-130.
Hebda, R.J. and Mathewes, R.W., 1986. Radiocarbon dates from Anthony Island, Queen Charlotte Islands, and their geological and archaeological significance. Can. J. Earth Sci., 23, 12: 2071-2076.
Hendry, M.D., 1982. Late Holocene sea-level changes in western Jamaica. In: D.J. Colquhoun (Editor), Holocene Sea Level Fluctuations, Magnitude and Causes. Dep. Geol., Univ. S. C., Columbia, pp. 71-80.
Henoch, W.E.S., 1964. Postglacial marine submergence and emergence of Melville Island, N.W.T. Geogr. Bull. 22: 105-126.
Héquette, A., 1988. Vues récentes sur l'évolution du Svalbard au Quaternaire. Rev. Geomorphol. Dyn., 37, 4: 129-141.
Héquette, A. and Ruz M.H., 1989. Les variations postglaciaires de la ligne de rivage au Spitsberg nord-occidental, Svalbard. Z. Geomorphol., 33, 3: 323-337.
Heron, S.D., Jr., Moslow, T.F., Berelson, W.M., Herbert, J.R., Steele, G.A., III and Susman, K.R., 1984. Holocene sedimentation of a wave-dominated barrier-island shoreline: Cape Lookout, North Carolina. Mar. Geol., 60: 413-434.
Heyworth, A., 1986. Submerged forests as sea-level indicators. In: O. van de Plassche (Editor), Sea-Level Research. Geo Books, Norwich, pp. 401-411.
Heyworth, A., 1986a. Dendrochronological dating. In: O. van de Plassche (Editor), Sea-Level Research. Geo Books, Norwich, pp. 561-566.
Heyworth, A. and Kidson, C., 1982. Sea-level change in southwest England and Wales. Proc. Geol. Ass., 93: 91-111.
Hill, P.R., Mudie, P.J., Moran, K. and Blasco, S.M., 1985. A sea-level curve for the Canadian Beaufort Shelf. Can. J. Earth Sci., 22: 1383-1393.
Hillaire-Marcel, C., 1976. La déglaciation et le relèvement isostatique sur la côte Est de la baie d'Hudson. Cah. Géogr. Qué., 20(50): 185-220.
Hillaire-Marcel, C., 1979. Les mers postglaciaires du Québec: quelques aspects. Thèse Univ. Paris VI, Vol. 1, 293 pp.; vol. 2, 241 fig. & 26 pl.

Hillaire-Marcel, C., 1980. Multiple component postglacial emergence, Eastern Hudson Bay, Canada. In: N.A. Mörner (Editor), Earth Rheology, Isostasy and Eustasy. Wiley, New York, pp. 215-230.
Hillaire-Marcel, C. and Vincent, J.S., 1980. Paleo-Quebec, 11, Montreal (quoted by Peltier and Andrews, 1983).
Hine, A.C., Snyder, S.W. and Neumann, A.C., 1979. Coastal plain and inner shelf structure, stratigraphy, and geologic history: Bogue Banks area, North Carolina. Final Rep. to N. C. Sci. Technol. Comm. (quoted by Belknap et al., 1987).
Hirai, Y., 1987. Lacustrine and sublacustrine microforms and deposits near the shoreline of Lake Saroma and the sea level changes in the Sea of Okhotsk in the late Holocene (in Japanese, with English abstract). Ann. Tohoku Geogr. Assoc., 39, 1: 1-15.
Hjort, C. and Funder, S., 1974. The subfossil occurence of *Mytilus edulis* L. in central East Greenland. Boreas, 3: 23-33.
Hoika, J., Schütrumpf, R. and Schwabedissen, H., 1972. Süssau, ein neolithischer Wohnplatz an der Ostsee. Archäol. Korrespondenzblatt, 2: 1-19 (quoted by Klug, 1980).
Hopley, D., 1978. Sea level change on the Great Barrier Reef: an introduction. Phil. Trans. R. Soc. Lond., A, 291: 159-166.
Hopley, D., 1982. The Geomorphology of the Great Barrier Reef: Quaternary Development of Coral Reef. John Wiley-Interscience, New York, 453 pp.
Hopley, D., 1983. Deformation of the north Queensland continental shelf in the late Quaternary. In: D.E. Smith and A.G. Dawson (Editors), Shorelines and Isostasy. Acad. Press, London, pp. 347-366.
Hopley, D., 1986a. Beachrock as a sea-level indicator. In: O. van de Plassche (Editor), Sea-Level Research. Geo Books, Norwich, pp. 157-173.
Hopley, D., 1986b. Corals and reefs as indicators of paleo-sea levels, with special reference to the Great Barrier Reef. In: O. van de Plassche (Editor), Sea-Level Research. Geo Books, Norwich, pp. 195-228.
Huang, Z., Li ,P., Zhang, Z. and Zong, Y., 1986. Changes of sea level in the late Pleistocene in South Sea coasts (in Chinese with English abstract). In: China Sea Level Changes. China Ocean Press, Beijing, pp. 178-194.
Huang, Z., Li, P., Zhang, Z. and Zong, Y., 1987. Sea-level changes along the coastal area of south China since late Pleistocene. In: Y. Qin and S. Zhao (Editors), Late Quaternary Sea-Level Changes. China Ocean Press, Beijing, pp. 142-154.
Hughes, T, Denton, G.H. and Grosswald, H.G., 1977. Was there a late-Würm Arctic ice sheet? Nature, 266: 596-602.
Hurtig, T. von, 1959. Das physisch-geographische Bild der Ostsee und ihrer Küstenabschnitte und das Problem der postdiluvialen Überflutung des Ostseebeckens. Geogr. Berichte, 10/11: 46-63.
Hyvärinen, H., 1987. Models of fluctuating eustatic sea level in the Holocene falsified by a Baltic sea-level record. Poster presented at the XII INQUA Congr., Ottawa.
Hyvärinen, H. and Eronen, M., 1979. The Quaternary history of the Baltic, The northern part. In: V. Gudelis and L.K. Königsson (Editors), The Quaternary History of the Baltic. Univ. Uppsala, pp. 7-27.
Ida, Y., Yonekura, N. and Kayanne, H., 1984. Holocene sea-level changes in the southern Marianas. In: Sea-Level Changes and Tectonics in the Middle Pacific. Rep. HIPAC Proj. in 1981, 1982 and 1983. Kobe Univ., pp. 187-204.
IGBP, 1988. A Plan for Action. IGBP Secr., R. Swed. Acad. Sci., Stockholm, Rep., 4, 200 p.

IGBP, 1989. Global Changes of the Past. IGBP Secr., R. Swed. Acad. Sci., Stockholm, Rep., 6, 39 p.

Iseki, H. and Moriyama, A., 1980. See: Y. Ota, Y. Matsushima and H. Moriwaki (Compilers), 1981. Atlas of Holocene Sea Level Records in Japan, pp. 67-68.

Iseki, H., Fujii, S. and Fuji, N., 1982. Holocene sea level changes based on the data from boring cores at the west part of Nagoya harbor, central Japan (in Japanese, with English abstract). Daiyonki Kenkyu (Quat. Res.), 21, 3: 179-182.

Ivanov, V.F., 1982. Sea level fluctuations near Eastern Chukotka coast during late Pleistocene and Holocene. In: P.A. Kaplin, R.K. Klige and A.L. Chepalyga (Editors), Sea and Oceanic Level Fluctuations for 15 000 Years. Nauka, Moscow, pp. 190-195.

Ives, J.D., 1964. Deglaciation and land emergence in northeastern Foxe Basin. Geogr. Bull. 21: 54-65. Quoted by Müller and Barr, 1965.

Izmailov, Y.A., 1982. Results of the Holocene terraces structure studies in the region of Adler and Lazarevsky seas coasts (the Black Sea coast of the Caucases) (in Russian). in: P.A. Kaplin, R.K. Klige and A.L. Chepalyga (Editors), Sea and Oceanic Level Fluctuations for 15000 Years. Nauka, Moskow, pp. 156-161.

Jardine, W.G., 1975. Chronology of Holocene marine transgression and regression in south-western Scotland. Boreas, 4: 173-196.

Jardine, W.G., 1986. Determination of altitude. In: O. van de Plassche (Editor), Sea-Level Research. Geo Books, Norwich, pp. 569-590.

Jaritz, W., Ruder, J. and Schlenker, B., 1977. Das Quartär im Küstengebiet von Moçambique und seine Schwermineralführung. Geol. Jahrb., B, 26: 3-93.

Jelgersma, S., 1961. Holocene Sea Level Changes in the Netherlands. Ph.D. Thesis, Leiden. Also in: Meded. Geol. Stichting, C-IV (7), 100 pp.

Jelgersma, S., 1966. Sea-level changes during the last 10,000 years. Proc. Int. Symp. World Climate from 8000-0 BC. R. Meteor. Soc. Lond.: 54-71.

Jelgersma, S., 1977. Zeespiegelbeweging en bodemdaling. In: C.J. van Staalduinen (Editor), Geologisch Onderzoek van het Nederlandse Waddengebied. Rijks Geol. Dienst, Haarlem, pp. 72-74.

Jelgersma, S., 1979. Sea-level changes in the North Sea basin. In: E. Oele et al. (Editors), The Quaternary History of the North Sea, Acta Univ. Ups. Symp. Univ. Ups. Annum Quingentesimum Celebrantis, 2: 233-248.

Jelgersma, S., Roep, T.B. and Beets D.J., 1975. New data on sea-level changes in the Netherlands. Guidebook INQUA Holocene & Shorelines Meeting, France, Belgium, The Netherlands and NW Germany.

Jo, W., 1980. Holocene sea-level changes on the east coast of Korea Peninsula (in Japanese with English abstract). Geogr. Rev. Japan, 53, 5: 317-328.

Julian M., Lilienberg D.A. and Nicod J., 1987. Rôle des modifications des niveaux quaternaires des mers Caspienne, Noire et Méditerranée dans la formation du relief de leurs marges montagnardes. Méditerranée, 2-3: 19-35.

Kaizuka, S. Matsuda, T., Nogami, M. and Yonekura, N., 1973. Quaternary Tectonic and Recent Seismic Crustal Movements in the Arauco Peninsula and its Environs, Central Chile. Geogr. Rep. Tokyo Metrop. Univ., 8: 1-50.

Kaizuka, S., Naruse, Y. and Matsuda, I., 1977. Recent formations and their basal topography in an around Tokyo Bay, central Japan. Quat. Res., 8: 32-50.

Kaizuka, S., Miyauchi, T. and Nagaoka, S., 1983. Marine terraces, active faults and tectonic history of Iwo-jima (in Japanese, with English abstract). Ogasawara Res., 9: 13-45.

Kaland, P.E., 1984. Holocene shore displacement and shorelines in Hordaland, western Norway. Boreas, 13: 203-242.

Kaland, P.E., Krzywinski, K. and Stabell, B., 1984. Radiocarbon-dating of transitions between marine and lacustrine sediments and their relation to the development of lakes. Boreas, 13: 243-258.

Kale, V.S. and Rajaguru, S.N., 1985. Neogene and Quaternary transgressional and regressional history on the west coast of India, an overview. Bull. Decc. College Res. Inst., 44: 153-165 (quoted by Merh, 1987).

Kambouroglou, E., 1989. Eretria: Paleogeographic and Geomorphological Evolution During the Holocene - Relationship Between Ancient Environment and Ancient Inhabitation (in Greek). Ph.D. Thesis, Univ. Athens. Publication of the Municipality of Eretria, Athens.

Kambouroglou E., Maroukian H. and Sampsos A., 1988. Coastal evolution and archaeology north and south of Khalkis (Euboea) in the last 5000 years. In: A. Raban (Editor), Archaeology of Coastal Changes. BAR Int. Ser., Oxford, 404: 71-79.

Katto, J. and Akojima, I., 1980. Rate and mode of Holocene crustal movement in the Muroto Peninsula, Shikoku (in Japanese, with English abstract). In: A. Taira and M. Tashiro (Editors), Selected Papers in Honor of Prof. Jiro Katto. Rinyakosaiki Press, Kochi, pp. 1-15.

Katupotha, J. and Fujiwara, K., 1988. Holocene sea-level change of the southwest and south coasts of Sri Lanka. Palaeogeogr., Palaeoclim., Palaeoecol., 68, 2-4: 189-203.

Kawana, T., 1982. Vertical distribution of several kinds of organisms attaching to notches along the coasts of Yoron, Okinawa, Miyako and Ishigaki islands, the Ryukyus. Geol. Studies Ryukyu Islands, 6: 107-113.

Kawana, T. and Pirazzoli, P.A., 1984. Late Holocene shorelines and sea level in Miyako Island, the Ryukyus, Japan. Geogr. Rev. Japan, 57, B, 2: 135-141.

Kayan I., 1988. Late Holocene sea-level changes on the western Anatolian coast. Palaeogeogr., Palaeoclim., Palaeoecol., 68: 205-218.

Kayanne, H., Yonekura, N., Ishii, T. and Matsumoto, E., 1988. Geomorphic and geologic development of Holocene emerged reefs in Rota and Guam, Mariana Islands. In: Sea-Level Changes and Tectonics in the Middle Pacific. Rep. HIPAC Proj. in 1986 and 1987. Univ. Tokyo, pp. 35-57.

Kaye, C.A. and Barghoon, E.S., 1964. Late Quaternary sea-level change and crustal rise at Boston, Massachusets, with notes on the autocompaction of peat. Geol. Soc. Am. Bull., 75: 63-80.

Keene, H.W., 1971. Postglacial submergence and salt marsh evolution in New Hampshire. Marit. Sediments, 7 (2): 64-68.

Kelletat D., 1975. Eine eustatische Kurve für das jüngere Holozän, konstruiert nach Zeugnissen früherer Meeresspiegelstände im östlichen Mittelmeegebiet. N. Jb Geol. Paläont. Mh., 6: 360-374.

Kelletat, D., 1985. Studien zur spät- und postglazialen Küstenentwicklung der Varanger-Halbinsel, Nord-Norwegen. Essen Geogr. Arb., 10: 1-111.

Kelletat, D., 1987. German contributions to IGCP-200, Late-Quaternary sea-level correlations and applications. Inst. Geogr. Univ. Essen-GHS, 19 pp.

Kessel, H. and Raukas, A., 1979. The Quaternary history of the Baltic, Estonia. In: V. Gudelis and L.K. Königsson (Editors),

The Quaternary History of the Baltic. Univ. Uppsala, pp. 127-146.

Kiden, P., 1989. Holocene water level movements in the lower Scheldt perimarine area. In: C. Baeteman (Editor), Quaternary Sea-Level Investigations from Belgium. Geol. Dienst Belg., Brussel, Prof. Pap. 1989/6 Nr 241, pp. 1-19.

Kidson, C. and Heyworth, A., 1978. Holocene eustatic sea level change. Nature, 273, 5665: 748-750.

King, C.A.M., 1969. Glacial geomorphology and chronology of Henry Kater Peninsula, East Baffin Island, N.W.T. Arct. & Alp. Res. 1 (3): 195-212.

Kirk, R.M., 1977. Rates and forms of erosion on intertidal platforms at Kaikoura Peninsula, South Island, New Zealand. N.Z. J. Geol. Geophys., 20, 3: 571-613.

Kjemperud, A., 1981. A shoreline displacement investigation from Frosta in Trondheimsfjorden, Nord-Trøndelag, Norway. Nor. Geol. Tidsskr., 61: 1-15.

Kjemperud, A., 1986. Late Weichselian and Holocene shoreline displacement in the Trondheimsfjord area, central Norway. Boreas, 15: 61-82.

Kliewe, H. and Janke, W., 1982. Der Holozäne Wasserspiegelanstieg der Ostsee im nordöstlichen Küstengebiet der DDR. Petermanns Geogr. Mitt., 2: 65-74.

Klige, R.K., 1980. The Level of the Ocean in the Geological Past (in Russian). Nauka, Moscow, 111 pp.

Klug, H., 1980. Der Anstieg des Ostseespiegels im deutschen Küstenraum seit dem Mittelatlantikum. Eiszetalt. Ggw., 30: 237-252.

Koba, M., Nakata, T. and Watabe, S., 1979. Late Quaternary reef caps of Takara and Kodakara Islands, Ryukyu Archipelago, and sea-level changes of late Holocene (in Japanese, with English abstract). Earth Sci., 33: 173-191.

Koba, M., Omoto, K. and Takahashi, T., 1980. Late Holocene higher sea level and its radiocarbon dates in Okierabu-jima, Ryukyus (in Japanese, with English abstract). Daiyonki Kenkyu (Quat. Res.), 19, 4: 317-320.

Koba, M., Nakata, T. and Takahashi, T., 1982. Late Holocene eustatic sea-level changes deduced from geomorphological features and their ^{14}C dates in the Ryukyu Islands, Japan. Palaeogeogr., Palaeoclim., Palaeoecol., 39: 231-260.

Kolp, O., 1979. The Quaternary history of the Baltic, The relation between the eustatic rise of the sea-level, submarine terraces and the stages of the Baltic Sea during Holocene. In: V. Gudelis and L.K. Königsson (Editors), The Quaternary History of the Baltic. Univ. Uppsala, pp. 261-274.

Konishi, K., Oshiro, I. and Tanaka, T., 1979. Holocene raised coral reef on Senkaku islands. Proc. Jpn. Acad., 55, B, 7: 335-340.

Korotkii, A.M., 1985. Qyaternary sea-level fluctuations on the northwestern shelf of the Japan Sea. J. Coastal Res. 1, (3): 293-298.

Korotkii, A.M., Karaulova, L.P. and Troitskaya, T.S., 1980. The Quaternary of Primorye: Stratigraphy and Paleogeography (in Russian). Novosibirsk (quoted by Selivanov and Stepanov, 1985).

Köster, R., 1961. Junge eustatische und tektonische Vorgänge im Küstenraum der südwestlichen Ostsee. Meyniana, 11: 23-81 (quoted by Klug, 1980).

Köster, R., 1968. Postglacial sea-level changes in the western Baltic region in relation to worldwide eustatic movements. In: Means of Correlation of Quaternary Successions. Proc. VII INQUA Congr., vol. 8, Univ. Utah Press, pp. 421-435.

Köster, R., 1971. Postglacial sea-level changes on the German Northsea and Baltic shorelines. Quaternaria, 14: 97-100.

Kraft, J.C., 1971. Sedimentary facies patterns and geologic history of a Holocene marine transgression. Geol. Soc. Am. Bull., 82: 2131-2158.

Kraft, J.C., 1976. Radiocarbon dates in the Delaware coastal zone (eastern Atlantic coast of North America). Delaware Sea Grant Tech. Rep. 19-76, Univ. Delaware, Newark, 20 pp.

Kraft, J.C. and John, C.J., 1976. The geological structure of the shorelines of Delaware. Delaware Sea Grant Tech. Rep. 14-76, Univ. Delaware, Newark, 106 pp.

Kraft J.C., Rapp G., Jr. and Aschenbrenner S.E., 1975. Late Holocene paleogeography of the coastal plain of the Gulf of Messenia, Greece, and its relationships to archaeological settings and coastal change. Geol. Soc. Am. Bull., 86: 1191-1208.

Kraft J.C., Aschenbrenner S.E. and Rapp G., Jr., 1977. Paleogeographic reconstructions of coastal Aegean archaeological sites. Science, 195 (4282): 941-947.

Kraft J.C., Rapp G.R., Jr. and Aschenbrenner S.E., 1980a. Late Holocene palaeogeomorphic reconstructions in the area of the Bay of Navarino: Sandy Pylos. J. Archaeol. Sci., 7: 187-210.

Kraft J.C., Kayan I. and Erol O., 1980b. Geomorphic reconstructions in the environs of ancient Troy. Science, 209, (4458): 776-782.

Krog, H., 1979. The Quaternary history of the Baltic, Denmark. In: V. Gudelis and L.K. Königsson (Editors), The Quaternary History of the Baltic. Univ. Uppsala, pp. 207-217.

Kvasov, D.D., 1979. The Late Quaternary history of large lakes and inland seas of eastern Europe. Ann. Acad. Sci. Fenn., A, 127, 71 pp.

Labeyrie J., Lalou C., Monaco A. and Thommeret J., 1976. Chronologie des niveaux eustatiques sur la côte du Roussillon de -33 000 ans BP à nos jours. C.R. Acad. Sci.,D, 282: 349-352.

Laborel, J., 1979-80. Les Gastéropodes Vermetidés: leur utilisation comme marqueurs biologiques fossiles. Oceanis, 5, Fasc. Hors-Sér.: 221-239.

Laborel, J., 1986. Vermetid gastropods as sea-level indicators. In: O. van de Plassche (Editor), Sea-Level Research. Geo Books, Norwich, pp. 281-310.

Laborel, J., 1987. Marine biogenic constructions in the Mediterranean: a review. Sci. Rep. Port-Cros Natl. Park, Fr, 13: 97-126.

Laborel, J., Pirazzoli, P.A., Thommeret, J. and Thommeret, Y., 1979. Holocene raised shorelines in western Crete (Greece). In: Proc. "1978 International Symposium on coastal evolution in the Quaternary", Sâo Paulo, Brazil, pp. 475-501

Laborel J., Delibrias G. and Boudouresque C.F., 1983. Variations récentes du niveau marin à Port-Cros (Var, France), mises en évidence par l'étude de la corniche littorale à *Lithophyllum tortuosum*. C.R. Acad. Sci., II, 297: 157-160.

Lambeck, K., 1990. Glacial rebound, sea-level change and mantle viscosity. Q. J. R. Astron. Soc., 31: 1-30.

Lan, D., Yu, Y., Chen C. and Xie, Z., 1986. Preliminary study on late Pleistocene transgression and Holocene sea-level fluctuation in Fuzhou basin (in Chinese with English abstract). Mar. Geol. Quat. Geol., 6, 3: 103-111.

Landvik, J.Y., Mangerud, J. and Salvigsen, O., 1987. The late Weichselian and Holocene shoreline displacement on the west-central coast of Svalbard. Polar Res., 5: 29-44.

Landvik, J.Y., Mangerud, J. and Salvigsen, O., 1988. Glacial history and permafrost in the Svalbard area. Proc. V Int. Conf. Permafrost, Tapir Publ., Trondheim, pp. 29-44.
Larsonneur, C., 1971. Manche centrale et Baie de Seine: géologie du substratum et des dépôts meubles. Thèse d'Etat, Univ. Caen, 394 pp., 16 pl.
Larsonneur, C., 1977. De la Baie de Seine à l'estuaire de la Seine, histoire du Quaternaire marin. Bull. Soc. Géol. Normandie et Amis du Museum du Havre, 64, 4: 9-19.
Lasca, N.P., 1966. Postglacial delevelling in Skeldal, northeast Greenland. Arctic, 19 (4): 349-353.
Lasca, N.P., 1969. The surficial geology of Skeldal, Mesters Vig, Northeast Greenland. Medd. om Grønl., 176 (3): 56 pp.
Lassalle, P. and Rondot, J., 1967. New ^{14}C dates from the Lac St-Jean area, Quebec. Can. J. Earth Sci., 4: 568-571.
Lebuis, J. and David, P.P., 1977. La stratigraphie et les événements du Quaternaire de la partie occidentale de la Gaspésie, Québec. Géogr. Phys. Quat., 31 (3-4): 275-296 (quoted by Hillaire-Marcel, 1979).
Lee, H.A., 1962. Surficial geology of Rivière-du-Loup-Trois-Pistoles area, Quebec, 22 c/3, 21 N/14, 21 N/13 (E/2); Geol. Surv. Can., Pap. 61-32 (quoted by Elson, 1969).
Le Fournier, J., 1974. La sédimentation holocène en bordure du littoral picard et sa signification dynamique. Bull. Centre Rech. Pau-SNPA, 8, 1: 327-349.
Lewis, C.F.M., 1970. Recent uplift of Manitoulin Island, Ontario. Can. J. Earth Sci. 7: 665-675.
Lézine, A.M., Bieda, S., Faure, H. and Saos, J.L., 1985. Etude palynologique et sédimentologique d'un milieu margino-littoral: la tourbière de Thiaye (Sénégal). Sci. Géol. Bull., 38: 79-89.
L'Homer A., Bazile F., Thommeret J. and Thommeret Y., 1981. Principales étapes de l'édification du delta du Rhone de 7000 BP à nos jours; variations du niveau marin. Oceanis, 7 (4): 389-408.
Lidén, R, 1938. Den senkvartära strandförskjutningens fötlopp och kronologi i Ångermanland. Geol. Fören Stockholm Förhandl., 60: 397-404 (quoted by Mörner, 1979b).
Lidz, B.H. and Shinn, E.A., 1991. Paleoshorelines, reefs, and a rising sea: South Florida, U.S.A. J. Coast Res., 7, (1): 203-229.
Lie, S.E., Stabell, B. and Mangerud, J., 1983. diatom stratigraphy related to late Weichselian sea-level changes in Sunnmøre, western Norway. Nor. Geol. Unders., 208: 416-429 (quoted by Svendsen and Mangerud, 1987).
Liew, P.M. and Lin, C.F., 1987. Holocene tectonic activity of the Hengchun Peninsula as evidenced by the deformation of marine terraces. Mem. Geol. Soc. China, 9: 241-259.
Lighty, R.G., Macintyre, I.G. and Stuckenrath, R., 1982. *Acropora palmata* reef framework: a reliable indicator of sea-level in the western Atlantic for the past 10,000 years. Coral Reefs, 1: 125-130.
Liljegren, R., 1982. Paleoekologi och strandförskjuntning i en littorinavik i mellersta Blekinge. Dep. Quat. Geol. Lund., Thesis 11, 95 pp.
Lin, C.C., 1969. Holocene geology of Taiwan. Acta Geol. Taiwan., 13: 83-126.
Linden, R.H. and Schurer, P.J., 1988. Sediment characteristics and sea-level history of Royal Roads Anchorage, Victoria, British Columbia. Can. J. Earth Sci., 25, 11: 1800-1810.

Lingle, C.S. and Clark, J.A., 1979. Antarctic ice-sheet volume at 18 000 years BP and Holocene sea-level changes at the West Antarctic margin. J. Glaciol., 24, 90: 213-230.
Linke, G., 1982. Der Ablauf der holozänen Transgression der Nordsee aufgrund von Ergebnissen aus dem Gebiet Neuwerk/Scharhörn. Probl. Küstenforsch. südl. Nordseegebiet, 14: 123-157 (quoted by Winn et al., 1986).
Locat, J., 1977. L'émersion des terres dans la région de Baie-des-Sables/Trois-Pistoles, Québec. Géogr. Phys. Quat., 31 (3-4): 297-306.
Løken, O.H., 1965. Postglacial emergence at the south end of Inugsuin Fiord, Baffin Island, N.W.T. Geogr. Bull., 7 (3-4): 243-258.
Long, D., Laban, C., Streif, H., Cameron, T.D.J. and Schüttenhelm, R.T.E., 1988. The sedimentary record of climatic variation in the southern North Sea. Phil. Trans. R. Soc. Lond., B, 318: 523-537.
Longva, O., Larsen, E. and Mangerud, J., 1983. Stad. Skildring av kvartaergeologisk kart 1019 II - M 1:50 000 (med fargetrykt kart). Nor. Geol. Unders., 393: 1-66.
Lortie, G. and Guilbault, J.P., 1984. Les diatomées et les foraminifères de sédiments marins post-glaciaires du Bas-Saint-Laurent (Québec): une analyse comparée des assemblages. Nat. Can. (Rev. Ecol. Syst.), 111: 297-310.
Louwe Kooijmans, L.P., 1971. Mesolithic bone and antler implements from the North Sea and from the Netherlands. Berichten van het R.O.B., 20-21: 27-35.
Louwe Kooijmans, L.P., 1974. The Rhine/Meuse Delta. Ph. D. Thesis, Leiden, 421 pp.
Louwe Kooijmans, L.P., 1976. Prähistorische Besiedlung im Rhein-Maas-Deltagebiet und die Bestimmung ehemaliger Wasserhöhen. Probleme der Kustenforschung im sudlichen Nordseegebiet 11: 119-143.
Ludwig, G., Müller, H. and Streif, H., 1979. Neuere Daten zum holozänen Meeresspielelanstieg im Bereich der Deutschen Bucht. Geol. Jb., D, 32: 3-22 (quoted by Winn et al., 1986).
Lumley H. de, 1976. Les lignes de rivage. In: La Préhistoire Française. CNRS, Paris, vol. 3: 24-26.
Luternauer, J.L., Clague, J.J., Conway, K.W., Barrie, J.V., Blaise, B. and Mathewes, R.W., 1989. Late Pleistocene terrestrial deposits on the continental shelf of western Canada: evidence for rapid sea-level change at the end of the last glaciation. Geol. 17: 357-360.
Machida, H., Nakagawa, H. and Pirazzoli, P.A., 1976. Preliminary study on the Holocene sea levels in the central Ryukyu Islands. Rev. Geomorphol. Dyn., 25: 49-62.
Macintyre, I.G., Pilkey, O.H. and Stuckenrath, R., 1978. Relict oysters on the United States Atlantic continental shelf: a reconsideration of their usefulness in understanding late Quaternary sea-level history. Geol. Soc. Am. Bull., 89: 277-282 (quoted by Matthews, 1990).
Maeda, Y., 1978. Holocene transgression in Osaka Bay. Environmental changes in the Osaka area, Part III. J. Geosci. Osaka City Univ., 21: 33-63.
Maeda, Y., 1980. Holocene transgression in Osaka Bay and Harima Nada (in Japanese, with English abstract). Umi to Sora, 56: 145-150
Maeda, Y., 1984. Paleoenvironmental change during the Holocene on the coast along Okhotsk, Hokkaido (in Japanese). Dohosha, pp. 430-440.

Maeda, Y., Yamashita, K., Matsushima, Y. and Watanabe, M., 1983. Marine transgression over Mazukari shell mound on the Chita peninsula, Aichi Prefecture, central Japan (in Japanese, with English abstract). Daiyonki Kenkyu (Quat. Res.), 22, 3: 213-222.

Magnusson, E., 1970. Beskrivning till geologiska kartbladet Örebro NV. Sver. Geol. Unders., Ae, 6: 64-92.

Malounguila-N'Ganga, D., 1983. Les environnements sédimentaires des plates-formes du Nord-Congo et du Sud Gabon au Quaternaire supérieur d'après les données des vibro-carottages. Thèse Univ. Toulouse, 180 pp. + annexes.

Marcus, L.F. and Newman, W.S., 1983. Hominid migrations and the eustatic sea level paradigm: a critique. In: P.M. Masters and N.C. Flemming (Editors), Quaternary Coastlines and Marine Archaeology. Academic Press, London, pp. 63-85.

Mariette, H., 1971. L'archéologie des dépôts flandriens du Boulonnais. Quaternaria, 14: 137-150.

Markov, Y.D. et al., 1979. Traces of glacio-eustatic changes of the Japan Sea level in the area of the Gulf of Peter the Great. In: Bottom Geology of the Japan and Philippine Seas (New Data). Vladivostok (quoted by Korotkii, 1985).

Marshall, J.F. and Jacobson, G., 1985. Holocene growth of a mid-Pacific atoll: Tarawa, Kiribati. Coral Reefs, 4: 11-17.

Marthinussen M., 1960. Coast- and fjord area of Finnmark, With remarks on some other districts. In: O. Holtedahl (Editor), Geology of Norway. Nor. Geol. Unders., 208: 416-429 (quoted by Corner, 1980).

Marthinussen, M., 1962. ^{14}C datings referring to shore lines, transgressions, and glacial substages in Northern Norway. Nor. Geol. Unders., 215: 37-67.

Martin, L., 1973. Morphologie, sédimentologie et paléogéographie au Quaternaire récent du plateau continental ivoirien. Thèse Doct. d'Etat Sci. Nat., Univ. Paris VI, 340 pp.

Martin, L. and Delibrias, G., 1972. Schéma des variations du niveau de la mer en Côte-d'Ivoire depuis 25 000 ans. C. R. Acad. Sci. Paris D, 274: 2848-2851.

Martin, L., Suguio, K., Flexor, J.-M., Bittencourt, A. and Vilas-Boas, G., 1979-1980. Le Quaternaire marin brésilien (littoral Pauliste, sud Fluminense et Bahianais). Cah. ORSTOM, Géol., 11, 1: 95-124.

Martin L., Flexor J.-M., Blitzkow, D. and Suguio, K., 1985. Geoid change indications along the Brazilian coast during the last 7000 years. Proc. 5th Int. Coral Reef Congr., Tahiti, vol. 3, pp. 85-90.

Martin, L., Suguio, K. and Flexor, J.M., 1986. Shell middens as a source for additional information in Holocene shoreline and sea-level reconstruction: examples from the coast of Brazil. In: O. van de Plassche (Editor), Sea-Level Research. Geo Books, Norwich, pp. 503-521.

Mathews, W.H., Fyles, J.G. and Nasmith, H.W., 1970. Postglacial crustal movements in southwestern British Columbia and adjacent Washington state. Can. J. Earth Sci., 7, 2: 690-702.

Matsumoto, H., 1981. Sea-level changes during the Holocene and geomorphic developments of the Sendai coastal plain, northeast Japan (in Japanese, with English abstract). Geogr. Rev. Japan, 54, 2: 72-85.

Matsumoto, E., Matsushima, Y. and Miyata, T., 1986. Holocene sea-level studies by swampy coastal plains in Truk and Ponape, Micronesia. In: Sea-Level Changes and Tectonics in the Middle Pacific. Rep. HIPAC Proj. in 1984 and 1985. Kobe Univ., pp. 95-110.

Matsushima, Y., 1982. Radiocarbon ages of the Holocene marine deposits along Kucharo Lake, northern Hokkaido (in Japanese, with English abstract). Bull. Kanagawa Pref. Mus., 13: 51-66.
Matsushima Y. (Editor), 1987. Report of Kawasaki City Mus., I (quoted by Ota et al., 1987b).
Matthews, B., 1967. Late Quaternary land emergence in northern Ungava, Quebec. Arctic, 20 (3): 176-202.
Matthews, R.K., 1990. Quaternary sea-level change. In: Sea-Level Change. Geophysics Study Committee, National Research Council, Natl. Acad. Press, Washington, D.C., pp. 88-103.
McFarlan, E., Jr., 1961. Radiocarbon dating of late Quaternary deposits, South Louisiana. Geol. Soc. Am. Bull., 72: 129-158.
McIntire, W.G. and Morgan, J.P., 1964. Recent geomorphic history of Plum Island, Massachusetts, and adjacent coasts. La. State Univ. Press, Baton Rouge, 44 pp.
McLaren, P. and Barnett, D.M., 1978. Holocene emergence of the south and east coasts of Melville Island, Queen Elizabeth Islands, Northwest Territories, Canada. Arct. 31 (4): 415-427.
McManus, D.A. and Creager, J.S., 1984. Sea-level data for parts of the Bering-Chukchi shelves of Beringia from 19 000 to 10 000 ^{14}C yr BP. Quat. Res., 21: 317-325.
McRoberts, J.H.E., 1968. Post-glacial history of Northumberland Strait based on benthic foraminifera. Marit. Sediments, 4: 88-95 (quoted by Scott et al., 1987a).
Mechetin, A.V., 1981. Quoted by Korotkii, 1985 (no bibliographic reference given).
Menke, B., 1976. Befunde und Überlegungen zum nacheiszeitlichen Meeresspiegelanstieg (Dithmarschen und Eiderstedt, Schleswig-Holstein. Eiszetalt. Ggw., 21: 5-21 (quoted by Behre et al., 1979).
Merh, S.S., 1987. Quaternary sea level changes: the present status vis-à-vis records along coasts of India. Indian J. Earth Sci., 14, 3-4: 235-251.
Mikkelsen, V.M., 1949. Praesto Fjord, The development of the postglacial vegetation and a contribution to the history of the Baltic Sea. Dansk Bot. Arkiv, 13, 5: 1-171.
Miller, A.A.L., Mudie, P.J. and Scott D.B., 1982. Holocene history of Bedford Basin, Nova Scotia: foraminifera, dinoflagellate and pollen records. Can. J. Earth Sci., 19 (12): 2342-2367 (quoted by Carrera and Vanicek, 1988).
Milliman, J.D. and Emery, K.O., 1968. Sea levels during the past 35,000 years. Science, 162: 1121-1123.
Milliman, J.D., Broadus, J.M. and Gable, F., 1989. Environmental and economic implications of rising sea level and subsiding deltas: the Nile and Bengal examples. Ambio, 18 (6): 340-345.
Miyata, T., Maeda, Y., Matsumoto, E., Matsushima, Y., Rodda, P. and Sugimura, A., 1988. Emerged notches and microatolls on Vanua Levu, Fiji. In: Sea-Level Changes and Tectonics in the Middle Pacific. Rep. HIPAC Proj. in 1986 and 1987. Univ. Tokyo, pp. 67-76.
Mobley, C.M., 1988. Holocene sea levels in southeast Alaska: preliminary results. Arctic, 41, 4: 261-266.
Mojski, J.E., 1982. Geological section across the Holocene sediments in the northern and eastern parts of the Vistula deltaic plain. Polish Acad. Sci., Inst. Geogr. Spatial Organization, Geogr. Stud. Special Issue 1, Warvaw, pp. 149-169.
Møller, J.J., 1987. Shoreline relation and prehistoric settlement in Northern Norway. Nor. Geogr. Tidsskr., 41-1: 45-60.
Møller, J.J., 1989. Geometric simulation and mapping of Holocene relative sea-level changes in Northern Norway. J. Coast. Res., 5, 3: 403-417.

Montaggioni, L.F., 1976. Holocene submergence on Réunion Island (Indian Ocean). Ann. S. Afr. Mus., 71: 69-75.
Montaggioni, L.F., 1978. Recherches géologiques sur les complexes récifaux de l'archipel des Mascareignes (Océan Indien occidental). Thèse Univ. Aix-Marseille, 2 vol., 524 pp.
Montaggioni, L.F., 1981. Les variations relatives du niveau marin à l'île Maurice (Océan Indien) depuis 7000 ans. C. R. Acad. Sci. Paris, II, 293: 833-836.
Montaggioni, L.F., 1988. Holocene reef growth history in mid-plate high volcanic islands. Proc. 6th Int. Coral Reef Symp., Australia, vol. 3, pp. 455-460.
Montaggioni, L.F. and Pirazzoli, P.A., 1984. The significance of exposed coral conglomerates from French Polynesia (Pacific Ocean) as indicators of recent relative sea-level changes. Coral Reefs, 3: 29-42.
Mook, W.G. and Plassche, O. van de, 1986. Radiocarbon dating. In: O. van de Plassche (Editor), Sea-Level Research. Geo Books, Norwich, pp. 525-560.
Moriwaki, H., 1978. Problems concerning Holocene sea-level changes. Geogr. Rep. Tokyo Metrop. Univ., 13: 49-64.
Moriwaki, H., 1979. The landform evolution of the Kujukuri coastal plain, central Japan (in Japanese, with English abstract). Daiyonki Kenkyu (Quat. Res.), 18: 1-16.
Mörner, N.A., 1969. The Late Quaternary history of the Kattegatt Sea and the Swedish West Coast, deglaciation, shorelevel, displacement, chronology, isostasy and eustasy. Sver. Geol. Unders., C, 640: 1-487.
Mörner, N.A., 1971. The Holocene eustatic sea level problem. Geol. Mjnb., 50: 699-702.
Mörner, N.A., 1976. Eustatic changes during the last 8,000 years in view of radiocarbon calibration and new information from the Kattegatt region and other northwestern European coastal areas. Palaeogeogr., Palaeoclim., Palaeoecol., 19: 63-85.
Mörner, N.A., 1976a. Eustasy and geoidal changes. J. Geol., 84: 123-151.
Mörner, N.A., 1979a. South scandinavian sea level records: a test of regional eustasy, regional paleoenvironmental changes and global paleogeoid changes. In: K. Suguio T.R. Fairchild, L. Martin and J.M. Flexor (Editors), Proc. 1978 Int. Symp. Coastal Evolution in the Quaternary. Univ. Sâo Paulo, pp. 77-103.
Mörner, N.A., 1979b. The Fennoscandian uplift and late Cenozoic geodynamics: geological evidence. GeoJournal, 3, 3: 287-318.
Mörner, N.A., 1980. The Northwest European "sea-level laboratory" and regional Holocene eustasy. Palaeogeogr., Palaeoclim., Palaeoecol., 29: 281-300.
Mörner, N.A., 1988. Terrestrial variations within given energy, mass and momentum budgets; paleoclimate, sea level, paleomagnetism, differential rotation and geodynamics. In: F.R. Stephenson and A.W. Wolfendale (Editors), Secular Solar and Geomagnetic Variations in the Last 10,000 Years. Kluwer Acad. Publ., Dordrecht, pp. 455-478.
Mörner, N.A., 1989. Holocene sea level changes in the Tierra del Fuego region. Bull. INQUA Neotecton. Comm., 12: 85-87.
Mörner, N.A. (in press). Holocene sea level changes in the Tierra del Fuego region. In: Proc. Int. Symp. Global Changes in South America During the Quaternary: Past-Present-Future. Sâo Paulo, 1989.
Morzadec-Kerfourn, M.T., 1974. Variations de la ligne de rivage armoricaine au Quaternaire. Mém. Soc. Géol., Minéral. Bretagne, 17: 1-208.

Morzadec-Kerfourn, M.T., 1979-80. Indicateurs écologiques du domaine littoral: végétation et plancton organique. In: (P.A. Pirazzoli, Editor), Les Indicateurs de Niveaux Marins. Oceanis, 5, Fasc. Hors-Série, pp. 207-213.

Moslow, T.F. and Colquhoun, D.J., 1981. Influence of sea level change on barrier island evolution. Oceanis, 7, (4): 439-454.

Moslow, T.F. and Heron, S.D., 1981. Holocene depositional history of a microtidal cuspate foreland cape: Cape Lookout, North Carolina. Mar. Geol., 41: 251-270.

Mottershead, D.N., 1977. The Quaternary evolution of the south coast of England. In: C. Kidson and M.J. Tooley (Editors), The Quaternary History of the Irish Sea. Seel House Press, Liverpool, pp.299-320.

Müller, F. and Barr, W., 1966. Postglacial isostatic movement in Northeastern Devon Island, Canadian Arctic Archipelago. Arct., 19 (3): 263-269.

Müller, W., 1962. Der Ablauf de holozänen Meerestransgression an der südlichen Nordseeküste und Folgerungen in bezug auf eine geochronologische Holozängliederung. Eiszetalt. Ggw., 13: 197-226.

Munk, W. and Revelle, R., 1952. Sea level and the rotation of the earth. Am. J. Sci., 250: 829-833.

Nagasawa, R., 1983. Late Holocene sea-level changes and coastal landform development in Tanabe Bay, southwestern Japan (in Japanese, with English abstract). Ann. Tohoku Geogr. Assoc., 35, 1: 11-19.

Nakada, M. and Lambeck, K., 1987. Glacial rebound and relative sea-level variations: a new appraisal. Geophys. J. R. Astron. Soc., 90: 171-224.

Nakada, M. and Lambeck, K., 1989. Late Pleistocene and Holocene sea-level change in the Australian region and mantle rheology. Geophys. J., 96: 497-517.

Nakata, T., Takahashi, T. and Koba, M., 1978. Holocene emerged coral reefs and sea-level changes in the Ryukyu Islands (in Japanese, with English abstract). Geogr. Rev. Japan, 51: 87-108.

Nakata, T., Koba, M., Jo, W., Imaizumi, T., Matsumoto, H. and Suganuma, T., 1979. Holocene marine terraces and seimic crustal movements. Sci. Rep. Tohoku Univ., 7th ser., 29, 2: 195-204.

Nakata, T., Koba, M., Imaizumi, T., Jo, W.R., Matsumoto, H. and Suganuma, T., 1980. Holocene marine terraces and seismic crustal movements in the southern part of Boso Peninsula, Kanto, Japan (in Japanese, with English abstract). Geogr. Rev. Japan, 53: 29-44.

Nakiboglu, S.M., Lambeck, K. and Aharon, P., 1983. Postglacial sealevels in the Pacific: implications with respect to deglaciation regime and local tectonics. Tectonophysics, 91: 335-358.

Nardin, T.R., Osborne, R.H., Bottjer, D.J. and Scheidemann, R.C.Jr., 1981. Holocene sea-level curves for Santa Monica shelf, California continental borderland. Science, 213: 331-333.

Naruse, T., Onoma, M. and Murakami, R.,1984. Data on late Holocene sea level changes in the coastal area along the Harimanada, Seto Inland Sea (in Japanese). Daiyonki Kenkyu (Quat. Res.), 22, 4: 327-331.

Neev D., Bakler N. and Emery K.O., 1987. Mediterranean coasts of Israel and Sinai. Taylor & Francis, New York, 130 pp.

Nelson, A.R., 1978. Quaternary glacial and marine stratigraphy of the Qivitu Peninsula, northern Cumberland Peninsula, Baffin Island, Canada. PhD dissertation, Univ. Colorado, Boulder (quoted by Andrews, 1987).

Nelson, H.F. and Bray, E.E., 1970. Stratigraphy and history of the Holocene sediments in the Sabine-High Island area, Gulf of Mexico. In: J.P. Morgan (Editor), Deltaic Sedimentation Modern and Ancient, Soc. Econ. Paleontol. Mineral. Spec. Publ. 15, pp. 48-77.

Nesteroff W.D., 1984. Etude de quelques grès de plage du sud de la Corse: datations ^{14}C et implications néotectoniques pour le bloc corso-sarde. In: Le Beach-rock. Travaux de la Maison de l'Orient, Lyon, 8: 99-111.

Neumann, C.A., 1971. Quaternary sea-level data from Bermuda. Quaternaria, 14: 41-43.

Nevessky E.N., 1970. Holocene history of the coastal shelf zone of the USSR in relation with processes of sedimentation and condition of concentration of useful minerals. Quaternaria, 12: 78-88.

Newell & Bloom, 1970. The reef flat and 'Two-meter terrace' of some Pacific atolls. Geol. Soc. Am. Bull., 81: 1881-1894.

Newman, W.S., 1966. Late Pleistocene environments of the western Long Island area. Ph.D. Dissertation, New York Univ. (quoted by Newman et al., 1971).

Newman, W.S., 1986. Palaeogeodesy data bank and palaeogeodesy. IGCP Project 200 Newletter and Annual Report, Univ. of Durham, p. 41.

Newman, W.S. and Baeteman, C., 1987. Holocene excursions of the northwest European geoid. Progr. Oceanogr., 18: 287-322.

Newman, W.S. and Munsart, C.A., 1968. Holocene geology of the Wachapreague lagoon, Eastern Shore Peninsula, Virginia. Mar. Geol., 6: 81-105.

Newman, W.S. and Rusnak, 1965. Holocene submergence of the eastern shore of Virginia. Science, 148: 1464-1466.

Newman, W.S., Thurber, D.H., Zeiss, H.S., Rokach, A. and Musich, L., 1969. Late Quaternary geology of the Hudson River estuary: a preliminary report. N.Y. Acad. Sci. Trans., 31: 548-570 (quoted by Newman et al., 1971).

Newman, W.S., Fairbridge, R.W. and March, S., 1971. Marginal subsidence of glaciated areas: United States, Baltic and North Seas. In: M. Ters (Editor), Etude sur le Quaternaire dans le Monde, VIII° Congr. INQUA Paris, 1969, pp. 795-801.

Newman, W.S., Cinquemani, L.J., Pardi, R.R. and Marcus, L.F., 1980. Holocene delevelling of the United States' East Coast. In N.A. Mörner: (Editor), Earth Rheology, Isostasy and Eustasy. Wiley, London, pp. 449-463.

Newman, W.S., Cinquemani, L.J., Sperling, I.A., Marcus, L.F. and Pardi, R.R., 1987. Holocene neotectonics and the Ramapo Fault Zone anomaly: a study of varying marine transgression rates in the lower Hudson estuary, New York and New Jersey. In: D. Nummedal, O.H. Pilkey and J.D. Howard (Editors), Sea-Level Fluctuation and Coastal Evolution. Soc. Econ. Paleontol. Mineral., Spec. Publ. 41, Tulsa, pp. 97-111.

Newman, W.S., Pardi, R.R. and Fairbridge, R.W., 1989. Some considerations of the compilation of Late Quaternary sea level curves: a North American perspective. In: D.B. Scott et al. (Editors), Late Quaternary Sea-Level Correlation and Applications. NATO ASI Ser., C, 256, pp. 207-228.

Nichols, R.L., 1968. Coastal geomorphology, McMurdo Sound, Antarctica. J. Glaciol., 7, 51: 449-478 (quoted by Lingle and Clark, 1979).

Nilsson, E., 1953. Om södra Sveriges senkvartära historia. Geol. Fören. Stockh. Förh., 75: 155-246 (quoted by Liljegren, 1982).

Nikonov, A.A., 1977 (quoted by Badyukov and Kaplin, 1979)

Nilsson, T., 1948. Versuch einer Anknüpfung der postglazialen Entwicklung des nordwestdeutschen und niederländischen Flachlandes an die pollenfloristische Zonengliederung Südskandinaviens. Medd. Lunds Geol. Min. Inst., 112: 1-79.

Nir Y. and Eldar I., 1987. Ancient wells and their geoarchaeological significance in dating tectonics of the Israel Mediterranean coastline region. Geology, 15: 3-6.

NIVMER (Editor), 1979-80. Les Indicateurs de Niveaux Marins. Oceanis, 5, Fasc. Hors-Sér.: 145-360.

NIVMER Informations: 2 (1978), 3 (1978), 4 (1979), 5 (1980), 7 (1981), 8 (1982).

Nunn, P.D., 1990. Coastal processes and landforms of Fiji: their bearing on Holocene sea-level changes in the south and west Pacific. J. Coast. Res., 6, 2: 279-310.

Oldale, R.N. and O'Hara, C.J., 1980. New radiocarbon dates from the inner continental shelf off southeastern Massachusets and a local sea- level-rise curve for the past 12,000 years. Geol., 8: 102-106.

Omoto, K., 1979. Holocene sea-level changes: A critical review. Sci. Rep. Tohoku Univ., 7th ser., 29: 205-222.

Ostrovsky A.B., Izmaylov Y.A., Balabanov I.P., Skiba S.I., Skryabina N.G., Arslanov S.A., Gey N.A. and Suprunova, N.I., 1977. New data on the paleohydrological regime of the Black Sea in the Upper Pleistocene and Holocene. In: Paleogeography and Deposits of the Pleistocene of the Southern Seas of the USSR. Nauka, Moscow, pp. 131-141 (quoted by Chepalyga, 1984).

Ota, Y., 1987. Sea-level changes during the Holocene: the northwest Pacific. In: R.J.N. Devoy (Editor), Sea Surface Studies. Croom Helm, London, pp. 348-374.

Ota, Y. (Editor), 1987a. Holocene Coastal Tectonics of Eastern North Island, New Zealand. Dep. Geogr., Yokohama Natl. Univ., Yokohama, 104 pp.

Ota, Y. (Editor), 1987b. Holocene coastal tectonics of eastern North Island, New Zealand. Bull. INQUA Neotectonics Comm., 10: 64-70.

Ota, Y., Machida, H., Hori, N., Konishi, K. and Omura, A., 1978. Holocene raised coral reefs of Kikai-jima (Ryukyu Islands) - An approch to Holocene sea level study (in Japanese, with English abstract). Geogr. Rev. Japan, 51: 109-130.

Ota, Y., Matsushima, Y. and Moriwaki, H. (Compilers), 1981. Atlas of Holocene Sea Level Records in Japan. Japanese Working Group of the IGCP Project 61, Dep. Geogr. Yokohama Natl. Univ., Yokohama, 195 pp.

Ota, Y., Ishibashi, K., Matsushima, Y., Matsuda, T., Miyoshi, M., Kashima, K. and Matsubara, A., 1986. Holocene relative sea-level change in the southern part of Izu peninsula, central Japan; data from subsurface investigation (in Japanese, with English abstract). Daiyonki Kenkyu (Quat. Res.), 25, 3: 203-223.

Ota, Y., Matsushima, Y. and Umitsu, M. (Compilers), 1987a. Atlas of Late Quaternary Sea Level Records in Japan. I: Review Papers and Holocene. Japanese Working Group of IGCP Project 200, Dep. Geogr. Yokohama Natl. Univ., Yokohama, 580 pp.

Ota, Y., Matsushima, Y., Umitsu, M. and Kawana, T., 1987b. Middle Holocene Shoreline Map of Japan. Japanese Working Group for IGCP Project 200. Dep. Geogr. Yokohama Natl. Univ., Yokohama.

Ota, Y., Berryman, K.R., Hull, A.G., Miyauchi, T. and Iso, N., 1988. Age and height distribution of Holocene transgressive deposits in eastern North Island, New Zealand. Palaeogeogr., Palaeoclim., Palaeoecol., 68: 135-151.

Palmer, A.J.M. and Abbott, W.H., 1986. Diatoms as indicators of sea-level change. In: O. van de Plassche (Editor), Sea-Level Research. Geo Books, Norwich, pp. 457-487.

Papageorgiou, S. and Stiros, S., 1990. Environmental changes, seismic activity and archaeological research in NW Greece (in Greek). In: Proc. First Historical and Archaeological Congr. of Aetoloakamania, Agrinio 21-23 Oct. 1988 (in press).

Paradis, G., 1976. Recherches sur le Quaternaire récent du Sud de la R.P. du Bénin (ex Dahomey): étude de thanatocénoses de mollusques. Thèse Univ. Paris-Sud, Orsay. Notes C.E.R.P.A.B., Contrib. 12, 173 pp. + 15 pl., 63 fig.

Pardi, R.R.,and Newman, W.S., 1987. Late Quaternary sea levels along the Atlantic coast of North America. J. Coast. Res., 3, (3): 325-330.

Paskoff, R. and Oueslati, A., 1991. Modifications of coastal conditions in the Gulf of Gabes (Southern Tunisia) since classical Antiquity. Z. Geomorphol., Suppl. 81: 149-162.

Paskoff R. and Sanlaville P., 1983. Les côtes de la Tunisie. Variations du niveau marin depuis le Tyrrhénien. Maison de l'Orient Méditerranéen, Lyon, 192 pp.

Påsse, T., 1987. Shore displacement during the late Weichselian and Holocene in the Sandsjöbacka area, SW Sweden. Geol. Fören. Stockholm Förh., 109-3: 197-210.

Peltier, W.R., 1974. The impulse response of a Maxwell earth. Rev. Geophys. Space Phys., 12: 649-669.

Peltier, W.R., 1976. Glacial isostatic adjustments - II: The inverse problem. Geophys. J. R. Astron. Soc., 46: 605-646.

Peltier, W.R., 1988. Lithospheric thickness, Antarctic deglaciation history, and ocean basin discretization effects in a global model of postglacial sea level change: a summary of some sources of nonuniqueness. Quat. Res., 29: 93-112.

Peltier, W.R., 1990. Glacial isostatic adjustment and relative sea-level change. In: Sea-Level Change. Geophysics Study Committee, National Research Council, Natl. Acad. Press, Washington, D.C., pp. 73-87.

Peltier, W.R., 1991. The ICE-3G model of late Pleistocene deglaciation: construction, verification, and applications. In: R. Sabadini et al. (Editors), Glacial Isostasy, Sea-Level and Mantle Rheology. NATO ASI Ser., C, 334, Kluwer, Dordrecht, pp. 95-119.

Peltier, W.R. and Andrews, J.T., 1976. Glacial-Isostatic adjustment - I: the forward problem. Geophys. J. R. Astron. Soc., 46: 605-646.

Peltier, W.R. and Andrews, J.T., 1983. Glacial geology and glacial isostasy of the Hudson Bay region. In: D.E. Smith and A.G. Dawson (Editors), Shorelines and Isostasy. Academic Press, London. pp. 285-319.

Peltier, W.R., Farrell, W.E. and Clark, J.A., 1978. Glacial isostasy and relative sea level: a global finite element model. Tectonophys., 50: 81-110.

Peng, T.H., Li, Y.H. and Wu, F.T., 1977. Tectonic uplift rates of the Taiwan Island since the early Holocene. Mem. Geol. Soc. China, 2: 57-69.

Petersen, K.S., 1981. The Holocene marine transgression and its molluscan fauna in the Skagerrak-Limfjord region, Denmark. Spec. Publ. Int. Assoc. Sedimentol., 5: 497-503.

Petersen, K.S., 1985. The late Quaternary history of Denmark, The Weichselian ice sheets and land/sea configuration in the late Pleistocene and Holocene. J. Danish Archaeol., 4: 7-22.

Petersen, K.S., 1986. Marine molluscs as indicators of former sea-level stands. In: O. van de Plassche (Editor), Sea-Level

Research: A Manual for the Collection and Evaluation of Data. Geo Books, Norwich, pp. 129-155.
Peterson, P.D., Scheidegger, K.F. and Schrader, H.J., 1984. Holocene depositional evolution of a small active-margin estuary of the northwestern United States. Mar. Geol., 59: 51-83.
Petit-Maire, N., 1986. Palaeoclimates in the Sahara of Mali. Episodes, 9 (1): 7-16.
Pickrill, R.A., 1976. The evolution of coastal landforms of the Wairau Valley. N.Z. Geogr., 32, 1: 17-29.
Pirazzoli P.A., 1974. Dati storici sul medio mare a Venezia. Atti Accad. Sci. Ist. Bologna, Rend. 13 (1): 125-148.
Pirazzoli P.A., 1976a. Sea level variations in the northwest Mediterranean during Roman times. Science, 194: 519-521.
Pirazzoli, P.A., 1976b. Les Variations du Niveau Marin Depuis 2000 Ans. Mem. Lab. Geomorphol. Ec. Prat. Ht. Etud., No. 30, Dinard, 421 pp.
Pirazzoli, P.A., 1977. Sea level relative variations in the world during the last 2000 years. Z. Geomorphol., 21, 3: 284-296.
Pirazzoli, P.A., 1978. High stands of Holocene sea levels in the northwest Pacific. Quat. Res., 10: 1-29.
Pirazzoli, P.A., 1979-80. Les viviers à poissons romains en Méditerranée. Oceanis, 5, Fasc. Hors-Sér.: 191-201.
Pirazzoli, P.A., 1979-80a. Encoches de corrosion marine dans l'arc Hellénique. Oceanis, 5, Fasc. Hors-Sér.: 327-334.
Pirazzoli, P.A., 1986a. The Early Byzantine Tectonic Paroxysm. Z. Geomorphol., Suppl. 62: 31-49.
Pirazzoli, P.A., 1986b. Secular trends of relative sea-level (RSL) changes indicated by tide-gauge records. J. Coast. Res., Spec. Issue 1: 1-26.
Pirazzoli, P.A., 1986c. Marine notches. In: O. van de Plassche (Editor), Sea-Level Research. Geo Books, Norwich, pp. 361-400.
Pirazzoli, P.A. and Delibrias, G., 1983. Late Holocene and Recent sea level changes and crustal movements in Kume Island, the Ryukyus, Japan. Bull. Dep. Geogr. Univ. Tokyo, 15: 63-76.
Pirazzoli, P.A. and Grant, D.R., 1987. Lithospheric deformation deduced from ancient shorelines. In: K. Kasahara (Editor), Recent Plate Movement and Deformation. Am. Geophys. Union, Geodynamics Ser., Washington, 20, pp. 67-72.
Pirazzoli, P.A. and Koba, M., 1989. Late Holocene sea-level changes in Iheya and Noho islands, the Ryukyus, Japan. Chikyu Kagaku (Earth Sci.), 43: 1-6.
Pirazzoli, P.A. and Montaggioni, L.F., 1986. Late Holocene sea-level changes in the northwest Tuamotu Islands, French Polynesia. Quat. Res., 25: 350-368.
Pirazzoli, P.A. and Montaggioni, L.F., 1988a. The 7000 yr sea-level curve in French Polynesia: geodynamic implications for mid-plate volcanic islands. Proc. 6th Int. Coral Reef Symp., Australia, vol. 3, pp. 467-472.
Pirazzoli, P.A. and Montaggioni, L.F., 1988b. Holocene sea-level changes in French Polynesia. Palaeogeogr., Palaeoclim., Palaeoecol., 68: 153-175.
Pirazzoli, P.A. and Pluet, J., 1991. Holocene changes in sea level as climate proxy data in Europe. In: Proc. Eur. Sci. Found. Workshop "Evaluation of Climate Proxy Data in Relation to the European Holocene", Arles, Dec. 1989. Paläoklimaforschung, Suppl.: 205-225 (in press).
Pirazzoli, P.A. and Thommeret, J., 1973. Une donnée nouvelle sur le niveau marin à Marseille à l'époque romaine. C. R. Acad. Sci. Paris, D, 277: 2125-2128.
Pirazzoli, P.A., Thommeret, J., Thommeret, Y., Laborel, J. and Montaggioni, L.F., 1981. Les rivages émergés d'Antikythira

(Cerigotto): corrélations avec la Crète occidentale et implications cinématiques et géodynamiques. In: Actes Colloque Niveaux Marins et Tectonique Quaternaires dans l'Aire Méditerranéenne. CNRS et Univ. Paris I, Paris, pp. 49-65.

Pirazzoli, P.A., Thommeret, J., Thommeret, Y., Laborel, J. and Montaggioni, L.F., 1982. Crustal block movements from Holocene shorelines: Crete and Antikythira (Greece). Tectonophysics, 86: 27-43.

Pirazzoli, P.A., Kawana, T. and Montaggioni, L.F., 1984. Late Holocene sea-level changes in Tarama Island, the Ryukyus, Japan. Chikyu Kagaku (Earth Sci.), 38, 2: 113-118.

Pirazzoli, P.A., Brousse, R., Delibrias, G., Montaggioni, L.F., Sachet, M.H., Salvat, B. and Sinoto, Y.H., 1985. Leeward Islands: Maupiti, Tupai, Bora Bora, Huahine, Society Archipelago. 5th Int. Coral Reef Congr., Tahiti, vol. 1, pp. 17-72.

Pirazzoli, P.A., Delibrias, G., Kawana, T. and Yamaguchi, T., 1985a. The use of barnacles to measure and date relative sea-level changes in the Ryukyu Islands, Japan. Palaeogeogr., Palaeoclimatol., Palaeoecol., 49: 161-174.

Pirazzoli, P.A., Montaggioni, L.F., Vergnaud-Grazzini, C. and Saliège, J.F., 1987. Late Holocene sea levels and coral reef development in Vahitahi Atoll, eastern Tuamotu Islands, Pacific Ocean. Mar. Geol., 76: 105-116.

Pirazzoli, P.A., Montaggioni, L.F., Saliège, J.F., Segonzac, G., Thommeret, Y. and Vergnaud-Grazzini, C., 1989. Crustal block movements from Holocene shorelines: Rhodes Island (Greece). Tectonophys., 170: 89-114.

Pirazzoli, P.A., Laborel, J., Saliège, J.F., Erol, O., Kayan, I. and Person, A., 1991. Holocene raised shorelines on the Hatay coasts (Turkey): paleoecological and tectonic implications. Mar. Geol., 96: 295-311.

Plafker, G. and Rubin, M., 1967. Vertical tectonic displacements in South-central Alaska during and prior the Great 1964 Earthquake. J. Geosci., Osaka City Univ., 10: 53-66.

Plassche, O. van de, 1982. Sea-level change and water-level movements in the Netherlands during the Holocene. Meded. Rijks Geol. Dienst, 36-1: 1-93.

Plassche, O. van de (Editor), 1986. Sea-Level Research. Geo Books, Norwich, 618 pp.

Plassche, O. van de, 1986a. Introduction. In: O. van de Plassche (Editor), Sea-Level Research. Geo Books, Norwich, pp. 1-26.

Plassche, O. van de, 1988. Stratigraphy, paleogeography and morphology of a macrotidal bay coastal plain: western Marais de Dol, Brittany, France. In: Extended Abstracts, Int. Symp. Theoretical and Applied Aspects of Coastal and Shelf Evolution, Past and Future, jointly with Inaugural Meeting of IGCP Project 274, Amsterdam, 19-24 Sept., 1988, pp. 132-135.

Plassche, O. van de, 1990. Mid-Holocene sea-level change on Virginia's eastern shore. Mar. Geol., (in press).

Plassche, O. van de, 1991. Coastal submergence of the Netherlands, NW Brittany (France), Delmarva Peninsula (Va, USA), and Connecticut (USA) during the last 5500 to 7500 sideral years. In: R. Sabadini et al. (Editors), Glacial Isostasy, Sea-Level and Mantle Rheology. NATO ASI Ser. C, 334, Kluwer, Dordrecht, pp. 285-300.

Plassche, O. van de and Roep, T.B., 1989. Sea-level changes in the Netherlands during the last 6500 years: basal peats vs. coastal barrier data. In: D.B. Scott, P.A. Pirazzoli and C.A. Honig (Editors), Late Quaternary Sea-Level Correlation and Applications. Kluwer, Dordrecht, NATO ASI Series C, 256: 41-56.

Plassche, O. van de, Mook, W.G. and Bloom, A.L., 1989. Submergence of coastal Connecticut 6000-3000 (^{14}C) years B.P. Mar. Geol., 86: 349-354.
Playford, P.E., 1988. Guidebook to the geology of Rottnest Island. Geol. Soc. Aust., West. Aust. Div., Excursion Guideb. No. 2. Perth, 67 pp.
Playford, P.E. and Leech, R.E.J., 1977. Geology and Hydrology of Rottnest Island. Geol. Surv. West. Aust., Rep. 6: 1-98.
Pomel, R., 1979. Géographie Physique de la Basse Côte d'Ivoire. Thèse 3e Cycle, Univ. Caen, 623 pp.
Porter, S.C., Stuiver, M. and Heusser, C.J., 1984. Holocene sea-level changes along the Strait of Magellan and Beagle Channel, southermost South America. Quat. Res., 22: 59-67.
Preuss, H., 1980. Computerauswertung von Seespiegeldaten für das IGCP-Projekt Nr. 61. Eiszetalt. Ggw., 30: 183-201.
Prigent, D., 1977. Contribution à l'Etude de la Transgression Flandrienne en Basse Loire, Apport de l'Archéologie. Thèse de 3e Cycle, Univ. Nantes, 177 pp.
Pugh, D., 1990. Sea-level: change and challenge. Nat. Resour., 26, 4: 36-46.
Pugh, D.T., Spencer, N.E. and Woodworth, P.L., 1987. Data Holding of the Permanent Service For Mean Sea Level. P.S.M.S.L., Birkenhead, 156 p.
Punning, Y.M., 1987. Holocene eustatic oscillations of the Baltic sea level. J. Coast. Res., 3, 4: 505-513.
Pushkar, V.S., 1979. Late Anthropogene Biostratigraphy of the Southern Far East. Moskow (quoted by Korotkii, 1985).
Qiu S., 1986. Development of reef flat and sea level changes. Geogr. Dep., South China Normal Univ., Guangzhou, 12 pp.
Quinlan, G., 1985. A numerical model of postglacial relative sea level change near Baffin Island. In: J.T. Andrews (Editor), Quaternary Environments, Eastern Canadian Arctic, Baffin Island and Western Greenland. Allen & Unwin, London, pp. 560-584.
Quinlan, G. and Beaumont, C., 1981. A comparison of observed and theoretical postglacial relative sea level in Atlantic Canada. Can. J. Earth Sci. 18 (2): 1146-1163.
Quinlan, G. and Beaumont, C., 1982. The deglaciation of Atlantic Canada as reconstructed from the postglacial relative sea-level record. Can. J. Earth Sci. 19: 2232-2246.
Raban A. and Galili, E., 1985. Recent maritime archaeological research in Israel - A preliminary report. Int. J. Naut. Archaeol. Underw. Explor. 14, (4): 321-356.
Rabassa, J., Heusser, C. and Stuckenrath, R., 1987. New data on Holocene sea transgression in the Beagle Channel: Tierra del Fuego, Argentina. Quat. South Am. Antarct. Penins., 4: 291-309.
Rabassa, J., Heusser, C.J. and Rutter, N., 1990. Late-Glacial and Holocene of Argentine Tierra del Fuego. Quat. South Am. Antarct. Penins., 7: 327-351.
Rampino, M.R., 1979. Holocene submergence of southern Long Island, New York. Nature, 280: 132-134.
Redfield, A.C., 1967. Postglacial change in sea level in the western North Atlantic Ocean. Science, 157: 687-692.
Redfield, A.C. and Rubin, M., 1962. Tha age of salt marsh peat and its relation to recent changes in sea level at Barnstable, Massachusets. Proc. Natl. Acad. Sci., 48 (10): 1728-1735.
Regrain, R., 1980. Geographie Physique et Télédétection des Marais Charentais. Thèse d'Etat de Géographie, Univ. Bretagne Occidentale, Brest, 512 pp.
Renberg, I. and Segerström, U., 1981. Appications of varved lake sediments in palaeoenvironmental studies. Wahlenbergia, 7: 125-133.

Revelle, R.R. et al., 1990. Overview and recommendations. In: Sea-Level Change. Geophysics Study Committee, National Research Council, Natl. Acad. Press, Washington, D.C., pp. 3-34.

Rhodes, E.G., 1980. Modes of Holocene Coastal Progradation, Gulf of Carpentaria. Unpublished PhD Thesis, Aust. Natl. Univ., Canberra (quoted by Chappell, 1987).

Ridley, A.P. and Seeley, M.W., 1979. Evidence for recent coastal uplift near Al Jubail, Saudi Arabia. Tectonophys., 52: 319-327.

Ristaniemi, O., 1984. Ancylusjärven aikainen rannansiirtyminen Salpausselkävyöhykkeessä Karjalohjan-Kiskon alueella Lounais-Suomessa. Publ. Dep. Quat. Geol. Univ. Turku, 53: 1-75.

Robbin, D.M., 1984. A new Holocene sea level curve for the upper Florida Keys and Florida reef tract. In: P.J. Gleason (Editor), Environments of South Florida: Present and Past. Miami Geol. Soc., Mem. 2, pp. 437-458.

Rodda, P., 1988. Visit to Western Samoa with the HIPAC Team. In: Sea-Level Changes and Tectonics in the Middle Pacific. Rep. HIPAC Proj. in 1986 and 1987. Univ. Tokyo, pp. 85-90.

Roeleveld, W., 1974. The Groningen coastal area: a study in Holocene geology and low-land physical geography. Bericht. Rijksdienst Oudheidk. Bodemonderz., 20-21 (1970-1971): 7-25 and 24 (1974), Suppl.

Roeleveld, W. and Van Loon, A.J., 1979. The Holocene development of the young coastal plain of Suriname. Geol. Mijnb., 58, 1: 21-28.

Roep, T.B., 1986. Sea-level markers in coastal barrier sands: examples from the North Sea coast. In: O. van de Plassche (Editor), Sea-Level Research. Geo Books, Norwich, pp. 97-128.

Roep, T.B. and Beets, D.J., 1988. Sea level rise and paleotidal levels from sedimentary structures in the coastal barriers in the western Netherlands since 5600 BP. Geol. Mijnb., 67: 53-60.

Rohde, H., 1984. New aspects concerning the increase of sea level on the German North Sea coast. 19th Coast. Engin. Conf., Houston, 1, pp. 899-911 (quoted by Kettetat, 1987).

Saarnisto, M., 1981. Holocene emergence history and stratigraphy in the area north of the Gulf of Bothnia. Ann. Acad. Sci. Fenn., A-III, 130: 1-42.

Sabadini, R., Doglioni, C. and Yuen, D.A., 1990. Eustatic sea level fluctuations induced by polar wander. Nature, 345: 708-710.

Sabadini, R., Lambeck, K. and Boschi, E. (Editors), 1991. Glacial Isostasy, Sea-Level and Mantle Rheology. NATO ASI Ser. C, 334, Kluwer, Dordrecht, pp. 1-708.

Sakaguchi, Y., Kashima, K. and Matsubara, A., 1985. Holocene marine deposits in Hokkaido and their sedimentary environments. Bull. Dep. Geogr. Univ. Tokyo, 17: 1-17.

Salomaa, R., 1982. Post-glacial shoreline displacement in the Lauhanvuori area, western Finland. Ann. Acad. Sci. Fenn., A-III, 134: 81-97.

Salvigsen, O., 1977. Holocene emergence and finds of pumice, whalebones, and driftwood at Snartknausflya, Nordaustlandet. Norsk Polarinstitutt Årbok 1977/1978, Oslo, pp. 217-228.

Salvigsen, O., 1981. Radiocarbon dated raised beaches in Kong Karls Land, Svalbard, and their consequences for the glacial history of the Barents Sea area. Geogr. Ann., 63, A, 3-4: 283-291.

Salvigsen, O. and Österholm, H., 1982. Radiocarbon dated raised beaches and glacial history of the northern coast of Spitsbergen, Svalbard. Polar Res., 1: 97-115.

Sandegren, R., 1940. Torvgeologisk och pollen analytisk undersökning av torvmarken N intill boplatskomplexet vid

Siretorp. In: Stenaldersboplatserma vid Siretorp i Blekinge. Kungl. Vitterhets Histor. Antik. Akad. Stockholm, Arkeol. Monogr., 3: 1-405.

Sandegren, R., 1952. Beskrivning till Kartbladet Onsala. Sver. Geol. Unders., Aa, 192: 1-99.

Sandegren, R. and Johansson, 1931. Beskrivning till kartbladet Göteborg. Sver. Geol. Unders.,Aa, 173: 50-141.

Sanlaville, P., 1973. L'utilisation des vermets dans la datation des changements récents du niveau de la mer. In: L'Archéologie Subaquatique: Une Discipline Naissante. Unesco, Paris, pp. 189-195.

Sanlaville P., 1977. Etude géomorphologique de la région littorale du Liban. Publ. Univ. Liban, Sect. Et. Géogr., Beyrouth, 2 vol.+ cartes.

Sanlaville, P., 1989. Considérations sur l'évolution de la basse-Mésopotamie au cours des derniers millénaires. Paléorient, 15 (2): 5-27.

Sanlaville, P. and Paskoff, R., 1986. Shoreline changes in Bahrain since the beginning of human occupation. In: S. Haya, A. Al Khalifa and M. Rice (Editors), Bahrain Through the Ages. KPI, London, pp. 15-24.

Sauramo, M., 1940. Suomen luonnon kehitys jääkaudesta nykyaikaan. WSOY, Porvoo, 286 pp. (quoted by Salomaa, 1982).

Savage J.C. and Plafker, G., 1991. Tide gauge measurements of uplift along the south coast of Alaska. J. Geophys. Res., 96, B: 4325-4335.

Scarre, C., 1984. Archaeology and sea-level in west-central France. World Archaeol., 16, 1: 98-107.

Schmitz, H., 1953. Die Waldgeschichte Ostholsteins und der zeitliche Verlauf der postglazialen Transgression an der holsteinischen Ostseeküste. Ber. dt. Bot. Ges., 66: 151-166 (quoted by Köster, 1968).

Schnack, E.J., Fasano, E.J. and Isla, F.I., 1982. The evolution of Mar Chiquita Lagoon coast, Buenos Aires Province, Argentina. In: Holocene Sea Level Fluctuations, Magnitude and Causes. Dep. Geol., Univ. S.C., Columbia, pp. 143-155.

Schnitker, D., 1974. Postglacial emergence of the Gulf of Maine. Geol. Soc. Am. Bull., 85: 491-494.

Schofield, J.C., 1960. Sea level fluctuations during the last 4000 years as recorded by a chenier plain, Firth of Thames, New Zealand. N.Z. J. Geol. Geophys., 3: 467-485.

Schofield, J.C., 1977. Late Holocene sea level, Gilbert and Ellice Islands, west central Pacific Ocean. N.Z. J. Geol. Geophys., 20, 3: 503-529.

Scholl, D.W., 1964. Recent sedimentary record in mangrove swamps and rise in sea level over the southwestern coast of Florida: Part I. Mar. geol., 1: 344-366.

Scholl, D.W. and Stuiver, M., 1967. Recent submergence of southern Florida: a comparison with adjacent coasts and other eustatic data. Geol. Soc. Am. Bull., 78: 437-454.

Scholl, D.W., Craighead, F.C., Sr. and Stuiver, M., 1969. Florida submergence curve revised: its relation to coastal sedimentation rates. Science, 163: 562-564.

Schütte, H, 1939. Sinkendes Land an der Nordsee? Schr. Dtsch. Naturkundever., N.F., 9 (quoted by Jelgersma, 1961).

Scott, D.B., 1977. Distributions and Population Dynamics of Marsh-Estuarine Foraminifera with Applications to Relocating Holocene Sea-Level. Ph.D. Dissertation, Geol. Dep., Dalhousie Univ., Halifax, 207 pp.

Scott, D.B. and Greenberg, D.A., 1983. Relative sea-level rise and tidal development in the Fundy tidal system. Can. J. Earth Sci., 20 (10): 1554-1564.

Scott, D.B. and Medioli, F.S., 1978. Studies of relative sea level changes in the Maritimes. Progr. Rep. Dep. Energy, Mines & Resour. 2239-4-31/78, 79 pp.

Scott, D.B. and Medioli, F.S., 1979. Marine Emergence and Submergence in the Maritimes. Progr. Rep. Dep. Energy, Mines & Resour., Dep. Geol., Dalhousie Univ., Halifax, 69 pp.

Scott, D.B. and Medioli, F.S., 1980. Post-glacial emergence curves in the Maritimes determined from marine sediments in raised basins. In: Proc. Can. Coast. Conf. 1980. Burlington, Ontario, pp. 428-446.

Scott, D.B. and Medioli, F.S., 1982. Micropaleontological documentation for early Holocene fall of relative sea level on the Atlantic coast of Nova Scotia. Geology, 10: 278-281 (quoted by Carrera and Vanicek, 1988).

Scott, D.B. and Medioli, F.S., 1986. Foraminifera as sea-level indicators. In: O. van de Plassche (Editor), Sea-Level Research. Geo Books, Norwich, pp. 435-456.

Scott, D.B., Medioli, F.S. and Duffett, T.E., 1984. Holocene rise of relative sea level at Sable Island, Nova Scotia, Canada. Geology, 12: 173-176.

Scott, D.B., Medioli, F.S. and Miller, A.A.L., 1987a. Holocene sea levels, paleoceanography, and late glacial ice configuration near the Northumberland Strait, Maritime Provinces. Can. J. Earth Sci. 24: 668-675.

Scott, D.B., Boyd, R. and Medioli, F.S., 1987b. Relative sea level changes in Atlantic Canada: observed level and sedimentological changes vs. theoretical models. In: D. Nummedal, O.H. Pilkey and J.D. Howard (Editors), Sea-Level Fluctuation and Coastal Evolution. Soc. Econ. Paleontol. Miner., Spec. Publ. n° 41, Tulsa, pp. 87-96.

Scott, D.B., Boyd, R., Douma, M., Medioli, F.S., Yuill, S., Leavitt, E. and Lewis, C.F.M., 1989. Holocene relative sea-level changes and Quaternary glacial events on a continental shelf edge: Sable Island bank. In: D.B. Scott, P.A. Pirazzoli and C.A. Honig (Editors), Late Quaternary Sea-Level Correlation and Applications. NATO ASI Ser., C, 256, Kluwer Academic Publishers, Dordrecht, pp. 105-119.

Searle, D.J. and Woods, P.J., 1986. Detailed documentation of a Holocene sea-level record in the Perth region, southern western Australia. Quat. Res., 26: 299-308.

Selivanov, A.O. and Stepanov, V.P., 1985. Geoarcheological investigations on the Soviet Primorye coast: their application to interpretations of paleoclimates and former sea levels. J. Coastal Res. 1, (2): 141-149.

Serebryanny L.R., 1982. Postglacial Black-Sea coast fluctuations and their comparison with the glacial history of the Caucasian high mountain region. In: P.A. Kaplin, R.K. Klige and A.L. Chepalyga (Editors), Sea and Oceanic Level Fluctuations for 15000 Years (in Russian). Nauka, Moscow, pp. 161-167.

Shaw, J. and Forbes, D.L., 1990. Relative sea-level change and coastal response, northeast Newfoundland. J. Coast. Res. 6 (3): 641-660.

Shennan, I., 1981. The nature, extent, and timing of marine deposits in the English Fenland during the Flandrian age. Striae, 14: 177-181.

Shennan, I, 1982. Interpretation of Flandrian sea-level data from the Fenland, England. Proc. Geol. Ass., 93, 1: 53-63.

Shennan, I., 1982b. Problems of correlating Flandrian sea-level changes and climate. In: A.F. Harding (Editor), Climatic Change in Later Prehistory. Edinburgh Univ. Press, pp. 52-67.
Shennan, I., 1986. Flandrian sea-level changes in the Fenland. II: tendencies of sea-level movement, altitudinal changes, and local and regional factors. J. Quat. Sci., 1, 2: 155-179.
Shennan, I., 1987. Holocene sea-level changes in the North Sea region. In: M.J. Tooley and I. Shennan (Editors), Sea-Level Changes. Basil Blackwell, Oxford, pp. 109-151.
Shennan, I., 1989. Holocene crustal movements and sea-level changes in Great Britain. J. Quat. Sci., 4: 77-89.
Shennan, I. and Tooley, M.J., 1987. Conspectus of fundamental and strategic research on sea-level changes. In: M.J. Tooley and I. Shennan (Editors), Sea-Level Changes. Basil Blackwell, Oxford, pp. 371-390.
Shennan, I., Tooley, M.J., Davis, M.J. and Haggart, B.A., 1983. Analysis and interpretation of Holocene sea-level data. Nature, 302, 5907: 404-406.
Shepard, F.P., 1960. Rise of sea level along northwest Gulf of Mexico. In: F.P. Shepard and T.H. van Andel (Editors), Recent Sediments, Northwest Gulf of Mexico. Am. Assoc. Pet. Geol., pp.338-344 (quoted by Bloom, 1963).
Shepard, F.P., 1963. Thirty-five thousand years of sea level. In: Essays in Marine Geology in Honor of K.O. Emery. Univ. South Calif. Press, Los Angeles, pp. 1-10.
Shepard, F.P., 1964. Sea level changes in the past 6000 years, Possible archaeological significance. Science, 143, 3606: 574-576.
Shepard, F.P. and Curray, J.R., 1967. Carbon-14 determination of sea level changes in stable areas. Progr. Oceanogr., 4: 283--291 (quoted by Newman et al., 1971).
Shepard, F.P. and Wanless, H.R., 1971. Our Changing Coastlines. McGraw-Hill, New York, 579 p.
Shepard, F.P., Curray, J.R., Newman, W.A., Bloom, A.L., Newell, N.D., Tracey, J.L. Jr. and Veeh, H.H., 1967. Holocene changes in sea level: evidence in Micronesia. Science, 157: 542-544.
Shilik K.K., 1975. The change in the level of the Black Sea in Late Holocene. Quoted by Badyukov and Kaplin (1979).
Shimoyama, S. and Shuto, T., 1978. Quaternary molluscan fossil assemblages from Arato, Fukuoka City (in Japanese, with English abstract). Sci. Rep. Dep. Geol. Kyushu Univ., 13: 47-59.
Siiriäinen, A., 1972. A gradient/time curve for dating Stone Age shorelines in Finland. Suomen Museo 79: 5-18.
Singh, L.J., 1973. Beach ridges of southeast Wellington, New Zealand. Ninth Congress INQUA, Abstracts. Christchurch, pp. 533-534.
Sissons, J.B., 1983. Shorelines and isostasy in Scotland. In: D.E. Smith and A.G. Dawson (Editors), Shorelines and Isostasy. Academic Press, London, pp. 209-225.
Sissons, J.B. and Brooks, C.L., 1971. Dating of early Postglacial land and sea level changes in the western Forth Valley. Nature, 234, 50: 124-127.
Sneh Y. and Klein M., 1984. Holocene sea level changes at the coast of Dor, southeast Mediterranean. Science, 226: 831-832.
Sørensen, R., 1979. Late Weichselian deglaciation in the Oslofjord area, South Norway. Boreas, 8: 241-246.
Spencer, N.E., Woodworth, P.L. and Pugh, D.T., 1988. Ancillary Time Series of Mean Sea Level Measurements. Permanent Service for Mean Sea Level, Birkenhead, Merseyside, 89 pp.
Stabell, B., 1980. Holocene shorelevel displacement in Telemark, southern Norway. Nor. Geol. Tidsskr., 60: 71-81.

Stea, R.R., Scott, D.B., Kelley, J.T., Kelley, A.R., Wightman, D.M., Finck, P.W., Seaman, A., Nicks, L., Bleakney, S., Boyd, R. and Douma, M., 1987. Quaternary Glaciations, Geomorphology, and Sea-Level Changes: Bay of Fundy Region. Centre for Marine Geology, Dalhousie Univ., Halifax, 79 pp.

Steele, G.A., 1980. Stratigraphy and depositional history of Bogue Banks, North Carolina. Thesis. Duke Univ., Durham, N.C., 201 pp. (unpublished, quoted by Moslow and Heron, 1981).

Straaten, L.M.J.U. van, 1954. Radiocarbon datings and changes of sea level at Velzen (Netherlands). Geol. Mijnb., 16: 247-253.

Street, F.A., 1977. Deglaciation and marine paleoclimates, Schuchert Dal, Scoresby Sund, East Greenland. Arct. & Alp. Res., 9: 421-426.

Streif, H., 1979. Holocene sea level changes in the Strait of Malacca. In: Proc. Int. Symp. Coastal Evol. Quat., Sâo Paulo 11-18 sept. 1978, pp. 552-572.

Stuiver, M. and Daddario, J.J., 1963. Submergence of the New Jersey coast. Science, 142: 951.

Stuiver, M. and Kra, R. (Editors), 1986. 12th Int. Radiocarbon Conf., Trondheim, Calibration Issue. Radiocarbon, 28, 2B, pp. 805-1030.

Stuiver, M., Denton, G.H. and Borns, H.W.Jr., 1976. Carbon-14 dates of *Adamussium colbecki* (Mollusca) in marine deposits at New Harbor, Taylor Valley. Antarct. U.S., 11, 2: 86-88 (quoted by Lingle and Clark, 1979).

Stuiver, M., Pearson, G.W. and Braziunas, T.F., 1986. Radiocarbon age calibration of marine samples back to 9000 cal yr BP. Radiocarbon, 28, 2B: 980-1021.

Suess, E, 1885. Das Antlitz der Erde. Wien.

Suggate, R.P., 1968. Post-glacial sea-level rise in the Christchurch metropolitan area, New Zealand. Geol. Mijnb., 47, 4: 291-297.

Sugimura, A., Maeda, Y., Matsushima, Y., Rodda, P. and Matsumoto, E., 1988a. Lobau lowland, Viti Levu, Fiji. In: Sea-Level Changes and Tectonics in the Middle Pacific. Rep. HIPAC Proj. in 1986 and 1987. Univ. Tokyo, pp. 59-65.

Sugimura, A., Maeda, Y., Matsushima, Y. and Rodda, P., 1988b. Further report on sea-level investigation in Western Samoa. In: Sea-Level Changes and Tectonics in the Middle Pacific. Rep. HIPAC Proj. in 1986 and 1987. Univ. Tokyo, pp. 77-84.

Sveian, H. and Olsen, L., 1984. A shoreline displacement curve from Verdalsøra, Nord-Trøndelag, Central Norway. Boreas, 8: 241-246 (quoted by Svendsen and Mangerud, 1987).

Svendsen, J.I. and Mangerud, J., 1987. Late Weichselian and Holocene sea-level history for a cross-section of Western Norway. J. Quat. Sci., 2: 113-132.

Svensson, N.O., 1985. Some preliminary results on the early Holocene shore displacement in the Oskarshamn area, south-eastern Sweden. Eiszetalt. Ggw., 35: 119-133.

Synge, F.M., 1977. Records of sea-levels during the late Devensian. Phil. Trans. R. Soc. Lond., B, 280: 211-228.

Synge, F.M., 1980. A morphometric comparison of raised shorelines in Fennoscandia, Scotland and Ireland. Geol. Fören. Stockholm Förh., 102, 3: 235-249.

Taira, K., 1975. Holocene crustal movements in Taiwan as indicated by radiocarbon dating of marine fossils and driftwood. Tectonophysics, 28: T1-T5.

Taira, K., 1980. Radiocarbon dating of shell middens and Holocene sea-level fluctuations in Japan. Palaeogeogr., Palaeoclim., Plaeoecol., 32: 79-87.

Takahashi, T., Koba, M. and Kan, H., 1988. Relationship between reef growth and sea level on the northwest coast of Kume Island, the Ryukyus: data from drill holes on the Holocene coral reef. Proc. 6th Int. Coral Reef Symp., Australia, vol. 3, pp. 491-496.

Takano, S., 1978. Development of the compound spit of Notsukezaki, Hokkaido (in Japanese, with English abstract). Ann. Tohoku Geogr. Assoc., 30: 82-90.

Tastet, J.P., 1981. Morphologie des littoraux sédimentaires liée aux variations du niveau de la mer: exemple du Golfe de Guinée. Oceanis, 7, 4: 455-472.

Taylor, R.B., Carter, R.W.G., Forbes, D.L. and Orford, J.D., 1986. Beach sedimentation in Ireland: contrasts and similarities with Atlantic Canada. Curr. Res. Geol. Surv. Can., 1986A: 55-64.

Teichert, C., 1950. Late Quaternary changes of sea level at Rottnest Island, Western Australia. Proc. R. Soc. Victoria, 59: 63-79.

Ten Brink, N.W., 1974. Glacio-isostasy: new data from west Greenland and geophysical implications. Geol. Soc. Am. Bull., 85: 219-228.

Ters, M., 1973. Les variations du niveau marin depuis 10 000 ans, le long du littoral atlantique français. In: Le Quaternaire: Géodynamique, Stratigraphie et Environnement. Suppl. Bull. Assoc. Fr. Et. Quat. 36: 114-135.

Thom, B.G. and Chappell, J., 1975. Holocene sea levels relative to Australia. Search, 6, 3: 90-93.

Thom, B.G. and Roy, P.S., 1983. Sea level change in New South Wales over the past 15,000 years. In: (D. Hopley, Editor), Australian Sea Levels in the Last 15 000 Years: A Review. Dep. Geogr., James Cook Univ. North Qld., Monogr. Ser., Occasion. Pap. No. 3. Townsville, pp. 64-84.

Thom, B.G. and Roy, P.S., 1985. Relative sea levels and coastal sedimentation in southeast Australia in the Holocene. J. Sediment. Petrol., 55 (2): 257-264.

Thommeret, J. and Thommeret, Y., 1978. ^{14}C datings of some Holocene sea levels on the North coast of the island of Java (Indonesia). Mod. Quat. Res. Southeast Asia, 4: 51-56.

Thommeret Y., Thommeret J., Laborel J., Montaggioni L.F. and Pirazzoli P.A., 1981. Late Holocene shoreline changes and seismo-tectonic displacements in western Crete (Greece). Z. Geomorphol., Suppl. 40: 127-149.

Tjia, H.D., Fujii, S., Kigoshi, K. and Sugimura, A., 1975. Additional dates on raised shorelines in Malaysia and Indonesia. Sains Malays., 4: 69-84.

Tjia, H.D., Fujii., S. and Kigoshi, K., 1977. Changes of sea-level in the southern South China Sea area during Quaternary times. In: Proc. Symp. Quat. Geol. Malay-Indon. Coast. Offshore Areas, CCOP/TP5, pp. 11-36.

Tjia, H.D., Fujii, S. and Kigoshi, K., 1983. Holocene shorelines of Tioman Island in the South China Sea. Geol. Mijnb., 62: 599-604.

Tooley, M.J., 1971. Changes in sea-level and the implications for coastal development. In: Association of River Authorities Year Book and Directory 1971, pp. 220-225.

Tooley, M.J., 1974. Sea-level changes during the last 9000 years in north-west England. Geogr. J., 140: 18-42.

Tooley, M.J., 1976. Flandrian sea-level changes in west Lancashire and their implications for the 'Hillhouse Coastline'. Geol. J., 11, 2: 137-152.

Toyoshima, Y., 1978. Postglacial sea level change along San'in district, Japan (in Japanese, with English abstract). Geogr. Rev. Japan, 51: 147-157.

Tracey, J.I. Jr. and Ladd, H.S., 1974. Quaternary history of Eniwetok and Bikini Atolls, Marshall Islands. Proc. Second Int. Coral Reef Symp., Brisbane, 2, pp. 537-550.

Umbgrove, J.H.F., 1947. Origin of the Dutch coast. Kon. Ned. Akad. Wet. Proc., Sect. Sci., 50: 227-237.

Umitsu, M., 1979. Geomorphic development of the Nobi plain in late Quaternary (in Japanese, with English abstract). Geogr. Rev. Japan, 52: 199-208.

Upson, J.E., Leopold, E.B. and Rubin, M., 1964. Postglacial change of sealevel in New Haven Harbor, Connecticut. Am. J. Sci., 262: 121-132.

Urien, C.M., 1970. Les rivages et le plateau continental du sud du Brésil, de l'Uruguay et de l'Argentine. Quaternaria 12: 57-69.

Veen, J. van, 1954. Tide gauges, subsidence -gauges and floodstones in the Netherlands. Geol. Mijnb. 16: 214-219.

Voskoboinikov V.M., Rotar M.F. and Konikov E.G., 1982. Relationship between rythmic composition of thick Holocene layers of the Black Sea region lagoons and oscillatory level regime of the Black Sea (in Russian). In: Sea Level Fluctuations. Moscow State University, pp. 264-274.

Voss, F. von, 1970. Der Einfluβ des jüngsten Transgressionsablaufes auf die Küstenenwichlung der Geltinger Birch im Nordteil der westlichen Ostsee. Die Küste, 20: 101-113.

Wada, M., 1972. Late Quaternary history of Niigata lowland (in Japanese, with English abstract). Mem. Geol. Soc. Japan, 7: 77-89.

Walcott, R.I., 1972a. Late Quaternary vertical movements in eastern North America: quantitative evidence of glacio-isostatic rebound. Rev. Geophys. Space Phys., 10, 4: 849-884.

Walcott, R.I., 1972b. Past sea levels, eustasy and deformation of the earth. Quat. Res., 2: 1-14.

Walcott, R.I., 1977. Rheological models and observational data of glacio-isostatic rebound. In: N.A. Mörner (Editor), Symposium Earth Rheology and Late Cenozoic Isostatic Movements, Abstracts. Geol. Inst., Stockh. Univ., p. 129.

Wang, J. and Wang, P., 1980. Relationship between sea-level changes and climatic fluctuation in east China since the Late Pleistocene. Acta Geogr. Sinica, 35: 299-312 (quoted by Ota, 1987).

Washburn, A.L. and Stuiver, M. 1962. Radiocarbon-dated postglacial delevelling in Northeast Greenland and its implications. Arct., 15 (1): 66-73.

Webber, P.J., Richardson, J.W. and Andrews, J.T., 1970. Post-glacial uplift and substrate age at Cape Henrietta Maria, southeastern Hudson Bay, Canada. Can. J. Earth Sci., 7: 317-325.

Weinsberg, I., Grinbergs, E., Danilans, I. and Ulst, A.I., 1974. Late- and post-Glacial history of the Baltic Sea according to the materials obtained by studying the Latvian coasts. Baltica Publishers, 5: 89-93 (quoted by Badyukov and Kaplin, 1979).

Wellman, H.W., 1962. Holocene of the North Island of New Zealand: a coastal reconnaissance. Trans. R. Soc. N.Z., Geol., 1, 5: 29-99.

Wellman, H.W., 1967. Tilted marine beach ridges at Cape Turakirae, N.Z. J. Geosci., Osaka City Univ., 10, 1-16: 123-129.

Wilkinson, T.J. and Murphy, P., 1986. Archaeological survey of an intertidal zone: the submerged landscape of the Essex coast, England. J. Field Archaeol., 13: 177-194.

Winn, K., Averdieck, F.R., Erlenkeuser, H. and Werner, F., 1986. Holocene sea level rise in the western Baltic and the question of isostatic subsidence. Meyniana, 38: 61-80.

Woodroffe, C.D., 1988. Mangroves and sedimentation in reef environments: indicators of past sea-level changes, and present sea-level trends? Proc. 6th Int. Coral Reef Symp., Australia, Vol. 3, pp. 535-539.
Woodroffe, C.D., Curtis, R.J. and McLean, R.F., 1983. Development of a chenier plain, Firth of Thames, New Zealand. Mar. Geol., 53: 1-22.
Woodroffe, C.D., Thom, B.G., Chappell, J., Wallensky, E., Grindrod, J. and Head, J., 1987. Relative sea level in the south Alligator River region, North Australia, during the Holocene. Search, 18, 4: 198-200.
Woodroffe, C., McLean, R., Polach, H. and Wallensky, E., 1990. Sea level and coral atolls: late Holocene emergence in the Indian Ocean. Geology, 18: 62-66.
Wu, X., 1987. Sea-level changes and stratigraphy during late Quaternary in the basins and plains along the coast of Fujian Province, southeastern China. In: Y. Qin and S. Zhao (Editors), Late Quaternary Sea-Level Changes. China Ocean Press, Beijing, pp. 223-238.
Wu, P. and Peltier, W.R., 1983. Glacial isostatic adjustment and the free air gravity anomaly as a constraint on deep mantle viscosity. Geophys. J. R. Astron. Soc., 74: 377-450.
Xie, Z., Shao, H., Chen, F., Chen, Z. and Dou, Y., 1986. Transgression since late Pleistocene in Fujian coast (in Chinese with English abstract). In: China Sea Level Changes. China Ocean Press, Beijing, pp. 156-165.
Xu, M., Ma, D., Zhou, Q., Zhang, G. and Lan, X., 1986. Quaternary sea-level fluctuation in Zhujiang River delta area (in Chinese with English abstract). Mar. Geol. Quat. Geol. (Qingdao & Beijing), 6, 3: 93-102.
Yang, D., 1986. Tidal level changes near the Changjiang estuary since Holocene (in Chinese with English abstract). In: China Sea Level Changes. China Ocean Press, Beijing, pp. 124-131.
Yang, H. and Chen, X., 1985. Quaternary transgressions, eustatic changes and shifting of shoreline in east China (in Chinese with English abstract). Mar. Geol. Quat. Geol. (Qingdao), 5, 4: 59-80.
Yang, H. and Xie, Z., 1984a. Sea-level changes in east China over the past 20 000 years. In: R.O. Whyte (Editor), The Evolution of Eastern Asian Environment. Univ. Hong Kong, pp. 288-308.
Yang, H. and Xie, Z., 1984b. A perspective on sea-level fluctuations and climatic variations (in Chinese with English abstract). Acta Geogr. Sinica, 39, 1: 20-39.
Yim, W.W.S., 1986a. A sea-level curve for Hong Kong during the last 40,000 years. In: Sea-level changes in Hong Kong during the last 40,000 years, Programme and Extended Abstracts. Univ. Hong Kong, pp. 23-30.
Yim, W.W.S., 1986b. Radiocarbon dates from Hong Kong and their geological implication. J. Hong Kong Archaeol. Soc., 11: 50-63.
Yokota, K., 1978. Holocene coastal terraces of the southeast coast of the Boso peninsula (in Japanese, with English abstract). Geogr. Rev. Japan, 51: 349-364.
Yonekura, N., 1975. Quaternary tectonic movements in the outer arc of southwest Japan, with special reference to seismic crustal deformation. Bull. Dep. Geogr. Univ. Tokyo, 7: 19-71.
Yonekura, N., Matsushima, Y., Maeda, Y. and Kayanne, H., 1984. Holocene sea-level changes in the southern Cook Islands. In: Sea-Level Changes and Tectonics in the Middle Pacific. Rep. HIPAC Proj. in 1981, 1982 and 1983. Kobe Univ., pp. 113-136.
Yonekura, N., Ishii, T., Saito, Y., Maeda, Y., Matsushima, Y., Matsumoto, E. and Kayanne, H., 1988. Holocene fringing reefs and

sea-level change in Mangaia Island, southern Cook Islands. Palaeogeogr., Palaeoclim., Palaeoecol., 68: 177-188.

Yoshikawa, T., Kaizuka, S. and Ota, Y., 1981. The Landforms of Japan. Univ. Tokyo Press, 222 p.

Zendrini, A., 1802. Sull'alzamento del livello del mare. Giornale dell'Italiana Letteratura (Padova), 2: 3-37.

Zhang M., 1987. Climate evolution and sealevel changes in the Xisha region since late Pleistocene. In: Y. Qin and S. Zhao (Editors), Late Quaternary Sea-Level Changes. China Ocean Press, Beijing, pp. 161-168.

Zhao, X., 1984. Study of Sea-level Changes in China (in Chinese). Fujian Sci. Press, Fuzhou, 194 pp.

Zhao, X., Geng, X. and Zhang, J., 1982. Sea level changes in eastern China during the past 20 000 years. Acta Oceanol. Sinica, 1, 2: 248-258.

Zwart, H.J., 1951. Postglaciale land- en zeeniveau-veranderingen. Kon. Ned. Akad. Wet. Proc. B, 14: 162-173.

LOCALITY INDEX

[Abbreviations: B. = Bay; L. = Lake; F. = Fjord (Fiord); Is. = Island(s); Pen. = Peninsula; R. = River]

AUTHOR INDEX